The Symbian OS
Ma Architecture
Te Sourcebook

-3

To be re,

The Symbian OS Architecture Sourcebook

Design and Evolution of a Mobile Phone OS

By

Ben Morris

Reviewed by

Chris Davies, Warren Day, Martin de Jode, Roy Hayun, Simon Higginson, Mark Jacobs, Andrew Langstaff, David Mery, Matthew O'Donnell, Kal Patel, Dominic Pinkman, Alan Robinson, Matthew Reynolds, Mark Shackman, Jo Stichbury, Jan van Bergen

Symbian Press

Head of Symbian Press
Freddie Gjertsen

Managing Editor
Satu McNabb

John Wiley & Sons, Ltd

Other Wiley Editorial Offices

Anniversary Logo Design: Richard J. Pacifico

Library of Congress Cataloging-in-Publication Data

Morris, Ben, 1958-
The Symbian OS architecture sourcebook : design and evolution of a
mobile phone OS / by Ben Morris.
    p. cm.
Includes bibliographical references.
ISBN-13: 978-0-470-01846-0
ISBN-10: 0-470-01846-1
1. Operating systems (Computers) 2. Symbian OS (Computer file) I.
Title.
QA76.76.O63M6835 2007
005.4'32 – dc22

2006103533

British Library Cataloguing in Publication Data

A catalogue record for this book is available from the British Library

ISBN: 978-0-470-01846-0

Typeset in 10/12pt Optima by Laserwords Private Limited, Chennai, India
Printed and bound in Great Britain by Bell & Bain, Glasgow
This book is printed on acid-free paper responsibly manufactured from sustainable
forestry in which at least two trees are planted for each one used for paper production.

To Philippa, with love.

Contents

About the Author

Ben Morris joined Psion Software in October 1997, working in the software development kit team on the production of the first C++ and Java SDKs for what was at that time still the EPOC32 operating system. He led the small team that produced the SDKs for the ER5 release of EPOC32 and, when Psion Software became Symbian, he took over responsibility for expanding and leading the company's system documentation team. In 2002, he joined the newly formed System Management Group in the Software Engineering organization of Symbian, with a brief to 'define the system'. He devised the original System Model for Symbian OS and currently leads the team responsible for its maintenance and evolution.

He can be found on the Internet at *www.benmorris.eu*

Acknowledgements

Some people told me it would be hard to write this book in and around my real job in the System Management Group at Symbian and a few promised me that it would be impossible. They were all right, of course, although none of them tried to stop me.

Many thanks to Wiley and Symbian Press therefore for their patience as I've stretched deadlines. Thanks to Fredrik Josephson for saying 'yes' to my starting the book as a 10% task and for turning a blind eye when it grew beyond that; and to Geert Bollen for being (almost) tolerant when he inherited the problem. Thanks to Freddie Gjertsen of Symbian Press for getting me to the end and to Phil Northam for his part in making it happen in the first place.

My biggest thanks, though, are due to those who took the time to talk to me, agreed to my using a recording device and let me use their words. They are: Geert Bollen, Martin Budden, Andy Cloke, Charles Davies, Bob Dewolf, Morgan Henry, Ian Hutton, Peter Jackson, Keith de Mendonca, Will Palmer, Howard Price, Murray Read, Martin Tasker, Andrew Thoelke and David Wood. I have done my best to make sure they are happy with the use to which I have put their words.

I am also very grateful to my technical reviewers from across the company (and, in a few cases, from outside it): Jan van Bergen, Chris Davies, Warren Day, Roy Hayun, Simon Higginson, Mark Jacobs, Martin de Jode, Andrew Langstaff, David Mery, Matthew O'Donnell, Kal Patel, Dominic Pinkman, Matt Reynolds, Alan Robinson, Mark Shackman, Phil Spencer, and Jo Stichbury. Jeff Lewis provided a final review from a commercial perspective.

Any errors which remain are mine, of course.

A special thanks to Jawad Arshad for his help in constructing the reference material in Appendix A, and for his careful review of what

I did with it, and to Bob Rosenberg for his great work on the System Model graphics (which is present in the book in the form of the color pull-out). Way back when, Martin Hardman was my original collaborator on early versions of the System Model, and I would like to acknowledge his contribution

Finally, my family have put up with this book for longer than was promised. Philippa, Nat, Jake and Henrietta – thanks.

Glossary of Terms

ABI	Application binary interface
ADT	Abstract data type
BAL	Bearer Abstraction Layer
BIO	Bearer-independent object
CDMA	Code Division Multiple Access
DFRD	Device family reference design
DRM	Digital rights management
DSP	Digital Signal Processor
EDGE	Enhanced Data Service for GSM Evolution
ETSI	European Telecommunications Standards Institute
FOMA	Freedom of Mobile Access
GPRS	General Packet Radio Service
IPC	Interprocess communication
MOAP	Mobile Application Platform
MTM	Message type module
MVC	Model–view–controller
OBEX	IrDA Object Exchange
OMA	Open Mobile Alliance
OTA	Over the air
PAN	Personal Area Networking
PIM	Personal information manager
PLP	Psion Link Protocol
QoS	Quality of Service
RTOS	Real-time operating system
RTP	Real-time transport protocol
SIP	Session initiation protocol

SMIL	Synchronized Multimedia Integration Language
UART	Universal Asynchronous Transmitter/Receiver
UMTS	Universal Mobile Telecommunications System
VoIP	Voice over IP
VPN	Virtual Private Network
WAP	Wireless Application Protocol
WDP	Wireless Datagram Protocol
XIP	Execute in place

Introduction

This book is part description, part reference, part case study and part history. My goal in writing it has been to try to make Symbian OS more accessible to a wider audience than has been catered for to date. I hope there is nothing dumbed-down about this book, but at the same time I have tried to make it accessible to those who are interested, but not expert, in the topics it covers, as well as useful to a more hands-on developer audience.

As Symbian OS becomes more mainstream – a volume product and not just a niche one – I hope this book will serve as a primer for the curious and a way in to a deeper understanding of what Symbian OS is, where it came from and why it is currently riding high.

Certainly there is material here which is useful to Symbian OS developers – both seasoned and novice – and which has previously been hard to find. However, this book takes a different approach to that of most Symbian Press books; it is not so much a 'how to' book as a 'what and why' book (and to some extent also a 'who and when' book).

Part 1 is a Symbian OS primer, a rapid introduction that sketches the background of the mobile telephony market, traces the emergence of Symbian OS and Symbian the company, conducts a rapid tour of the architecture of Symbian OS, and provides a refresher – or introduction – to the key ideas of object orientation (OO) in software.

Part 2 begins the more detailed exploration of the architecture of Symbian OS, following the Symbian OS System Model layering to provide a complete, high-level, architectural description of Symbian OS.

Part 3 returns to the historical approach of the primer chapters, and presents five case studies, each exploring some aspect of Symbian OS, or of its history and evolution, in depth. Drawing on the insights – and the recollections – of those who were involved, these studies trace and try to understand the forces that have shaped the operating system.

Appendix A contains a component by component reference, ordered alphabetically by component name, which is definitely intended for a developer audience. It also includes a color pull-out of the System Model for the current public release, Symbian OS v9.3.

Who This Book Is For

This book is for anyone who wants to understand Symbian OS better – what Symbian OS is, why it is what it is, and how it got to be that way. If you work with Symbian OS, or intend to, this book is for you. If you want to get under the skin of the OS and understand it more deeply, this book is very definitely for you. This book is for you too if you are interested in the software or mobile phone industries more generally, or in the perennial themes of software development, or are merely curious about how real systems get made and evolve.

A reasonable degree of software technical literacy is assumed, but not so much that the more casual reader should shy away. There are no exercises. And there is no sample code.

How to Use This Book

This book calls itself a *sourcebook* and it is intended to be used both as a primer and as a reference. Its different sections are useful in their different ways as reference material. Both Part 1 and Part 3 are structured as a straight-through read and, I hope, they offer a good starting point from which to come to Symbian OS for the first time. The material in Part 2 is probably deeper than a non-developer audience needs. And while this is not (strictly) a programming book, I hope that Symbian OS developers find its reference material useful, or better.

Telling Stories

Someone else wrote the phrase before I did: "In every great software product is a great story" [McCarthy 1995]. I think it's true. So while this book is aimed at a technically aware audience, it is not addressed exclusively to an audience of programmers. I hope programmers and, more generally, software developers, designers and architects will find it useful, especially those coming new to the OS and trying to understand it. But I hope it will be just as useful to academics and students, marketeers, technical decision makers and managers seeking to evaluate and understand Symbian OS, and indeed anyone else who is broadly in the business of software or phones or who is just interested in such things, and who is encountering Symbian OS (or its close competitors) for the first time and needs to understand it. Speaking personally, I have long been something of an operating system junkie; to some extent, therefore, this book attempts to scratch that itch. (You can't work for an operating system company and not have a bit of the operating system junkie in you.)

I hope that understanding the deeper story behind Symbian OS will help those who want to (or have to) work with it to understand it better and more deeply. Above all, I hope it will help them work better with Symbian OS than would be the case without this book.

I have another purpose too. One of the things which appealed to me most in my early days in the company (which became Symbian a few months after I joined) was the degree to which everyone involved in creating the system shared the sense that making software is a visionary activity and that making good software, indeed the best possible software, is as much a moral imperative as a business one. For an activity which likes to count itself as a branch of engineering, the number, and variety, of value words which cropped up in any daily conversation could be surprising. Making software, which is to say making this software in particular, is value-laden. 'Delight', 'elegance', 'trust', 'integrity', 'robustness', 'reliability', 'economy' and 'parsimony' were all among the company buzzwords and very much part of the fabric of the effort, and give a flavor of those times. Above all, to be part of the effort to create Symbian OS was to be part of the revolution, no less. The truly personal, individual, pocketable, always-on, human-scaled device you could trust your data to, and to some extent therefore also your identity, and your heart as well as your head, was not yet the commonplace thing which the mobile phone revolution has made of it. Symbian – the operating system and the company – has played its part, too, in that revolution.

Symbian is currently riding high. Symbian OS has done more than find a niche; it has found (and, indeed, it has founded) a global market and has led that market from its inception. To make that point more concretely, consider this: when I was starting work on this book, I drafted a paragraph about 2005 being a watershed year for Symbian OS, potentially its breakout year. Between then and now, as I write this at the end of 2006, the number of shipped Symbian OS phones has doubled from 50 million to 100 million, and counting.

Way back when, the company was a company of individuals – who could be opiniated, strident and arrogant but could just as quickly switch to humility in the face of a powerful intellectual argument. Inevitably, some of that individuality has been lost with success and growth. I hope that by capturing some of the flavor of those times, that particular flame can be kept burning.

I have been mindful both of commercial and personal confidences and I believe that nothing I have written (or quoted) breaches either. (Any instances of 'Don't quote me!' which appear in the text have been carefully approved.)

I have tried everywhere to observe the mantra 'Tell no lies', which is not always the case in books such as this, and which here and there has not been easy. Let me quote Bjarne Stroustrup as one inspiration for honesty, 'I abhor revisionist history and try to avoid it'.[1] I have done my best to follow that example.

Getting Symbian OS

Anyone, anywhere, can download Symbian OS in a form in which they can learn to program it, work with it, explore it and experiment with it. Anyone can learn to write Symbian OS applications: development kits are free, and easily available, for UIQ and S60 platforms; development tools (GCC and Eclipse) are free; the Symbian Press programming books are widely available; and the possible languages range from OPL (which began life as the Psion Organiser Language and is now an open-source, rapid application development language for phones based on Symbian OS) and Visual Basic (available from AppForge), through Java and Python, to full-on native Symbian OS C++. The range is covered, in other words, for everyone from the hobbyist to the enterprise developer to phone manufacturers and commercial developers.

---

[1] In [Stroustrup 1994, p2].

Part 1

The Background to Symbian OS

1

Why Phones Are Different

1.1 The Origins of Mobile Phones

The first mobile phone networks evolved from the technologies used in specialist mobile phone radio systems, such as train cab and taxi radios, and the closed networks used by emergency and police services and similar military systems.

The first ever open, public network (i.e., open to subscribing customers rather than restricted to a dedicated group of private users) was the Autoradiopuhelin (ARP, or car radio phone) network in Finland. It was a car-based system, inaugurated in 1971 by the Finnish state telephone company, that peaked at around 35 000 subscribers [Haikio 2002, p. 158].

A more advanced system, the Nordic Mobile Telephone (NMT) network, was opened a decade later in 1981 as a partnership between the Nordic state telecommunications monopolies (of Denmark, Finland, Norway and Sweden), achieving 440 000 subscribers by the mid-1990s, that is, more than a ten-fold increase on ARP [Haikio 2002, p. 158]. Unlike ARP, a car boot was no longer required to house the radio hardware. Ericsson, and later Nokia, were primary suppliers of infrastructure and phones, helping to give both companies an early edge in commercial mobile phone systems.

Elsewhere, Motorola and AT&T competed to introduce mobile phone services in the Americas, with the first Advanced Mobile Phone System (AMPS) network from AT&T going public in 1984. European networks based on an AMPS derivative (Total Access Communication System, TACS) were opened in 1985 in the UK (Vodafone), Italy, Spain and France.[1] Germany had already introduced its own system in 1981. In

---

[1] See for example the company history at **www.vodafone.com**.

Japan, a limited car-based mobile phone service was introduced in 1979[2] by NTT, the not-yet privatized telecommunications monopoly, but wider roll-out was held back until 1984. A TACS-derived system was inaugurated in Japan in 1991.

All these systems were cellular-based, analog networks, so-called first-generation (1G) mobile phone networks (ARP is sometimes described as zeroth-generation).

The history of the second-generation (2G) networks begins in 1982 when the Groupe Speciale Mobile (GSM) project was initiated by ETSI, the European telecommunications standards body, to define and standardize a next-generation mobile phone technology,[3] setting 1991 for the inauguration of the first system with a target of 10 million subscribers by 2000. GSM was endorsed by the European Commission in 1984; spectrum agreements followed in 1986; and development began in earnest in 1987. GSM reflected a deliberate social as well as economic goal, that of enabling seamless communications for an increasingly mobile phone world as part of the wider project to create a unified Europe. The politics of deregulation was also an important factor in the emergence of new mobile phone networks as rivals to the traditional monopoly telecommunications providers.[4]

The first GSM call was made, on schedule, in Finland on 1 July 1991, inaugurating the world's first GSM network, Radiolinja. By 1999, the network had achieved three million subscribers, a ten-fold increase on first-generation NMT and a hundred-fold increase on ARP.

GSM rapidly expanded in Europe, with new networks opening in the UK (Vodafone, Cellnet, One2One and Orange), Denmark, Sweden and Holland, followed by Asia, including Hong Kong, Australia and New Zealand. By the mid-1990s, new GSM networks had sprung up globally from the Philippines and Thailand to Iran, Morocco, Latvia and Russia, as well as in the Americas and to a lesser extent the USA, making GSM the dominant global mobile phone network technology.

Through the 1990s, GSM penetration rose from a typical 10% after three years to 50% and then 90% and more in most markets (all of Europe, for example, with the Nordic countries leading the way, but with Italy

[2] A useful history appears at *www2.sims.berkeley.edu/courses/is224/s99/GroupD/project1/paper.1.html*.

[3] For a history of GSM see *www.gsmworld.com/about/history.shtml*, as well as [Haikio 2002, p. 128].

[4] Political events unfolding between 1988 and 1992, such as the pulling down of the Berlin Wall, German unification and the collapse of the Soviet Union, were also indirectly significant, for example in causing Nokia to refocus on the mobile phone market [Haikio 2002, Chapters 5 and 7].

and the UK not far behind). By the end of the decade, the USA and Japan were atypical, with the USA opting for a different technology (CDMA[5]) and Japan languishing at less than 50% GSM penetration.[6]

1.2   From 2G to 3G

Famously, 3G is the technology that the network operators are most frequently said to have overpaid for, in terms of their spectrum licenses. (Auctions of the 3G spectrum raised hundreds of billions various currencies globally in the first years of the 21st century.)

In the GSM world, 3G means UMTS, the third-generation standard designed as the next step beyond GSM, with a few half-steps defined in between including GPRS, EDGE (see [Wilkinson 2002]), and other '2.5G' technologies. In the CDMA world, 3G means CDMA2000. (In other words the division between the USA and the rest of the world persists from 2G into 3G.)

The significant jump that 3G makes from 2G is to introduce fully packetized mobile phone networks. (GPRS, for example, is a 'halfway' technology that adds packet data to otherwise circuit-switched systems.) The significance of packetization is that it unifies the mobile phone networks, in principle, with IP-based (Internet technology) data networks. Japan has led the field since a large-scale 3G trial in 2001 but, as of the last quarter of 2005, it seems that 3G has arrived 'for the rest of us', with the introduction (finally) of competitively priced 3G networks from the likes of Vodafone and Orange in Europe, opening the way for competition to improve the 3G network offering.

Disappointingly, in terms of services 3G has not yet found a distinct identity. But from the phone and software perspective, the story is rather different. Early problems with the greater power drain compared to GSM, for example, made for clunky phones and poor battery life. Those problems have been solved and 3G phones are now interchangeable with any others. From a software perspective, there are no longer particular issues. Symbian OS has been 3G-ready for several releases. (From a user perspective, of course, 3G is different because it is 'always on'.)

---

[5] CDMA, also known as 'spread spectrum' transmission, was famously co-invented in a previous career by Hedy Lamarr, the Hollywood actress. [Shepard 2002] provides a very approachable survey of telecommunications technologies. [Wilkinson 2002] is an excellent, mobile phone-centric survey.

[6] [Haikio 2002, p. 157] presents figures for mobile phone network penetration for 20 countries between 1991 and 2001.

1.3   Mobile Phone Evolution

Mobile phones for the early analog networks were expensive, almost exclusively car-mounted devices selling to a niche market. Equipment vendors sold direct to customers. Network operators had no retail presence and generated cash flow solely from call revenues. As the analog networks evolved into GSM networks, mobile phones were liberated from the car and the early car phones evolved into personal portable phones and then began to shrink until they fitted, firstly, into briefcases and, finally, into pockets. From around 1994, when GSM started to boom, mobile phones and perhaps even more importantly mobile phone network services began to emerge as potential mass-market products.

The iconic Mobira Cityman, introduced by Nokia in 1986, was the size of a small suitcase and, with its power pack, weighed in at nearly 800 grams [Haikio 2002, p. 69]. By 1990, phones had halved in size and weight and they had halved again by 1994, when the Nokia 2100 was released. It was the first ever mass-market mobile phone and weighed in at 200 grams [Haikio 2002, p. 160]. (It is credited with selling 20 million units, against an initial target of 400 000.)

As it happens, 1998, the year that Symbian was created, saw a temporary market reversal[7] but mobile phone uptake boomed again towards the turn of the millennium.[8]

The PC and mobile phone trend lines crossed in 2000 when mobile phones outsold personal computers globally for the first time[9] (by a factor approaching four: 450 million phones to 120 million PCs). This was also the year in which the first Symbian OS phone shipped, the Ericsson R380, followed in 2001 by the Nokia 9210. Neither were volume successes but both products were seminal. In particular, the Nokia 9210 instantly put Nokia at the top of the sales league for PDAs, ahead of Palm, Compaq and Sharp. (The Communicator was classified by market analysts as a PDA, partly because it had a keyboard, but also partly because Symbian phones really were a new category, and analysts didn't quite know what to do with them.) The death of the PDA, much trumpeted since (and real enough, if Microsoft's Windows CE sales numbers and the demise of Palm OS are indicators), probably dates from that point.[10]

---

[7] Nokia failed to meet sales targets; Motorola issued a profits warning and cut jobs; Philips canceled joint ventures with Lucent; Siemens cut jobs; and Ericsson issued profits warnings.

[8] Mobile phone telephony thus acquires something of a millennial flavor, see [Myerson 2001, p. 7].

[9] Market data for the period can still be found on the websites of market analysis companies such as Canalysis, Gartner, IDC and others, as can the subsequent wider coverage from news sites ranging from the BBC and Reuters to The Register.

[10] In Q3 2005, for example, PDA shipments fell 18% while smartphone shipments rose 75% year on year. See, for example, commentary at The Register, ***www.theregister.co. uk***.

Although in 1997 Nokia shipped just over 20 million mobile phones, in 2001 it shipped 140 million and the trends were broadly similar for other vendors. (Nokia was the clear leader with over 30% of the market in 2001, compared to second placed Motorola with closer to 14%.) Even so, numbers which looked astonishing in 2001 [Myerson 2001] look decidedly tame today. In 2005, global mobile phone sales broke through the barrier of 200 million phones per quarter, with year-end shipments of 810 million, close to 20% and shipments for 2006 rising a further 21%, almost touching the 1 billion mark.[11] Close to 40% of the sales growth in 2005 came from Eastern Europe, Africa and Latin America.

Against these totals, annual sales of smartphones at closer to 50 million in 2005 look small (which is why Symbian has begun to chase the mid-range market). Nonetheless, Symbian OS still leads the market, having doubled its shipments in pretty much every year since the company's creation. Thus, shipments in 2003 more than doubled from 2 million to 6.7 million; in 2004, they doubled again to 14.4 million; and in 2005 they more than doubled again, with almost 34 million Symbian OS phones shipped in the year (see *www.symbian.com/news/pr/2006/pr20063419.html*).

1.4 Technology and Soft Effects

Almost as astonishing as the raw numbers are the social and technology changes packed into little more than half a decade. The Nokia 7650, introduced in spring 2002, was a breakthrough product. The first camera phone in Europe [Haikio 2002, p. 240] with MMS, email, a color display and a joystick, the Nokia 7650 introduced the Series 60 (now rebranded as S60) user interface and was the first Symbian phone to sell in significant volume. Looking back, it is easy to forget how novel its camera was.

Not even five years on, the mobile phone seems well on the way to subsuming digital photography (the digital camera market began to shrink for the first time in 2005, although arguably that may indicate saturation as much as competition). It is an open question whether mobile phones will do the same to the personal music-player market.[12] Phones seem already to have subsumed PDAs. This is the principle of convergence; on the evidence of the market to date, given the choice between multiple dedicated single function devices or multifunction mobile phone terminals (as mobile phones are increasingly described), the market is choosing the latter.

---

[11] See *www.gartner.com/it/page.jsp?id=501734*

[12] Apple's Quarter 1 2006 sales numbers, for example, show a decline in iPod sales at the same time as the Nokia 3250 'music player' phone has hit 'triple platinum' (i.e. 1 million units shipped) within a single quarter.

It may not even matter what impact convergence has on existing markets. Broadcast TV, Wi-Fi, and VoIP[13] are queuing up for the role of newest hot mobile phone technology and seem likely to sustain continued growth, with or without markets such as the personal music player. (Digital terrestrial broadcast TV may yet prove to be the 'killer app' for the mobile phone.) What seems certain is that personalization has worked. Whatever the market drivers (and they are not necessarily the same in all markets), person-to-person communications have moved from their Victorian origins in fixed lines anchored to fixed locations, to what used to be the distinctly science-fictional model of ubiquitous mobile phone personal communications (something rather more like the Star Trek model).

Genuine culture shock accompanied the emergence of the mass mobile phone market, with its new habits and behaviors: people chatting into their phones in the street and breezily answering their phones in restaurants and trains, breaking the unwritten rules of public–private spaces and frequently meeting hostility in consequence. Similarly, the rapid rise of a 'texting' culture produced a predictable gap between those who did (typically young users) and those who didn't, with an equally predictable spate of newspaper scare stories. Today, these seem like reports from a world long gone. Looking back at the vision for the future mobile phone information society that Nokia began promoting from around 1999, it is remarkable how much of it has come to pass. The vision is spelled out in detail in [Kivimaki 2001].

Telephony has always had a sociological dimension, ever since the fixed-line phone shrank the world and collapsed time, making two-way communications between remote locations instantaneous. This is even more striking for mobile telephony. Again, it is easy to forget how completely in the UK, for example, the first brick-like mobile phones became the personification of the London 'Big Bang' deregulation of the City, of the Thatcher era and the Lawson boom, every bit as much as red Ferraris. (Local TV news reported at the time that motorists were buying dummy mobile phones, simply to be seen talking into them while waiting at the traffic lights, thus catching some of the Big Bang glow.) Again, the curious notion of the 'car phone' has left its legacy in the name of one of the UK's larger mobile phone retailers, Carphone Warehouse. (Elsewhere in Europe, where the sociology presumably was different, the brand is simply Phone House.)

The mobile phone is an astonishing product phenomenon. Not just the businesses of the phone vendors, but completely new operator businesses too have been built on the back of selling and serving the mobile phone. New business models have been invented from subsidies for phones

---

[13] Voice Over IP (VoIP) telephony uses non-dedicated IP networks to carry voice telephony traffic. Internet phone services such as Skype are VoIP-based as, increasingly, are discount packages offered by mainstream phone providers.

to pre-payment and the marketing of intangibles such as 'airtime' and 'messages'. Meanwhile some old business models have collapsed under pressure from the cannibalization of neighboring markets including fixed-line telephony.

It is easy to underestimate the depth of these 'soft' effects. The PC brought about several social revolutions: as the visible embodiment of the ubiquitous microprocessor, as the medium for the Internet, including email, and most recently as the medium for the web. Arguably, the mobile phone transformation runs even deeper, because it impacts public and not just private behavior. It has both caused and enabled new social uses (it has changed family relationships, enabled 'remote mothering' [Ling 2004, p. 43] and so on), as well as new patterns of behavior which have rapidly become the norm (it has changed the way much business is done, changed the way people set up meetings, and melted private–public distinctions). The mobile phone 'fits into the folds of everyday life' (L. Fortunati quoted in [Ling 2004, p. 51]) in a way that few other technologies have and the effect has been extremely powerful.

1.5   Disruption and Complexity

A strong theme of this book is that mobile phones are uniquely complex, both as devices and as products, and are therefore uniquely challenging from a software perspective. Of course many things are complex. Rockets are complex and so is the Internet, and so are corporate services, battleships and submarines. But mobile phones outdo them all in the complexity of the package.

Mobile phones are complex packages of multiple software functions (computing, communications and multimedia), hardware technologies (battery and power, radio, displays, optics (lenses), and audio), and fabrication and manufacturing technologies (miniaturization, online customization and localization, global procurement, and sourcing and distribution) which are sold globally in unprecedented volumes. They have moved from a niche market to the mainstream in two decades, with much of that growth in the last five years. They have been technologically, commercially, and socially disruptive.

The typical pattern of a disruptive technology is that it succeeds not by outperforming existing technologies (many disruptive technologies have in fact failed first time around as direct challengers), but by subtly shifting the ground on which it competes. Instead of competing like-for-like, it outperforms the incumbents on shifted ground, in effect skewing the existing market and creating a new, related and overlapping but essentially different market. It removes the ground beneath the old technology not by replacing it directly but by sidelining it, often by moving the market in an unprecedented direction. It is rather like adaptive evolution, in which an unproductive mutation becomes unexpectedly relevant and therefore successful because of a shift in the external context.[14]

Disruption is part of what makes it so hard to predict the future. WAP failed dismally in one market whereas i-Mode was a runaway success in another, but on the face of it both offer essentially the same service. The missing ingredient for success in the case of WAP was not a technology ingredient but a market or social one. Andrew Seybold says that i-Mode 'is a cultural success – not a wireless success' (quoted in [Funk 2004, p. 13]). While the analysis is probably only half true, it does make the point that the social and cultural dimensions of technologies cannot be ignored.

Arguably, convergence is itself a form of disruption. One reason to believe that the mobile phone will dominate at the expense of laptops, PDAs, digital cameras or dedicated music players, all of which are objectively fitter for a single purpose, is that while these devices may score higher on function (in their niche), they score lower on personalization and value as an accessory. Symbian OS does not itself count as a disruptive technology, but it is a vehicle for the disruptive effects of convergence.[15]

1.6   The Thing About Mobile Phones

Mobile phones are different from other devices for many reasons and most of those reasons make them more complex too.

- Mobile phones are multi-function devices.

- Mobile phone functionality is expanding at an exponential rate.

- Phone-related technologies are evolving at an exponential rate.

- Mobile phones are enmeshed in a complex and still evolving business model.

- Mobile phones are highly personal consumer devices (even when someone else pays for them).

In a word, the mobile phone difference is 'complexity' and the trend towards complexity appears to be growing at an exponential pace.

---

[14] Disruption is a widely discussed (and fashionable) concept, first identified by Christensen [1997] as innovative change for which the market is the trigger point (see [Tidd 2005, p. 29]. [Funk 2004, p. 4] has a simple definition. [Davila 2006] defines it neatly as 'semi-radical technology innovation'.

[15] Symbian OS is, of course, itself at risk from the disruptive business model offered by Linux.

Mobile Phone Hardware and Software

Baseband (radio 'modem') hardware is complex. In effect, the baseband hardware is a complete package in its own right, consisting of CPU, data bus, dedicated memory, memory controller, digital signal processors (DSPs), radio hardware, and so on.

The baseband software stack is complex too. Mobile phone protocols are complex and require real-time systems to support their signaling timing tolerances. Real-time support cannot be faked. A real-time operating system is required at the bottom of the stack to manage the hardware and support the layers of software protocols all the way up to the phone-signaling stack.

Treating the phone as a black box encapsulated by a communications protocol simplifies the software problem but has drawbacks in terms of both speed and capability. The power and speed requirements of the phone's hardware cannot be ignored.

Mobile Phone Applications

A typical Symbian OS phone has a complete application suite: phone book application, email and messaging clients, jotter, clock and alarm applications, connection and network setup utilities (not to mention web browser, camera support and photo album applications, video clip player and editor, and music player).

The application layer requires a full function graphical user interface (GUI) framework to support it, from widget set to full application lifecycle. Most (and probably all) of the expected applications also demand fairly deep system support from the operating system.

While Symbian OS staked its initial claim at the high end of the market, partly on the strength of its application support, the downward push towards the mass-market volumes of mid-range phones does not mean it has to do less. There are persuasive arguments that the mid-range is not defined by functional breadth (the range of available applications) so much as by functional depth (the size of the mailbox, the number of fields in a contact, and so on). Equally, critical factors such as performance are typically more demanding in the mid-range, where users have higher expectations that things 'just work' and lower tolerance for failure.

Convergence and Commoditization

Phone functionality is extending in every direction. Two-camera phones are becoming commonplace, true optical cameras have arrived (with Zeiss lenses, for example), as have phones with boom-box stereo speakers, a gigabyte of RAM and a built-in global positioning system (GPS).[16]

---

[16] Siemens and Mitac for example have both announced GPS-enabled GSM phones.

Device convergence is not a hypothesis, it is the reality. As discussed above, mobile phones have cannibalized the PDA market, appear to have eroded the digital camera market, and threaten other markets including the personal music-player market.

At the same time, new technologies and advances in existing technologies continue to be relentlessly absorbed into and commoditized by the mobile phone market. For example, Wi-Fi is causing the connection model for mobile phones to be reinvented, with hot-spot connectivity offering alternative network options. Meanwhile advances in storage media, from flash drive densities to micro hard-drives, challenge the use-case assumptions for mobile phones. From being the equivalent of snapshot cameras, they have become full video-recording devices; with internal memories of several gigabytes, they now compete with dedicated music players.

While the PC market, for example, has been essentially mature for a decade and now exhibits little more exciting than consolidation, the mobile phone market continues to be transformed by convergence and commoditization.

Services

Possibly the biggest difference between phones and other mobile devices is the integration of uniquely complex technology with uniquely complex business models. Phone services are almost as important in the product offering as immediate phone functionality.

Everything about the mobile-phone-network business model is complex, from spectrum licenses to roaming, to network subsidies for phones, to packetization of data and the interaction with legacy technology models, be they fixed-line telephony or radio and TV broadcasting or the Internet.

This complexity has its impact on the software in mobile phones, whether it is the requirement to support custom network services, to enable customized applications, or to be invisible beneath the top-line branding of networks and vendors (which leads, for example, to the demand to support custom user interfaces).

Open Platforms

Symbian OS sets out to be an open application platform, in other words a platform for which anyone can write and sell (or share, or simply give away), installable software, whether end-user applications and utilities or service and feature extensions.

Symbian therefore must provide the development tools and support (tool chains, support programs, compatibility guarantees and documentation, including books) needed by external developers to understand and

use the system, and to design and write stable and secure appl
run over various releases of the operating system and on vari
models, including phones from different vendors.

Open platforms are easy to promise and hard to deliver. Success can
present acute problems of scaling. Thus, for example, while vendors were
bringing to market only one or two models per year, it was possible for
third parties to test their applications on all available phones. Those days
are long gone, with the biggest Symbian licensees sometimes bringing out
a dozen or more Symbian-based models in a single quarter. Managing the
success of the platform means managing compatibility better; adopting
and adhering to open standards including tool and language standards
(standard C++, the ARM EABI, and so on); producing more and better
documentation; providing more developer services such as the Symbian
Signed program; the list could go on. In turn, these things can only be
achieved by creating a healthy ecosystem around the platform to increase
the overall pool of available resources and maximize the community
contribution.

User Expectations

Users expect and demand rock-solid stability and performance from their
phones; desktop computer performance standards are not acceptable.

At the same time, users are fickle, tending either to be infinitely happy
or infinitely unhappy.[17] When they are infinitely unhappy, they return
the phone. However, it is not always easy to understand precisely what
triggers happiness or unhappiness (the trigger often seems removed from
ordinary measures of good, bad and defective behavior). Desktop PC
users seem more likely to be either *infinitesimally* happy (the machine
has not crashed) or unhappy (it crashed but they did not lose much data).

The conclusion is that phones really are different from other systems
and they are complex.

---

[17] Thanks to Phil McKerracher for this idea.

2

The History and Prehistory of Symbian OS

2.1 The State of the Art

Symbian OS reached market for the first time towards the end of 2000, with the release of the Ericsson R380 mobile phone in November and the announcement almost immediately afterwards of the Nokia 9210 Communicator, which came to market in June 2001. Both phones were based on versions of what had previously been known as Psion's EPOC operating system. The final EPOC release was EPOC32 Release 5 (strictly speaking, the final version was the full Unicode build, designated ER5u). The first release of Symbian OS was therefore designated v6.0.

Since then, well over a hundred phone models later (the 100th model[1] shipped in early Q2, 2006) and with more than 100 million (and rising) cumulative unit sales, Symbian OS has undergone continuous evolution to keep pace with the rapidly changing technology in the market it targets: communications-enabled mobile terminals including, of course, mobile phones.

The latest release of Symbian OS is v9. In v9, and its precursor v8, dozens of new APIs offer access to services and technologies which in many cases simply did not exist when Symbian OS first launched.[2] Bluetooth support was one of the earliest additions (v6); Wi-Fi is one of the most recent (v9). Telephony support, meanwhile, has evolved from basic GSM and GPRS (in v6) to include EDGE (v7), CDMA (v8) and 3G (v8). Networking support including IPSec has been integral from

---

[1] The Nokia 3250 (also, as it happens, the first Symbian OS v9.1 phone to market) was the 100th model, reaching the shops in April 2006, soon followed by the Sony Ericsson P990, also based on v9.1.

[2] To name just the three most obvious examples, Java ME, Bluetooth and 3G networks did not exist when Symbian OS was first launched.

the beginning, evolving to a dual IPv4/v6 stack in v7 and enabling full Internet browsing on Symbian phones, with recent additions including support for VPN clients. New multimedia APIs (v8) support the high data rates required for two-way streaming and high definition interactive TV (DVB-H). The graphics system supports vector graphics (OpenGL ES in v8), with direct screen access and double resolution displays. The new platform security model (v9) enables the platform to remain open, but safe, with a signing service to support trusted application download. The list goes on.

The foundation for these latest services is the new real-time kernel (available in v8.1b and from v9), supporting the multiple fast interrupts needed for high data throughput, the latest generation of ARM processor architectures (ARMv6 is supported in v9) and single core phone designs.

The latest Symbian OS phones are full multimedia devices, including multimegapixel cameras with integrated flash and optical zoom, support for hot-swappable media cards up to 2 GB, MP4 (video) and MP3 (audio) players (supporting WMA and AAC too), 24-bit color (16.7 million colors), not to mention Wi-Fi, and Universal Plug and Play (UPnP, which enables remote control of compatible PCs, audio systems and TVs from a Symbian phone).

Having achieved its first '1 million phones shipped' year in 2002 (2.1 million Symbian OS phones were shipped that year, compared with 0.5 million the year before), Symbian OS achieved 1 million phones shipped in one quarter in Q1 2003, and 1 million phones shipped in one month in December 2003. Volumes have continued to rise steadily since then, and Symbian OS passed the 100 million phones shipped milestone in Q4 2006. Those numbers translate into close to 70% of the high-end, or smartphone, market according to independent sources.[3]

At the same time, competition has probably never been greater. In 2006, Linux phones are shipping in substantial numbers in Japan and China. Microsoft launched the latest version of its mobile phone platform, Windows Mobile 5, in late 2005 and phones based on it are now shipping.[4] Qualcomm has signed up European networks for the first time to support its (previously CDMA-only) Brew platform. And while the future of the Java-based SavaJe platform is uncertain, 'all-Java' phones remain a possibility.[5]

But, at the time of writing, the biggest volume of phones (i.e. across the whole market) are based on none of these platforms at all and remain the mid- to low-end phones based on vendors' own proprietary operating systems. In 2006, Symbian set its sights on addressing this market and its

---

[3] In fact, Canalys puts Symbian's market share at nearer 80% based on data for Q3 2006 (see **www.canalys.com/pr/2006/r2006102.htm**).

[4] Windows Mobile 5.0 is based on WinCE 5.1.

[5] To split hairs a little, SavaJe is not in fact 'all Java': the kernel and low-level system is written in C and the system layers are a mix of C/C++ and Java.

v9 releases will increasingly be aimed at scaling not just for the high end, where it is a proven platform for the latest feature-laden phones, but for the mid-range, mass-volume consumer market.

2.2 In the Beginning

In the summer of 1994, Psion was a company of perhaps 40 software engineers and as many hardware engineers, with a product line of handheld organizers that was highly profitable. The most recent was the Psion Series 3a, the second in the Series 3 family, a pocket-sized phone with a clamshell design sporting a letterbox format grayscale display hinged over a QWERTY keyboard, with an x86-family processor inside, up to 2 MB of RAM, removable flash memory cards and a ROM-based 16-bit operating system (named SIBO) for an all solid-state design. Its hardware design was not revolutionary but it was striking. Even more so was its built-in set of easy-to-use productivity applications. Supported by a dedicated, BASIC-like programming language called OPL, a thriving hobbyist community had established itself, self-organized (in pre-World Wide Web style; the first release of the Netscape browser appeared that same year) around bulletin boards and news groups and writing add-on software.

OPL was in fact a carry-over from Psion's original Organiser product line, which was also doing nicely, having been enthusiastically adopted as a stock control tool by UK high-street retailers such as Marks & Spencer.

That particular summer, the big project was a true Visual Basic clone (called OVAL) for the Series 3a, intended not just to increase the capabilities of the machine, but to open a bridge to the programming mainstream and tap the rich potential of the hobbyist programming market in BASIC for DOS and the Macintosh.

At the same time a much smaller project was also kicked off to create a next-generation operating system for the 32-bit devices which the company was already planning as replacements for the 16-bit Series 3 range as part of its strategy for retaining its lead in the handheld market. (In 1994, Palm had yet to release the Pilot; indeed it was still a software house, writing connectivity software for Psion's Series 3, among other things. Apple's Newton was a year old and genuinely innovative but had failed to find much of a market. Microsoft had not yet released Windows CE and the Hewlett Packard machines which were the nearest competitors to the Series 3 were based on MS-DOS and primitive in comparison.[6])

---

[6] In 1991, Hewlett Packard introduced the HP-95LX palmtop running MS-DOS and applications such as Lotus 1-2-3, with a 16x40 text display. It was improved to an 80x25 display on the HP-100LX in 1993 and upgraded again with the HP-200LX in 1994. Devices based on Windows CE, starting with the HP-300LX, did not appear until 1997.

The follow-on to the Series 3 was codenamed Protea,[7] and over the next year the project continued to grow. By the end of 1995 it was driving a rapid expansion of the company and in particular the project to create the new operating system (which was eventually named EPOC) was consuming the lion's share of the company's software development budget, although the Series 3 software remained in active development. For example, email and Internet extensions, in particular, were being prepared as it became increasingly clear that accessing Internet services from handheld devices was likely to become a significant market driver.

The Protea story has been told before [Tasker 2000, p. 14]. The brief was simple enough – create the next-generation successor to the Series 3, a more sophisticated 32-bit handheld to be called the Psion Series 5.[8] In this sense, then, the project was quite narrowly focused on creating the next successful product. But from the software perspective, the longer term vision for EPOC was explicit. The design brief called for it to support not just the explicit requirements for Protea applications, but the as-yet unidentified requirements for other future products. While there was as yet no talk of licensing the operating system, there was a long-term vision. The next generation, like the current generation, would be a family of products and there was an explicit intention that the software should aim for a design life of perhaps ten to fifteen years.

The Protea project delivered in the early summer of 1997. Like many complex software projects, it was late but not excessively so. However, somewhere along the way an interesting shift had occurred. By the time the all-new Psion Series 5 shipped in June 1997, the software side of the company had been spun out (as Psion Software, in late 1996) and the first licensee software projects had started.

The Series 5 was an outstanding industrial design, with a true tactile keyboard (on which you really could touch type) and a backlit touch-screen with an ingenious hinge that ensured the device remained stable when used with a pen in touchscreen mode.[9] (Competing, non-Psion products tended to fall over backwards when the screen was pressed.) A CF card slot was provided for expandability and, best of all, the Series 5 seemed to run forever on two AA batteries. As for the software, it rapidly acquired a reputation for extreme usability and legendary robustness (after some natural early teething troubles).

The Series 5 was a best-seller though, quite probably, it did not sell as well as its predecessor, the Series 3. (In its lifetime of five years of

---

[7] Protea is the name of a flower native to South Africa. As it happens, Psion founder, David Potter, and the first two CEOs of Symbian, Colly Myers and David Levin, all share a connection with Southern Africa.

[8] Actually, as David Wood recalls, for a long time it was assumed it would be called the Series 4.

[9] Credit for the Series 5's famous hinging clamshell case goes to Martin Riddiford of the Therefore design consultancy.

Figure 2.1   The Psion Series 5 MX

production, the Series 3 is thought to have sold more than 1.5 million units.[10] The Series 5 and its immediate successors including the Revo, had a lifetime of four years of production, during which it sold probably around a million units (see Figure 2.1).

The EPOC team had started with a clean slate, but the operating system did not come out of nowhere. Many of the ideas had been tried, tuned and proven in one or more, sometimes all, of the previous systems. Clean and 'from the ground up' it may have been but it was nonetheless a from-the-ground-up rewrite of the 16-bit operating system for the Series 3, which in turn was a from-the-ground-up rewrite of the second-generation 8-bit operating system for the Organiser II. (The first-generation 8-bit system for the Organiser I had only rudimentary operating system features and was, in effect, written straight to the metal.)

While, by any measure, the new operating system was written remarkably quickly,[11] the fact remains that operating systems gestate slowly and cost years of effort to create.[12] Counting from the first Organiser systems, Psion had already invested a dozen years in operating system development when the Protea project began. Planning for a design life of at least as many years for the new operating system was a matter of basic commercial common sense.

It is likely that, had Psion had been a pure software company (or just a larger and more mature company), a from-the-ground-up rewrite,

---

[10] See ***http://3lib.ukonline.co.uk/historyofpsion.htm***.

[11] Martin Tasker puts its development time at 3.5 years and its cost at £6 million [Tasker 2000, p. 15].

[12] There have been some interesting attempts to quantify the development cost of Linux (see for example the article by David Wheeler at ***www.dwheeler.com/sloc/redhat71-v1/redhat71sloc.html***).

let alone one using a new and unfamiliar, object-oriented language, would not even have been considered, let alone allowed to complete. The business logic would almost certainly have favored extending the existing system and Psion very likely would have missed its moment. But Psion was not a software company, nor did it really think of itself as a computer hardware company; it was a product company, driven by a whole-product vision. And what's more, it had enough cash in the bank to do what it liked in pursuit of that vision. Which is just what it did.

2.3   The Prehistory of Psion

Psion started life distributing computer hardware but moved quickly into software, capitalizing on the pre-PC microcomputer boom, writing games and then office applications for machines such as the Sinclair ZX81 and Spectrum. It was a small-scale operation in a mews behind Gloucester Place in Marylebone, London. Early hits included a flight simulator for the ZX81 and a spreadsheet application for the Spectrum. The flight simulator was written by Charles Davies, an early director of Psion, later the Chief Technical Officer of Psion and now of Symbian. The spreadsheet application was written by Colly Myers, another long time Psion director, later CEO of Psion Software and Symbian's first CEO. The legend has it that Myers wrote the complete application from scratch in the course of a single flight from Johannesburg to London.

Few people still in Symbian can trace their roots back quite so far, but someone who can is Howard Price, now a senior system architect at Symbian, who joined Psion in 1983 with a math degree, having settled in London after traveling, largely to avoid returning to military service in (pre-democratic) South Africa.

Howard Price:

In 1983 we were only doing Spectrum work, mainly games. We all sat in a row along a workbench, about eight of us, with Charles Davies at a desk near the stairs and David Potter in a little office at the end. And everybody programming in assembler, pretty much, Z80 assembler for the Spectrum, which we wrote on some HP-type machine. Downstairs were all the boxes containing our programs on cassette that had come back from Ablex, the mastering company. Once a week, a truck would arrive and everybody would line up to throw the boxes onto the truck.

Charles Davies had joined the company at the invitation of its founder, David Potter, after completing a PhD in computational plasma physics. Potter had been his thesis supervisor.

Charles Davies:

I was programming 3D models in Fortran and then I left and joined him. There were Fortran programmers who didn't know the length of a word on the CDC machines we were using, but I always had an interest in programming, so I learned assembler. And then I went to microprocessors and I programmed a lot, first Pascal and then C. I learned C on the job at Psion.

Psion bet heavily on the success of Sinclair's follow-on to the Spectrum, the Sinclair QL, developing an office suite application. It was badly jolted when the machine flopped and the software didn't sell. The surprise success which emerged at around this time was not software at all but hardware: the original Psion Organiser, the pet project of the company's single hardware engineer.

The Organiser launched Psion as a product company and manifested what became its signature traits: carefully designed hardware products whose very modest means were maximized by great software. The games had made money but the devices rapidly became the soul of the company.

Charles Davies:

We had a product vision for the Organiser. It was an 8-bit device and the software was written straight above the bare metal, so there was no operating system. The first ROM was 4 KB, subsequently 16 KB and we had a programming language in there called Forth, even in the 4 KB, because Forth is that tight that you can do that.

The success of the Organiser put enough money in the bank to fund development of a second-generation device.

Charles Davies:

The Organiser II had the luxury of a 16 KB ROM. The first product didn't have any serial port and people wanted to add barcode readers and things, so we added a serial port. But then you had to write add-on software to talk to the serial port and Forth wasn't up to it. So OPL was invented at that time, a BASIC-like language. And because of that we ended up having to document certain library routines for extending the software in the ROM. That introduced us to the idea that when we go to the next generation we need an operating system proper. We need to separate the applications from the system, because we had library routines, but no operating system separation.

A few months after Howard Price joined, the original HP machine on which Z80 software was developed was replaced by a VAX, at the time a huge investment for what was still a very small company. When the VAX arrived, development moved from assembler to C. It was a big and exciting new language for the Psion programmers. Programs were written and cross-compiled on the VAX for the Z80 chip. But C had not been the first choice of programming language when the VAX arrived.

Charles Davies:

When we first got the VAX and decided to go from assembler to a high-level language we chose Pascal, which was the system language for the Apple Macintosh at that time. That choice lasted a few months. But then we read Kernighan and Ritchie and it was just obvious that this is a whole load better than Pascal for what we wanted to do. We recognized that C was the right language for us, because with C, you know, 'how low can you go?'. We recognized that in C, we could do the sorts of things that using Pascal we had to do in assembler. But that switch wasted money. At that time, a VAX cost £100 000 and you work out what that is in real terms now, it was a big investment and the compilers cost thousands. So we bought Pascal compilers and wasted a whole load of money and had to re-buy C cross-compilers.

Despite its expense, the VAX turned out to be a fortuitous choice. The VAX ran DEC's VMS operating system. That the early influence of VMS should eventually show up in an operating system which has become best known as a mobile phone operating system is startling at first sight. VAX, after all, was the dominant mini-computer of the late 1970s and pre-PC 1980s and VMS is in many ways a dinosaur of the big-metal era. But Symbian OS traces a very specific legacy to VMS, which indeed goes all the way back to the first Psion operating systems for the Organiser products.

2.4   The Beginnings of Symbian OS

When Geert Bollen joined Psion in May 1995, the 32-bit operating system project (EPOC) was well under way. Bollen had been in the UK for just six months, after moving on from a Belgian startup which had folded. A Macintosh-only shop with strong university links and specializing in document management systems, Bollen found Psion instantly quite different, 'a little bit homegrown', as he puts it.

The EPOC project was dominated by a few key personalities and largely divided up between the teams they had gathered around them. Colly Myers was responsible for the kernel and base layers, Charles Davies for the middleware and David Wood for the user interface. All were directors of Psion, and later Psion Software and Symbian, as was Bill Batchelor who had taken over the running of the overall Protea project.

Geert Bollen:

Most of the architecture came ultimately from the interaction between Charles Davies and Colly Myers and the creative tension between them. And sometimes it could take a long time to settle something. So they would have an argument and then out of that they would come up with something sufficiently rich for Charles, sufficiently doable for Colly and then an implementation would appear.

And once the implementation had appeared, that was very much that. As Bollen puts it, it could be extremely difficult to influence the implementation after the fact. Myers and Davies had styles which were as different as were their personalities.

Geert Bollen:

Charles Davies the cerebral purist, Colly Myers the bull-dog-like pragmatist, 'We are building the system and I have to know what to build!' So to give you an idea, Colly was off implementing the system, meanwhile Charles was masterminding a complete Rational Rose model for it.

Martin Tasker was another early recruit, joining a few weeks after Bollen when the software team was around 30 strong. Tasker had been at IBM, working on System 370 mainframes, having studied computing at Cambridge.

Martin Tasker:

I joined on 19 June 1995, 180 years and 1 day after the battle of Waterloo and 6 weeks before my wedding! There was fantastic intellect and purity of design. I think the atmosphere was really quite frontier. You got a senior position, I say 'senior' but this was a very small team, but you got authority within that team by basically stating your opinions and being seen to have good ones.

There was Colly Myers, in those days with a beard, who would sit there talking and he just couldn't keep still, he would be twitching.

And there was Charles Davies, face raised at a Victorian angle.

Commitment was total and the dynamic could be abrasive. Strongly held opinions, strongly defended, were part of the culture.

Martin Tasker:

Colly Myers was a combination of charismatic leadership and utter frustration!

Tasker indeed recalls an incident from the early days of the project, a small but significant difference of views between Myers and Davies which turned, as such differences usually did, on weighing the balance between purity and pragmatism.

Martin Tasker:

Colly Myers had a theory that array arguments should be unsigned, which meant that a descriptor length should be an unsigned value, in other words a `TUint` argument. And I well remember a meeting in the first floor corner office of Sentinel House, which was the operations room for the project and Charles Davies' exact words were, 'I'm having a bit of trouble with the troops'. They were unhappy about using unsigneds, because there are all kinds of things you might legitimately do when manipulating descriptors which would result in a signed value as an index, for instance you multiply by some signed value. So that plays havoc with the array, because you're getting all `fffffs` as an index.

And what this triggered in Colly! 'Are you mad?!' You know, 'Who are these programmers of yours?' So the abstraction is that an array has only a zero or a positive index, it cannot have a negative index. Therefore, what possible advantage could there be in allowing a signed index? So Colly's line was, 'You just need to teach your programmers how to program!'

Well I remember four weeks of totally polarized debate and heated argument. Of course Colly was right, but the fact is that it's just too hard to do the right thing. So I walked in one Monday morning and checked out a release note, this would have been October 1995, it just read 'As agreed, changed all `TUint` to `TInt`.' Of course we kept `TUints` for flags and such – but as for numbers, that was how that debate got closed. And Colly's 'agreement', when it came, was characteristically unilateral, announced through the release note after a gruelling weekend's work which couldn't really be automated – Colly really had to check every change manually.

The relentless development pace and constant project pressure were hard, too. Peter Jackson, who these days is responsible for Symbian's software configuration management systems, remembers the approach to project management, as directed by Bill Batchelor, with mixed feelings.

Peter Jackson:

Bill Batchelor liked getting his hands dirty, but then he became project manager for Protea and so he had a dual nature. He was passionate about the right things, but at the same time he didn't like it when you told him how long it would take to do 'the right thing'.

But the company was vibrant and small and people didn't actually mind hacking away for all hours of the day and night to achieve the end goal. Bill's over-optimistic project plans were just part of that mix. He would cajole you into committing to something that was impossible, and you'd do your best to achieve it, and then eventually there would be this undercurrent where everybody knew it was totally absurd, but no one was going to say it. Eventually it would all come out. And then there would be another planning iteration and the same thing would happen again.

The big practical problem with the Protea project, one which caused a succession of headaches for Geert Bollen, was the lack of real hardware. Software development started well ahead of the availability of any prototype hardware but even by mid-1995, when the software project was in full swing, the device prototypes were still not ready.

Geert Bollen:

The on-the-metal version of the kernel was started and delivered after I arrived and Colly Myers assembled a team for that. Before that Colly had been a one-man band. The GNU tools at that time were coming on. I had some involvement in that but they were still a long way from being rolled out.

Andrew Thoelke is another veteran who joined in March 1994 and is now the Chief Technology Architect at Symbian for the base services and kernel layers of the system. In the absence of hardware prototypes, built code was run on PCs using an emulator layer which mimicked a full system by mapping low-level operating system calls to their Windows equivalents, essentially the same approach used by Symbian OS developers today in the first stages of development, before moving to hardware targets.

Andrew Thoelke:

Down in the base team, not having hardware was a problem, so the system was first brought up on x86 as a hardware port before it was ever brought up on ARM. In the original kernel architecture, probably 40% or 50% of code is shared with the target, but there's still vast amounts of kernel code which is target only, all of the scheduling and threading, the interrupt model, the device-driver model, so all of that needed to be done with a real target. So they used a 486, they basically built an 8386 port of the system first, because that brought online another 40% of the kernel code. Obviously there was still ARM-specific hardware code and a different MMU and all that sort of thing. But it was actually much less work when hardware did become available because they had already got a generic kernel mostly working.

Whatever the problems, there was no doubt in anyone's mind that what they were creating was special. Martin Budden, now Chief System

Architect at Symbian, was a veteran of the two 16-bit projects before moving onto the EPOC project. He puts it very simply.

> **Martin Budden:**
>
> I came to the company because I wanted to do something that was exciting. As soon as I saw what Psion was doing, I just knew that was what I wanted to do.

Looking back, Psion's timing was good; it had judged the moment perfectly.

> **Martin Tasker:**
>
> Psion, like many companies then and not just in Britain but elsewhere, had achieved success by innovating according to rules which nobody had ever written. It just did its own thing and it found a niche in the market.

2.5   The Mobile Opportunity

When the Psion board decided to spin off its software division, which at that time numbered around 70 engineers, it was effectively a public commitment to a software-licensing strategy.

It is clear that a number of different options were considered. There were rumors at about that time that Psion had considered buying Palm. A possible purchase of Amstrad got as far as due diligence. The background is revealing. For the Psion board, the real target seems to have been a Danish phone-making company, Dancall, which Amstrad had previously bought and absorbed into its empire. Thus, buying Amstrad would have enabled Psion to become a phone manufacturer. This indicates very clearly the direction in which Psion was pressing at that time. In the event, Psion did not buy Amstrad and Dancall was eventually sold to Bosch, before being sold on to Siemens. Much later, it was the formerly Dancall site at which the Siemens Symbian OS phone, the SX1, was developed. Still more recently, the site has been sold on to Motorola.

Psion, of course, did make its move into the phone market, but in a quite different direction. It was a visionary move and one for which the company founder David Potter deserves enormous credit. There were other visionaries too. In particular, Juha Christensen, Psion Software's bravura marketing director,[13] had assiduously begun to cultivate mobile phone manufacturers, Nokia included. Psion Software was certainly not their only choice of partner for collaboration at the time (just as Symbian is by no means the only choice today). However, the company was perfectly positioned, with just the right product at just the right time in the evolution of the mobile phone market. It has succeeded remarkably well in extending that early lead into a commanding position in the market.

2.6   Background to the First Licensee Projects

The first Organiser shipped in 1984. Over more than ten years, Psion honed its hardware and software skills and learned through three complete iterations (Organiser, Organiser II and Series 3) what it took to create a complete software system for mobile, battery-powered, small-footprint, ROM-based systems, before embarking on the 32-bit EPOC operating system from scratch. The Series 5 shipped in June 1997. Almost exactly a year before, Psion's software division had been spun off into a separate company, Psion Software. Almost exactly a year afterwards, in June 1998, Symbian was created as a joint venture aimed at bringing EPOC as a new operating system to mobile phones.[14]

Even before the Series 5 project completed, licensees of Symbian OS from at least three companies were waiting in the wings. There are different versions of the story, but they all agree on the main points.

Martin Budden:

As I heard the story, Nokia were in the market for a new operating system for their Communicator and they approached us. I know that Juha was instrumental in brokering the deal, but it was Nokia's idea and I remember there was a time when we were told Nokia were coming to see us. It wasn't exactly 'smarten up the office', but you know, 'if they ask questions, give good answers'. It was Nokia that was strongly in favor of bringing in other phone manufacturers to form a consortium, or that's what I understand. They fairly quickly brought Ericsson on board and then Motorola got on board at the last minute, and that was also quite significant.

---

[13] The legend within the company when I joined in 1997 featured Juha going off to cold Northern climes, sharing saunas and vodka with Important People and coming home with a Nokia deal in his pocket. Juha was later tempted away to Microsoft to lead the Windows Smartphone effort.

[14] [Tasker 2000, Chapter 1] provides a definitive history.

'Symbian Day' was June 24th 1998 and the Psion share price rocketed on the news (causing much excitement in the office over the next few days). A few days before, we had delivered the first free SDK for what was still Release 2 of EPOC.[15] Version 5 was still a whole year away and the first Unicode release, ER5u as it was known, was a step further still.[16] This, arguably, was Symbian OS 'version 5' (although it was never called that), the first operating system release that was 'fit for phones', although even then the first phones were still a year away from market. The first designated Symbian OS release, v6, appeared in Spring 2001.

From a public perspective much was made of the Nokia versus Microsoft angle and some commentary viewed the creation of Symbian as an attempt by Nokia to build an alliance against Microsoft. But it seems just as meaningful (and more useful) to view it more as a case of Nokia making a shrewd move to work with competitors to consolidate and grow a new market (that of highly capable, multi-function, phone-enabled terminals: 'smartphones' in other words), at the same time enabling Nokia to focus on what it clearly saw as its strength, the user interface. The evidence [Lindholm 2003] is that Nokia viewed the user interface as the critical software design factor for the phone market – if not the key determiner of success then at least a critical one – and also that it viewed the user interface as its critical strength.

However, even before the Nokia approach, Psion had been actively evolving its own strategy and there is no doubt that a fundamental shift occurred after the start of the Protea project, leading to the spinning off of the software division to open the way for software licensing. The team led by Howard Price moved across quite late to the EPOC project to start what turned out to be the last rewrite of OPL, this time for the 32-bit platform. By then the company's focus had shifted quite noticeably.

> **Howard Price:**
>
> The big thing in every team meeting was, 'Where are we with licensees?' So we'd go to the senior team brief and a lot of the talk would be about winning another licensee, or that the licensees were getting unhappy because of this delay or that delay, or that they were worried we were delaying their products to concentrate on Psion work – the Series 5 project was running worryingly

---

[15] Giving SDKs away to developers was still considered controversial within the company at that time.

[16] ER5u was an interesting experience to live through, a complete rebuild of a system which still, at that time, did not routinely build from source (as it does these days, with nightly builds of multiple variants from a single master codeline), with the 'wide' flag set for all components in the system so that all descriptors, text data and resource strings (anything with text, in other words) built 'wide' using multibyte (UTF-8) Unicode text encoding. A complicated system of 'baton passing' was evolved to follow the dependency graph up through the system and ensure that for every component, all dependencies built first; not trivial in a system which still harbored some awkward circular dependencies.

late. It took a year longer than planned for us to ship and the licensees were
waiting.

The licensee strategy was squarely pitched at the phone market.

Andrew Thoelke:

The Series 3 family had been Intel based, but even at that point in '94, '95
the view was to migrate towards mobile and cellular applications. The jump
to ARM was intentional, because Intel was clearly not a player in that field
and ARM was already doing well and had ambitions to become far more
important in that space, so it was quite a strategic move. And part of the
mindset behind the next generation operating system was to target ARM. Even
at that point David Potter could see that handheld computers and PDAs and
cell phones would converge. And that's why in '96, before the Series 5 was
actually shipped, Psion put its software division out into a separate company,
specifically so that it could look at licensing its software externally.

The very first of those early collaborations was a project to create
the software for a mobile companion device for the Philips Ilium phone.
The companion and phone clipped together back to back and connected
through a hardware slot on the back of the phone, turning it into a
PDA/Communicator with 4 MB RAM, a Series 5-sized landscape-mode
touch screen, a choice of soft (on-screen) keyboard or handwriting
recognition and a full PDA application suite including calendar, organizer
and contacts book. Communications functions included email, web, fax,
SMS and full voice calling.[17]

The software was based on the November 1997 Message Suite release
of EPOC (also used in the Series 5 mx), which added email and web
applications, dial-up networking and TCP/IP, the C Standard Library and
the Message Suite itself. The project was publicly announced as the
Philips Ilium/Accent and showed at CeBIT at the end of 1998, but it never
came to market.

Martin Budden was the technical lead on the project, which involved
writing not just a bespoke user interface, but also a complete applications
suite including messaging and contacts applications. As he says, it was
a significant amount of work. However, compared with later projects,
these were very much toe-in-the-water exercises, both for Symbian and
its licensees. On the Symbian side, the team was relatively small, perhaps
a dozen developers working on the user interface and applications and
half as many again working on the software port to new hardware.

---

[17] The Ilium is described at ***www.noodlebug.demon.co.uk/goingmob/spphiila.htm***.

The Philips device was the first licensee project (unless Psion itself is counted as a licensee) but, most significantly, it was the first project to generate licensing revenue in the form of pre-paid royalties.

Martin Budden:

The Philips project was the first bit of money that we ever got in from licensing; the first licensed product we ever got and made some money from.

Other licensee projects followed, including the Series 5 look-alike Osiris from Oregon Scientific and the Geofox One, a keyboard-based PDA with a larger keyboard and screen than the Series 5 and with a laptop-style touchpad instead of a touchscreen. It also added a built-in modem and a standard Type II PCMCIA slot.

While they demonstrate the enthusiasm with which Psion Software set out to develop a licensing model, all of these projects were ultimately false starts, failing to capture much attention from the market. Straight after the Philips project, Budden moved onto another small licensee project, working on behalf of Ericsson, and stayed as technical lead through the project startup. While the biggest problem in the Philips project had been trying to work around the limitations of the hardware design, which had been more or less fixed before the start of the project, the Ericsson project was a true phone design. In particular, the hardware design provided a robust solution to the problem of communications between the phone hardware and the application processor. As Budden says, the feeling on the team was that the hardware design was right from the start.

The fact that it was another phone project indicates where Psion saw the market opportunities, but it also indicates the direction in which the licensees saw the phone market moving. The goal of the Ericsson project was to create a mobile phone with full PDA functions, as full as was then possible. The result was the Ericsson R380, a breakthrough product not because it sold particularly well (it was probably too far ahead of both the market and the current state of technology) but because it rehearsed key principles which led the way to later successful Sony Ericsson Symbian phones, starting with the P800 and followed up by the highly successful P900 family of phones.

Biggest by far of all these projects was the collaboration with Nokia to create the Nokia 9210 Communicator, which started while the Philips project was still running. While the Ericsson R380 team had roughly double the numbers of the Philips project, the Nokia 9210 project eventually involved probably half the company; by the time it completed, the company had grown from 70 to over 200.

The Nokia 9210 project completed after that of the Ericsson R380, but began before it. Earlier projects and, to some extent the Ericsson

Figure 2.2   The Ericsson R380

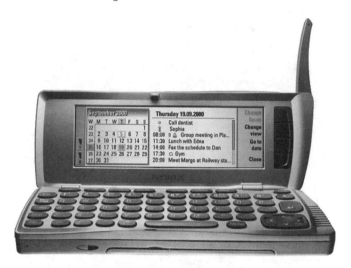

Figure 2.3   The first Symbian phone, the Nokia 9210 Communicator

R380, had been based on snapshots of the evolving operating system (which was still named EPOC), with the deepest changes concerned with the adaptation to new hardware and bespoke customization of the user interface. In contrast, the Nokia 9210 project drove a complete iteration of the operating system from the ER5 baseline to what became known, finally, as Symbian OS v6.0. Conceptually, the Nokia 9210 was the first Symbian phone, even though it wasn't the first to market (see Figure 2.3).

The transition from the Series 5 to the Nokia 9210 was less a series of steps than a route march, four years of hard work (from inception to completion). Symbian OS has been (and will no doubt remain) a continuous evolution towards a destination which is always one twist of the road away.

2.7   Device Families

The native EPOC graphical user interface (GUI), which defined the look, feel and interaction style of the device software, was known as Eikon. Eikon was designed for extensibility and customization. However, the extent of the variations required by different customers, driven by the needs of devices that, increasingly, were not PDAs but phones with PDA functions, significantly exceeded the assumptions of the original design. Each project effectively created a complete bespoke user interface, albeit from the common starting point of the Eikon code. Not only did this level of customization not scale, it was clearly threatening to fragment the platform.

> **Martin Budden:**
>
> The model of doing a bespoke user interface was already there. We did a bespoke user interface for Philips and for the Ericsson R380. And for the Nokia 9210 Communicator, again there was a new user interface to Nokia's specifications. But there were fundamental conflicts between these user interfaces, in practical terms of 'Did they have pens?' or 'Were they keyboard based?' and 'What was the screen size?', but also in deeper terms of the whole user interface philosophy and what you expose to the user. And it just became clear that if we did a user interface for every single phone, that wasn't going to be sustainable.

Symbian's solution was the so-called reference design strategy. The specific phone types were genericized to reference specifications: in practice, that meant the keyboard-based Communicator-style device and the pen-based 'smartphone' equipped with PDA functions and based loosely on the Ericsson R380. As well as a form-factor definition specifying the essential features of the physical design and therefore, in effect, parameterizing each design to a particular market point (in terms of features, size and key use cases), Symbian would supply a generic user interface for each form-factor, which licensees would then customize.

As realized in Symbian OS v6.0, devices were identified as 'smartphones' (phone form-factor devices) and 'communicators' (PDA form-factor devices). Communicators were further divided into keyboard-based (the Nokia 9210) and tablet-based devices (both Ericsson and Sanyo showed off prototypes broadly similar to Palm or Windows CE devices such as the Pilot and iPaq). Two reference user interfaces were included in v6.0 as 'device-ready' designs: Crystal, which shipped with the Nokia 9210 and which eventually became Nokia's Series 80 user interface; and Quartz, which eventually evolved into UIQ.

A number of other device family reference designs (DFRDs) were proposed and several proceeded to reasonably advanced specification, including Sapphire which was split into Red and Blue variants, depending on screen size and interaction mode (pen or keypad); Ruby, which evolved from Red Sapphire; and Emerald, which encapsulated the original smartphone concept as realized in the Ericsson R380. Neither Ruby nor Emerald were announced or came to market. Blue Sapphire eventually evolved into the Pearl DFRD and finally reached market branded as the Nokia Series 60 user interface (see Figure 2.4). Pearl had first been defined as a 'headless' DFRD (without a user interface), before acquiring code branched from Crystal and eventually unifying with the work which had been going on independently within Nokia to develop what was known as the 'square' user interface.[18]

Pearl in effect became the first true smartphone platform (defined as a phone with information capabilities) and was realized in the first Series 60 device, the Nokia 7650.

Quartz, meanwhile, never came to market in its original form, that of a tablet-style device most closely resembling a phone-enabled, pen-orientated PDA, which was dubbed the Mediaphone reference design when prototypes were shown at CeBIT in 2001. The Quartz design had originated at Symbian's Ronneby site in Sweden. Originally an Ericsson development laboratory specializing in Windows CE devices, the site had been transferred to Symbian as part of the original Ericsson investment in the consortium. Quartz quite clearly inherited Ronneby's design legacy. However, the device format with which Quartz did eventually come to market, by this time rebranded UIQ, was the one pioneered by Ericsson with the R380: pen-operated, a screen that could switch between portrait and landscape modes and with a key-pad flip. In, first, the P800 and then the P900 (see Figure 2.5), this format has become a signature design of Sony Ericsson's high-end, business-orientated range and has been extremely successful.

The Crystal user interface of the Nokia 9210 Communicator was eventually rebranded Series 80 (see Figure 2.6) and remains the basis for the product line which continues to evolve and innovate. (A Communicator was the first Symbian phone to offer Wi-Fi connectivity, for example.)

Figure 2.4   The Series 60 user interface, as used on the Nokia 7650

---

[18] David Wood believes that Nokia's work on 'square' had been in progress for at least two years before the formation of Symbian.

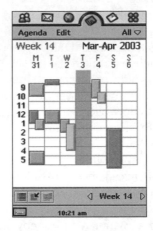

Figure 2.5   The UIQ user interface, as used on the Sony Ericsson P900

Figure 2.6   The Series 80 user interface, as used on the Nokia 9500

While a number of licensees worked on Quartz devices and others (besides Nokia) expressed interest in Crystal-based devices, the DFRD strategy eventually stalled. The reality is that, although they aimed to be generic, the designs could not escape the pull of licensees. In the end, they were more licensee-specific than generic, reflecting particular licensee's views about what a phone should be. Nokia drove Crystal; Ericsson and then Sony Ericsson drove Quartz; and Sapphire seemed to split and split again, until there was a one-to-one mapping between DFRDs and licensees.

Martin Budden's view then and now is that the problem was essentially not resolvable. It was not possible to agree on a Symbian-based user interface, in other words one evolved from the original Eikon GUI of EPOC and the Series 5, which was suitable for the different device visions of Nokia, Ericsson and Motorola.

And this was the problem. Each phone vendor had different, deeply held views about what makes a phone. Symbian was trying to create a software platform that would satisfy them all. Motorola wanted a pen-based user interface. Nokia wanted a keyboard-based user interface. Nokia did not place much value in the power of the pen and to date have only ever released one pen-based range of Symbian OS phones, the Series 90 user interface on the 7700 and 7710. Quartz, coming from a design unit which had started out working on Microsoft Windows CE devices, came up with a pen-based tablet format, such as the Compaq iPaq or the classic Palm devices.

Not everyone in the company was convinced by the DFRD strategy. To some it seemed more like an attempt to paint over the deeper problem, that conflicting licensee user interface requirements were irreconcilable.

Martin Budden:

In my view, we just could not resolve the issue. We couldn't come to an agreement on what the Symbian-based user interface would be that was suitable for all licensees, which I think was ultimately why Nokia went off to do their own.

The designs were not so much generic as licensee-specific, reflecting each licensee's views about what a phone should be.

Martin Budden:

The DFRD idea was to have families of user interfaces, so there would be one family for devices like the Nokia 9210 Communicator and the Series 5; there would be a family that was based on the Ericsson R380 which was a phone; and at about this time, Quartz started up and that was for another Ericsson phone with quarter-VGA tablet form-factor. There was a basic conflict there between Nokia and the Series 5 user interfaces, because what Nokia wanted went further than the customizations the user interface could easily accommodate and then the other conflict that started to manifest itself was a user interface for a phone form-factor smaller than even the Ericsson R380.

After his time on the Ericsson R380 project, Budden had moved across to work on Quartz as technical lead and spent the best part of a year commuting between London and Ronneby. He witnessed the difficulties of implementing Quartz as a product-ready, concrete instance of a DFRD at first hand. Partly the problem was one of resourcing, with the company's main focus dedicated to the underlying features of the operating system, many of which were driven by the ambitious requirements of the Nokia 9210 project. The fact that Quartz was being developed at a remote site did not help. Nor was it necessarily easy for the Ronneby engineering team to adapt itself to the very different style of Symbian, of which it was newly a part, compared with its previous parent, Ericsson. For its part, Symbian probably found integration of a new remote site just as painful. Beneath it all, were the basic engineering problems.

Martin Budden:

The Quartz team had very great difficulty getting anything done which would support their work and that led to a lot of fragmentation and reimplementation and it also highlighted that there was a lot of code that was not easily separable. So, for example, the messaging code had user interfaces in the engine layer, which meant that to change the user interface we had to redo a lot of the engine code as well. So there were a lot of things that made it difficult to separate out the bits. And this is when the idea of resolving all these conflicts by defining DFRDs emerged. I thought at the time that it was never going to fly.

However, the strategy served its purpose as, between 2000 and 2002, it enabled the important focus to become that of developing the underlying operating system. It also underwrote the splitting of generic from specific functionality in Eikon, the original EPOC GUI, as a necessary engineering step to enable the creation of the DFRD variant implementations.

When it abandoned the DFRD strategy in 2002, Symbian made a tactical retreat out of the user interface space altogether. The Pearl DFRD which was being actively developed in collaboration with Nokia, was taken over entirely by Nokia to become Series 60. Quartz, which by this time was the basis for projects with both Sony Ericsson and Motorola, was spun out into a new Symbian subsidiary, UIQ Technology AB, based at the Ronneby development site. UIQ became the name for the user interface.[19] Japanese licensees working under the FOMA umbrella, as DoCoMo's new 3G network was branded, went their own way and collaborated on a common user interface known as MOAP.[20]

Symbian's strategy since 2002 has been based on the concept of a 'headless' delivery to its customers with a custom user interface integrated as part of the product creation lifecycle, either by the phone vendor (in the case of the FOMA licensees and Nokia) or by a user interface vendor

---

[19] In late 2006, Sony Ericsson announced that agreement had been reached for it to acquire UIQ Technology AB from Symbian.

[20] FOMA is DoCoMo's 3G network, which launched in 2001. Mobile Application Platform (MOAP) was originally an internal designation which has now begun to appear in public forums.

(UIQ licenses its own user interface pre-integrated with Symbian OS to customers such as Sony Ericsson and Motorola; similarly, Nokia's independent Series 60 business unit licenses S60 preintegrated with Symbian OS to customers such as Samsung and LG).

Symbian OS is GUI-centric in the sense that the user model is exposed only through the GUI, while being designed into the operating system at a deep level; nonetheless, Symbian does not provide its own shippable GUI. This model has its challenges (as any model does) but, with the sales record as it stands, it can be considered proven. It is driven by the recognition that the mobile phone market is quite different from the desktop computing market, for example, in which users (as well as vendors) seem reconciled to the greater than 90% domination of a single user interface, Microsoft Windows. There are alternatives, in the form of Macintosh and Unix/Linux, but these are not mass-market alternatives. The attempts to create mass-market alternatives (BeOS, for example, in the late 1990s) have been spectacularly unsuccessful.

Personal mobile devices, from phones to music players to cameras, are very different. Consumer devices, from TVs and DVD players to Hi-Fi and radio to car dashboard controls, are very different too.[21] What users want from them and how they wish to interact with them, is quite different from what they want from the beige box underneath the desk.

2.8 Operating System Influences

While the user interface defines a system from the perspective of end-users and translates the design philosophy of the system into tangible behavior accessible to users, the real character of an operating system is defined at a deeper level, by the fundamental choices its designers make – typically in the form of trade-offs between aspirations and limitations, whether of performance against price, features against time-to-market or innovation against familiarity. As a company of engineers rather than computer scientists, it seems that Psion absorbed multiple influences, but then proceeded from immediate practicalities, almost as if no theory existed.

Its first true operating system was written for the 8-bit Organiser and then re-written for the next generation of 16-bit Organisers. Since these were x86-based, DOS was a possible alternative candidate. The company however decided to follow its own path.

---

[21] Christian Lindholm provides a fascinating public insight into the issues driving user interface design for phones as we have known them, and for the multi-function mobile terminals they are becoming. These devices, as he says, are evolving from impersonal objects to intimate possessions 'containing one's most important data and thoughts' [Lindholm 2003, p. 154].

> **Charles Davies:**
>
> We considered using DOS and rejected it, which was controversial at the time because there was a view that DOS was the only kind of operating system for PCs downwards and of course it was also what the HP200 used, which was our big competitor at that time.

But DOS is a strictly synchronous system, single-user and single-tasking and it also provides an interface which insists on dragging the user down to the machine level, which was not the vision at all. In fact Psion did later release a DOS-based version (MC600) of the MC400 laptop, precursor to the Series 3, just as it had released a DOS-based version of the HC handheld. As David Wood remembers, after that first-hand experience of MS-DOS, it became known in the company as 'MS-dogs'.

In contrast, the exposure to VMS was critical because it showed that there was another way.

> **Charles Davies:**
>
> From that experience of cross-compiling on VAX for the ZX80 we got to know VMS pretty well. It was a multitasking operating system with asynchronous services. Colly Myers was the architect and we decided to make our system pre-emptive multitasking.

While these ideas began to influence the Organiser operating systems, they became central to the design of the next iteration 16-bit system. This was the operating system which eventually made its way into the Series 3, from its first iteration for the MC400 laptop.

> **Charles Davies:**
>
> The strong drivers were firstly, 'always on', in other words, no bootup – the idea that the operating system ran forever – and, secondly, that you could switch from one application to another without waiting, which was the multitasking. One of the early design principles was that we started to play with servers and that was a direct VMS influence. So if you have a multitasking operating system and if applications are executables in that operating system, then the number of applications is extensible and you can switch between those applications without exiting the current application. We take multitasking for granted now, but DOS PCs at that time were single tasking, you ran an application and you had to exit it to run another application. Remember TSRs, Terminate And Stay Resident programs which allowed you to switch in and out? We thought that was crazy! We thought the main system should be like that.

The VMS influence was sufficiently visible in the final system to be noticeable, for those who cared to look. Peter Jackson had worked with VMS while at BP Exploration and recognized the influence even in the first iteration 8-bit Organiser system. He quickly became a Psion fan.

Peter Jackson:

I was attracted to Psion in the first place because I got hold of the internal documentation for the Organiser. This was the 8-bit operating system and I read the documentation and thought, 'That looks familiar. This has been influenced by VMS'. And I also thought, 'This is very clever. This is a company that could be worth paying attention to'.

The Organiser operating system displayed many of the properties of consistent and elegant design that Jackson had admired in VMS and had found wanting in other systems, for example Unix.

Peter Jackson:

I would characterize Unix as being something that wasn't really designed, it evolved. So, by contrast, VMS was a system that was much more carefully designed from the beginning and it was carefully designed in parallel with the emergence of the VAX hardware architecture. So when you look at the design of VMS, it's quite rigorous in terms of API definition and quite consistent in terms of patterns of use of those APIs. For example, typically in VMS all APIs are asynchronous and they all have ways of monitoring the outcome of a request made to the operating system or to the I/O system and so on. So once you've learnt a corner of the VMS system and you want to explore another corner of it, you don't have to climb the same learning curve all over again. And that level of consistency applies all the way up to the command interface.

By the time Psion produced its first 16-bit systems, the key VMS-inspired patterns were well entrenched.

Peter Jackson:

It was harder to get into the 16-bit system, but there was still that consistency and elegance in how they had implemented things and I would say that the attention being paid to asynchronous interfaces was a good example of that. The whole event-driven programming model was very strong in Psion in those days.

Jackson attributes those patterns not just to VMS, but also to a more general mainframe influence. While the hobbyist culture typified by

CPM and early DOS gravitated upwards from micros to PCs, the more sophisticated professional programming culture of multiple-peripheral, multiuser, 'big-iron' computing began to drift down towards smaller systems, first to minis (including VAX, as well as the PDP family on which Unix first evolved), micros and then PCs, but also to new classes of device such as those being pioneered by Psion.

Peter Jackson:

On mainframes and mini-computers, such as VAXs, where everything is event driven, you don't put in a synchronous I/O request. You don't say 'write this data to disk' and have the call return when the write is completed. You issue an instruction to write data to the disk and you get on with the rest of your life and at some point you'll get notified that the write is actually complete.

The asynchronous, server-based model has evolved to be one of the primary patterns in Symbian OS. Servers provide for serialized access to shared resources and are used throughout the system, wherever multiple users (client programs, including other system services) require access to a system resource, whether a logical resource such as a process or a physical resource such as the physical device screen (or screens).

For Jackson, the visible influence of VMS on the Psion operating system seemed to present a perfect opportunity.

Peter Jackson:

People who learned how to program on PCs, specifically on DOS, came from a different culture where that asynchronous, event-driven model did not hold. There were people that knew about event-driven programming and there were people that didn't. So I had this vision that said, all my mainframe expertise now applies here, I understand stuff that these PC programmers aren't at all familiar with. And the way I saw it then, when I was coming into the company, is that there's a whole set of software idioms to do with mainframe computing, and the way technology was evolving meant that you could now apply those idioms to smaller and smaller devices. And I was able to capitalize on my knowledge of mainframe software architecture.

The 16-bit system was a classic C API operating system, exposing a small number of system calls as C functions. Higher up in the system, object-oriented ideas had been widely applied. Initially, the operating system appeared in the new laptop-like product. Jackson recalls it with the enthusiasm of a convert. The MC400 in his view was way ahead of its time in terms of its hardware design, a perfect match for an operating system which was also distinctive and innovative.

Peter Jackson:

It was a really nice machine. It looked like a laptop computer, only slightly thicker than laptops are today, but on almost every face of this thing, every way you turned it you'd find an interface that you could plug something into. It was way ahead of its time. There was a speech synthesizer sound module that was quite sophisticated; it had a module that was a built-in modem; it had a superb keyboard and a long battery life. You could put batteries in it and it would be good for 90 hours and this was in the era when for laptops, three hours was your limit.

Unfortunately, the MC400 didn't sell, which was a substantial set-back for the company. However, Psion responded with a complete software overhaul (to reduce ROM size and improve all-round performance) and with a new hardware product, the Series 3. Significantly, the Series 3 played on the company's core competence in creating and marketing compelling small systems, which also avoided the DOS (and soon-to-be Windows) mainstream.

In software terms, the Series 3 gave Psion a second chance to prove the merits of its new 16-bit operating system. The most obvious conclusion to draw from its subsequent success seems to be that, while the market was prepared to embrace a novel design in a new device space, it rejected innovation when it conflicted with the incumbent standard, which, in the case of the laptop-like MC400, was DOS.

David Wood also believes that the MC failed because it simply had too many flaws, both hardware and software. It was less a question of standards than of quality and fitness for purpose. The digitizer was awkward to use, for example and the machine would sometimes reset as a result of electrostatic discharge when the user touched the removable media door.

Whatever the reason, its failure was compensated for by the huge success of the Series 3.

Peter Jackson:

The Series 3 used the same basic software technology and it was fantastically successful. Without the Series 3, none of us would be where we are today. So it was all to do with packaging, what device the software was packaged in.

The architecture of the early Psion operating systems and the patterns which have evolved from them and in their turn influence Symbian OS, can be traced back to a few key principles.

Charles Davies:

If you start with the idea of going from one application to another and having processes and tasks, processes more than tasks in fact, then you say, well those processes are running concurrently, they're running all the time, even though only one of them may be on the screen. And by the way we did produce devices like the MC400 where multiple applications were visible on the screen at the same time, though we ended up going toward smaller devices. But the vision was that the user could see multiple applications at the same time. We recognized that all of those applications would need to access a file system or whatever, that if you were running multiple applications at the same time and they compete for the same resources then you need to sort that out. One way of sorting that out, which is the design pattern we adopted, was to use a server to serialize access to a shared resource. So that was the slogan, 'servers serialize access to a shared resource'.

  And the mass-storage media card on the devices at that time was a shared resource too, so you couldn't just have applications writing to mass storage, you had to have something in between that was sorting out that access. VMS used servers for that, a file server, so we made a file server and we also had a supervisory process server. So the design principle of the 16-bit system was that you had client–server and fast context switching, so there's no penalty for using servers and you get a clean architectural way of serializing access to shared resources, including memory, which I suppose is what you could say the system supervisor was.

The server principle runs deep and important design consequences follow. For example, the need for fast context switching is what determines the process and thread architecture of the operating system.

Charles Davies:

So to grow the heap dynamically, well you had a server process to do that, which is the modern pre-emptive multitasking kernel approach. And a file server and the idea of servers for other things. We knew we wanted to do graphics and we wanted to have a windowing system. The competition was still doing character-mode graphics at that time and we wanted a true graphics mode, with variable fonts and more than one application drawing to the screen at the same time. And so we had to have a model for doing that. So we said, 'Okay, the file server shares access to mass storage and the window server is the right architecture for serializing shared access to the screen.' And you need a windowing system. Even on phones, other processes can pop up a notification at any time and it basically handles that. So that was an advanced attribute of the design.

The design principles were not necessarily novel but their application to the class of small device that Psion was pioneering certainly was, the

Apple Newton notwithstanding. The vision that Psion was pursuing so hungrily was of sophisticated pocket computers aimed at an audience of consumer users rather than technical wizards – mobile and pocket computing for all.

Charles Davies:

There were GUIs around at that time, Amiga and Macintosh, that did clipping. Windows was tiled at that time, but we said we wanted overlapping windows. We couldn't afford a hardware solution. We had been involved in doing an abortive piece of work for Thompson, which we called a Thompsonitosh, which was Macintosh-like with hardware support for overlapping windows, so we knew about those things. We had also worked on software for the Sinclair QL and we'd worked on PC software. Remember, this was the age of integrated suites, which at that time were still character-mode-based and came with their own windowing APIs. IBM, for example, had something called TopView at the time. So that was the kind of environment, but we wanted to do graphics and we wanted a contemporary, modern way of doing it.

So we had pre-emptive multitasking for windows and the Window Server was born and lots of things got done in servers. The idea of fast lightweight client–server internals and servers managing clients were early design principles. The other part of client–server architecture, I guess, was the idea that this is an operating system that needed to run for years at a time without a reboot and that meant you had to have system software that looked after badly behaving applications and so that led to the idea that servers managed their clients if the clients didn't do the right thing, so that the servers didn't get left with memory leakage or data from long-gone clients. Servers knew about client processes and managed their data on behalf of client processes, even if they terminated, so there are services to let you know if a client dies and also to clean up server-side resources so that you could run for a long time, because for sure you were going to have many dying applications panicking over time.

Another design principle that came in the early days was asynchronicity. We learned that from VMS, the idea that you had event signaling and not polling. So part of that was that we were designing for battery-powered devices and ROM-based devices, which is why we thought DOS was not appropriate because it wasn't designed for either ROM-based or battery-powered systems. For us, 'execute in place' was the norm – the idea that you executed in place in ROM but you could add applications that loaded – that was part of the design. The idea of dynamic libraries was an early part of the design and I guess we were aware that Windows had dynamic libraries. We knew we had to have shared libraries and we didn't want to be loading multiple copies of the same code, which by the way was the norm if you just used compilers and linkers in the usual way. I mean these were times when people had overlays, where the code got loaded in at the same addresses, right? That was when Bill Gates famously said 640 KB should be enough for everybody. And having written software with the overlay model, we figured we didn't want that. Overlays were impossible to manage. They fell over under their own complexity after a time. So we wanted libraries that were loaded once.

These principles were rehearsed through the three generations of Psion operating systems leading up to the creation of EPOC and eventually of Symbian OS. But they were driven also by a product vision. The company was driven by the vision of creating products aimed squarely at ordinary users, which would entice them and charm them and become indispensable pocket companions.

Charles Davies:

We were building products. We were working from an idea of the user experience that we wanted. So we didn't just do pre-emptive multitasking because we thought we wanted to do an operating system and that was the fun thing to do, although there was that element to it too, if we're being honest. But we had a vision that you shouldn't have to wait for boot up, that this would be an instantly available, instant-on device and one where you didn't have to exit one application before you could run another one, because that wasn't an appropriate user experience for a handheld device.

We also thought that multitasking was a good thing for writing robust software. We had this ethic of robustness, that the product didn't go wrong and that you didn't have to be a techie to use it. Because in those days you know, I remember the first 5 MB hard disk we bought for £6000. £6000! And you went on a training course to learn how to use it! And that was not the vision of the product that we had. We had a vision of a product used by somebody who wasn't stupid, but who wasn't going to read the manual, a device where the operating system did the work for you rather than the other way around.

So it was based from the user experience backwards; the technology was in support of the user experience. That was in the bones of the product vision. We didn't think of ourselves as producing an operating system and an application suite. We thought of ourselves as producing a product that would sell. It would walk off the shelves because people wanted it and it would be hard to imitate because we'd put some good technology in it.

With hindsight, the prehistory of the company looks very much like a dress rehearsal for a category of device which did not then exist – the mobile phone.

3

Introduction to the Architecture of Symbian OS

3.1   Design Goals and Architecture

Architecture is goal driven. The architecture of a system is the vehicle through which its design goals are realized. Even systems with relatively little formal architecture, such as Unix,[1] evolve according to more or less well-understood principles, to meet more or less well-understood goals. And while not all systems are 'architected', all systems have an architecture.

Symbian OS follows a small number of strong design principles. Many of these principles evolved as responses to the product ethos that was dominant when the system was first being designed.[2] That ethos can be summarized in a few simple rules.

- User data is sacred.

- User time is precious.

- All resources are scarce.

And perhaps this one too, 'while beauty is in the eye of the beholder, elegance springs from deep within a system'.

In Symbian OS, that mantra is taken seriously. What results is a handful of key design principles:

- ubiquitous use of servers: typically, resources are brokered by servers; since the kernel itself is a server, this includes kernel-owned resources represented by R classes

---

[1] 'Bottom up' and 'informal' typify the Unix design approach, see [Raymond 2004, p. 11].

[2] That is, the ethos which characterized Psion in the early-to-mid 1990s. By then, the company was a leader in the palmtop computer market. It was a product company.

- pervasive asynchronous services: all resources are available to multiple simultaneous clients; in other words, it is a service request and callback model rather than a blocking model

- rigorous separation of user interfaces from services

- rigorous separation of application user interfaces from engines

- engine reuse and openness of engine APIs.

Two further principles follow from specific product requirements:

- pervasive support for instant availability and instant switching of applications

- always-on systems, capable of running forever: robust management and reclaiming of system resources.

Symbian OS certainly aims at unequaled robustness, making strong guarantees about the integrity and safety (security) of user data and the ability of the system to run without failure (to be crash-proof, in other words). From the beginning, it has also aimed to be easy and intuitive to use and fully driven by a graphical user interface (GUI). (The original conception included a full set of integrated applications and an attractive, intuitive and usable GUI; 'charming the user' is an early Symbian OS slogan.[3])

Perhaps as important as anything else, the operating system set out from the beginning to be extensible, providing open application programming interfaces (APIs), including native APIs as well as support for the Visual Basic-like OPL language and Java, and easy access to Software Development Kits (SDKs)[4] and development tools.

However, systems do not stand still; architectures are dynamic and evolve. Symbian OS has been in a state of continuous evolution since it first reached market in late 2000; and for the three years before that it had been evolving from a PDA operating system to one specifically targeting the emerging market for mobile phones equipped with PDA functions. In view of this, it may seem remarkable that the operating system exhibits as much clarity and consistency in design as it does.

---

[3] For example, see almost anything written by David Wood. Today, the GUI is no longer supplied by Symbian, however GUI operation remains intrinsic to the system design. The original integrated applications survive in the form of common application engines across multiple GUIs, although their inclusion is a licensee option.

[4] Symbian no longer directly supplies SDKs, since these are GUI-dependent. Symbian provides significant 'precursor' content to licensees for inclusion in SDKs, including the standard documentation set for Symbian OS APIs.

Architectures evolve partly driven by pressures from within the system and partly they evolve under external pressures, such as pressures from the broad market, from customers and from competition.

Recent major releases of Symbian OS have introduced some radical changes, in particular:

- a real-time kernel, driven by evolving future market needs, in particular, phone vendors chasing new designs (for example, 'single core' phones) and new features (for example, multimedia)

- platform security, driven by broader market needs including operator, user and licensee needs for a secure software platform.

While both are significant (and profound) changes, from a system perspective they have had a relatively small impact on the overall shape of the system. Interestingly, in both cases the pressure to address these particular market needs arose internally in Symbian in anticipation of the future market and ahead of demand from customers.

It is tempting to idealize architecture. In the real world, all software architecture is a combination of principle and expediency, purity and pragmatism. Through the system lifecycle, for anything but the shortest-lived systems, it is part genuine, forward-looking design and part retrofitting; in other words, part architecture and part re-architecture.

Some of the patterns that are present in Symbian OS were also present (or, in any case, had been tried out) in its immediate precursors, the earlier Psion operating systems. The 16-bit operating system (SIBO) had extended the basic server-based, asynchronous, multitasking model of previous Psion products and re-engineered it using object-oriented techniques. SIBO also pioneered the approach to GUI design, designed communications services into the system at a deep level, and experimented with some idioms which have since become strongly identified with Symbian OS (active objects, for example).

In fact, surprisingly many features of Symbian OS have evolved from features of the earlier system:

- the fully integrated application suite: even though Symbian OS no longer includes a user interface or applications, it remains strongly application-centric

- ubiquitous asynchronous services

- optimization for battery-based devices

- optimization for a ROM-based design: unlike other common operating systems, SIBO used strategies such as 'execute-in-place' (XIP) (compare this with MS-DOS, which assumes it is loaded into RAM to execute) and re-entrancy[5] (MS-DOS is non-re-entrant), as well as a design for devices with only solid-state disks

- sophisticated graphical design: from the beginning, SIBO supported reactive repainting of windows and overlapping windows, in an age of tiled interfaces (for example, Windows 2.0 and the character-mode multitasking user interfaces of the day, such as TopView and DesqView)

- an event-driven programming model

- cross-platform development: the developers' mindset was more that of embedded systems engineering than the standard micro-computer or PC model.[6]

SIBO also introduced some of the programming constraints which show up in Symbian OS, for example forbidding global static variables in DLLs (because the compilers of the day could not support re-entrant DLLs), an early example of using the language and tools to constrain developer choices and enforce design and implementation choices, a consistent theme in Symbian's approach to development.

Symbian OS, or EPOC as it was then, was able to benefit from the experience of the earlier implementation in SIBO. The 16-bit system was, in effect, an advanced prototype for EPOC.

Meanwhile, of course, Symbian OS has continued to evolve. In particular, some crucial market assumptions have changed. Symbian OS no longer includes its own GUI, for example; instead it supplies the framework from which custom, product-ready GUIs such as S60, MOAP and UIQ are built. Hardware assumptions have changed quite radically too. Execute-in-place ROMs, for example, depend on byte-addressable flash silicon (so-called NOR flash); more recently, non-byte-addressable NAND flash has almost wholly superseded NOR flash, making execute-in-place a redundant strategy. Other technology areas, for example display technologies, have evolved almost beyond recognition compared to the 4-bit and 8-bit grayscale displays of earlier times. Not least, the telephony standards that drive the market have evolved significantly since the creation of the first mobile phone networks.

Despite sometimes radical re-invention and change, the original design conception of Symbian OS is remarkably intact.

---

[5] In designing for re-entrant DLLs (that is, re-entrant shared libraries), SIBO was significantly in advance of the available tools. For example, C compilers were poor in this area. Geert Bollen makes the point that it is not just language features that determine whether a given language is suitable for a particular project; the tools infrastructure that supports the language is equally important.

[6] It is interesting to note that Bill Gates has identified as one of Microsoft's key strengths (and, indeed, a key competitive advantage), that it develops all of its systems on its own systems. The advantage breaks down completely in the mobile phone context.

3.2 Basic Design Patterns of Symbian OS

The design principles of a system derive from its design goals and are realized in the concrete design patterns of the system. The key design patterns of Symbian OS include the following:

- the microkernel pattern: kernel responsibilities are reduced to an essential minimum

- the client–server pattern: resources are shared between multiple users, whether system services or applications

- frameworks: design patterns are used at all levels, from applications (plug-ins to the application framework) to device drivers (plug-ins to the kernel-side device-driver framework) and at all levels in between, but especially for hardware adaptation-level interfaces

- the graphical application model: all applications are GUI and only servers have no user interface

- an event-based application model: all user interaction is captured as events that are made available to applications through the event queue

- specific idioms aimed at improving robustness: for example, active objects manage asynchronous services (in preference, for example, to explicit multi-threading) and descriptors are used for type-safe and memory-safe strings

- streams and stores for persistent data storage: the natural 'document' model for applications (although conventional file-based application idioms are supported)

- the class library (the User Library) providing other user services and access to kernel services.

3.3 Why Architecture Matters

'Doing architecture' in a complex commercial context is not easy. Arguably all commercial contexts are complex (certainly they are all different), in which case architecture will never be easy. However, the business model for Symbian OS is particularly complex. While it must be counted as part of Symbian's success, it also creates a unique set of problems to overcome and work around, and to some extent those problems are then manifested as problems for software architecture.

Architecture serves a concrete purpose; it makes management of the system easier or more difficult, in particular:

- managing the functional behavior and supported technologies
- managing the size and performance
- retaining the ability to evolve the system.

Elegance, consistency, and transparency were all early design drivers in the creation of the system. Charles Davies, now Symbian CTO, was the early architect of the higher layers of the operating system.

> **Charles Davies:**
>
> I remember looking at Windows at that time and thinking that this is all very well, here is this Windows API, but just look what's happening underneath it, it was ugly. I wanted something that you could look inside of.

The early 'ethic of robustness', to use his phrase, came straight from the product vision.

Managing the Bounds of the System

In some ways, the hardest thing of all for Symbian is managing the impact of its business model on the properties of the system and, in particular, the problem that Charles Davies calls 'defining the skin' – understanding, maintaining, and managing the bounds of the system under the impact of the business model. As well as originating the requirements push and feeding the requirements pipeline, generating almost all of the external pressure on the system to evolve and grow, licensees and partners also create their own extensions to the system. (S60, arguably, is the most extreme example, constituting a complete system in itself, at around twice the size of the operating system.)

Being clear where to draw the boundary between the responsibilities of Symbian OS and the responsibilities of GUIs, in terms of who makes what and where the results fit into the architecture, becomes difficult. Charles Davies is eloquent on the subject.

> **Charles Davies:**
>
> One of the things I've done since being here is to try and identify where the skin of Symbian OS is, to define the skin. When I was at Psion and we were building a PDA, I understood where the PDA ended and where the things outside the PDA began, and so I knew the boundaries of the product. And then I came to Symbian and Symbian OS, and I thought, where are the boundaries? It's really tough to know where the boundaries are, and I still sometimes wonder if we really know that. That's debilitating from the point of view of knowing what to do. In reality we're trying to fit some kind of rational

boundary to our throughput, because you can't do everything. We've got, say, 750 people in software engineering working on Symbian OS, and we can't make that 1500 and we don't want to make that 200. So with 750 people, what boundary can we draw that matches a decent product?

In one sense the problem is particular to the business model that Symbian has evolved, and is less a question of pure technology management, which to some extent takes care of itself (or should, with a little help to balance the sometimes competing needs of different licensees), than of driving the operating system vision in the absence of a wider product vision. In that wider sense, the licensees have products; Symbian OS has technologies and it is harder to say what the source of the technology vision for the operating system itself should be. To remain at the front of the field, Symbian OS must lead, but on the whole, customers would prefer that the operating system simply maps the needs of their own leading products. The problem is that by the time the customer product need emerges, the operating system is going to be too late if it cannot already support it. (At least, in the case of complex technologies and, increasingly, all new mobile technologies are complex.) Customers therefore build their own extensions or license them from elsewhere, and the operating system potentially fails under the weight of incompatibilities or missing technologies).

Product companies are easier to manage than technology companies because it is clear what product needs are; either they align with the market needs or the product fails in the market. The Symbian model is harder and forever raises the question of whether Symbian is simply a supplier or integrator to its customers, or an innovator. Is Symbian a product company, where the operating system is the product, or does it merely provide a useful 'bag of bits'?

Architecture is at the heart of the answer. If there is an architecture to describe, then there is more to the operating system than the sum of its parts.

Managing Competitive Threats

There are many external threats to Symbian OS. Some of the threats are household names. Microsoft is an obvious threat, but the likelihood is that Microsoft itself will always be unacceptable to some part of the market, whatever the quality of its technology offering. (It is hard to see Nokia phones, for example, sharing branding with Microsoft Windows, and the same issues no doubt apply to some network operators, but clearly not to all of them.) It is almost as certain that Nokia in turn is unacceptable to some other parts of the market. S60 aims at building a stable of licensees, vendors for whom the advantages of adopting a proven, market-ready user interface outweigh the possible disadvantages of licensing a solution from a competitor, or the costs of developing an in-house solution. There will always be vendors, though, for whom licensing from Nokia is likely to be unacceptable. Interestingly, the more Microsoft resorts to branding its own phones, in order to increase market share, the more it competes with those it is seeking to license to. It is hard to see any scenario in which the phone market could become as homogeneous as the PC market.

Linux is also a clear and visible threat, even though again there are natural pockets of resistance. Linux, for example, is viral. Linux does not just take out competitors, it takes out whole parts of the software economy, and it is not yet clear what it replaces them with.[7] To put Linux in a phone, for example, seems to require just the same 'old' software economy as to put any other operating system into a phone, dedicated software development divisions which do the same things that other software development divisions do: write code, miss deadlines, fix defects, pay salaries. Linux may be royalty-free, but that translates into 'not-free-at-all' if you have to bring it inside your own dedicated software division. Nonetheless, to ignore Linux would be a (possibly fatal) mistake.

Architecture is part of the answer. If Symbian OS is a better solution, it is because its architecture is more fit for purpose than that of its competitors, not because its implementation is better. Implementation is a second-order property, easy to replace or improve. Architecture, in contrast, is a deep property.

3.4   Symbian OS Layer by Layer

The simplest architectural view of Symbian OS is the layered view given by the Symbian OS System Model.[8]

UI Framework Layer

The topmost layer of Symbian OS, the UI Framework layer provides the frameworks and libraries for constructing a user interface, including the basic class hierarchies for user interface controls and other frameworks and utilities used by user interface components.

The UI Framework layer also includes a number of specialist, graphics-based frameworks which are used by the user interface but which are also available to applications, including the Animation framework, the Front End Processor (FEP) base framework and Grid.

The user interface architecture in Symbian OS is based on a core framework called Uikon and a class hierarchy for user interface controls

---

[7] Where it is clear, it is not clear how to make a profit from what it replaces them with.
[8] The System Model (see Chapter 5) is relatively constant across different releases, although its details evolve to track the evolution of the architecture.

called the control environment. Together, they provide the framework which defines basic GUI behavior, which is specialized by a concrete GUI implementation (for example, S60, UIQ or MOAP), and the internal plumbing which integrates the GUI with the underlying graphics architecture.

Uikon was originally created as a refactoring of the Eikon user interface library, which was part of the earliest versions of the operating system. Uikon was created to support easier user interface customization, including 'pluggable' look-and-feel modules.

The Application Services Layer

The Application Services layer provides support independent of the user interface for applications on Symbian OS. These services divide into three broad groupings:

- system-level services used by all applications, for example the Application Architecture or Text Handling

- services that support generic types of application and application-like services, for example personal productivity applications (vCard and vCal, Alarm Server) and data synchronization services (OMA Data Sync, for example); also included are a number of key application engines which are used and extended by licensees (Calendar and Agenda Model), as well as legacy engines which licensees may choose to retain (Data Engine)

- services based on more generic but application-centric technologies, for example mail, messaging and browsing (Messaging Store, MIME Recognition Framework, HTTP Transport Framework).

Applications in Symbian OS broadly follow the classic object-oriented Model–Viewer–Controller (MVC) pattern. The framework level support encapsulates the essential relationships between the main application classes (representing the application data model, the views onto it, and the document and document user interface that allow it to be manipulated and persisted) and abstracts all of the necessary underlying system-level behavior. In principle, a complete application can be written without any further direct dependencies (with the exception of the User Library).

The Application Services layer reflects the way that the system as a whole has evolved. On the one hand, it contains essential application engines that almost no device can do without (the Contacts Model for example), as well as a small number of application engines that are mostly now considered legacy (e.g. the WYSIWYG printing services and the office application engines, including Sheet Engine, a full spreadsheet engine more appropriate for PDA-style devices). On the other hand, it contains (from Symbian OS v9.3) the SIP Framework, which provides the foundation for the next generation of mobile applications and services.

Java ME

In some senses, Java does not fit neatly into the layered operating system model. Symbian's Java implementation is based around:

- a virtual machine (VM) and layered support for the Java system which complements it, based on the MIDP 2.0 Profile

- a set of standard MIDP 2.0 Packages

- an implementation of the CLDC 1.1 language, I/O, and utilities services

- a number of low-level plug-ins which implement the interface between CLDC, the supported packages, and the native system.

Java support has been included in Symbian OS from the beginning, but the early Java system was based on pJava and JavaPhone. A standard system based on Java ME first appeared in Symbian OS v7.0s. Since Symbian OS v8, the Java VM has been a port of Sun's CLDC HI.

The OS Services Layer

The OS Services layer is, in effect, the 'middleware' layer of Symbian OS, providing the servers, frameworks, and libraries that extend the bare system below it into a complete operating system.

The services are divided into four major blocks, by broad functional area:

- generic operating system services

- communications services

- multimedia and graphics services

- connectivity services.

Together, these provide technology-specific but application-independent services in the operating system. In particular, the following servers are found here:

- communications framework: the Comms Root Server and ESock (Sockets) Server provide the foundation for all communications services

- telephony: ETel (Telephony) Server, Fax Server and the principal servers for all telephony-based services

- networking: the TCP/IPv4/v6 networking stack implementation

- serial communications: the C32 (Serial) Server, providing standard serial communications support

- graphics and event handling: the Window Server and Font and Bitmap Server provide all screen-drawing and font support, as well as system- and application-event handling

- connectivity: the Software Install Server, Remote File Server and Secure Backup Socket Server provide the foundation for connectivity services

- generic: the Task Scheduler provides scheduled task launching.

Among the other important frameworks and libraries found in this layer is the Multimedia Framework (providing framework support for cameras, still- and moving-image recording, replay and manipulation, and audio players) and the C Standard Library, an important support library for software porting.

The Base Services Layer

The foundational layer of Symbian OS, the Base Services layer provides the lowest level of user-side services. In particular, the Base Services layer includes the File Server and the User Library. The microkernel architecture of Symbian OS places them outside the kernel in user space. (This is in contrast to monolithic system architectures, such as both Linux and Microsoft Windows, in which file system services and User Library equivalents are provided as kernel services.)

Other important system frameworks provided by this layer include the ECom Plug-in Framework, which implements the standard management interface used by all Symbian OS framework plug-ins; Store, which provides the persistence model; the Central Repository, the DBMS framework; and the Cryptography Library.

The Base Services layer also includes the additional components which are needed to create a fully functioning base port without requiring any further high-level services: the Text Window Server and the Text Shell.

The Kernel Services and Hardware Interface Layer

The lowest layer of Symbian OS, the Kernel Services and Hardware Interface layer contains the operating system kernel itself, and the supporting components which abstract the interfaces to the underlying hardware, including logical and physical device drivers and 'variant support', which implements pre-packaged support for the standard, supported platforms (including the Emulator and reference hardware boards).

In releases up to Symbian OS v8, the kernel was the EKA1 (Kernel Architecture 1) kernel, the original Symbian OS kernel. In Symbian OS v8, the EKA2 (Kernel Architecture 2) real-time kernel shipped for the first time as an option. (It was designated Symbian OS v8.1b; Symbian OS v8.1a is the Symbian OS v8.1 release with the original kernel architecture.) From Symbian OS v9, EKA1 no longer ships and all systems are based on the real-time EKA2 kernel.[9]

3.5   The Key Design Patterns

Probably the most pervasive architectural pattern in Symbian OS is the structuring client–server relationship between collaborating parts of the system. Clients wanting services request them from servers, which own and share all system resources between their clients.

Another widely used pattern is the use of asynchronous methods in client–server communications. Together, these two patterns impose their shape on the system. Like any good architecture, the patterns repeat at multiple levels of abstraction and in all corners of the system.

A third pervasive pattern is the use of a framework plug-in model to structure the internal relationships within complex parts of the system, to enable flexibility and extensibility. Flexibility in this context means run-time flexibility and is particularly important when resources are constrained. The ability to load the requested functionality on demand enables more efficient use of constrained resources (objects which are not used are not created and loaded). Extensibility is important too in a broader sense. The use of plug-ins enables the addition of behavior over a longer timescale without re-architecting or re-engineering the basic design. An example is the structure of the telephony system which encapsulates generic phone concepts which are then extended, for example for GSM- or CDMA-specific behaviors, by extension frameworks. The use of plug-ins also enables licensees to limit or extend functionality by removing or replacing plug-in implementations.

At a lower level, Symbian OS makes much use of specific, local idioms. For example, active objects are the design idiom which makes asynchronicity easy and are widely used. ('Asynchronicity' here means the ability to issue a service request without having to wait for the result before the thread of execution can continue.) Encapsulating asynchronicity into active objects is an elegant object-oriented design. (Active objects are examples of cooperative multitasking: multiple active objects execute in effect within the context of a single thread. Explicit multithreading is an example of non-cooperative multitasking, that allows pre-emption.)

Symbian OS has also evolved a number of implementation patterns, including 'leaving' functions and the cleanup stack, descriptors for safe strings, local class and member naming conventions and the use of manifest constants for some basic types.

---

[9] This history is described in detail in [Sales 2005], the in-depth, authoritative reference.

Symbian's microkernel design dates back to its original conception, but becomes even more significant in the context of the new real-time kernel architecture. The real-time architecture is essential for a system implementing a telephony stack, which depends on critical timing issues, and is also becoming increasingly important for fast, complex multimedia functionality. Together, phone and multimedia are arguably the most fundamental drivers for any contemporary operating system. As mobile phones, in particular, reach new levels of multimedia capability, to become fully functional converged multimedia devices (supporting streamed and broadcast images and sound, e.g. music streaming, two-way streaming for video phone conferencing and interactive broadcast TV), achieving true real-time performance has become an essential requirement for a phone operating system. The real-time kernel allows Symbian OS to meet that requirement, making it a suitable candidate for directly hosting a 3G telephony stack.

The real-time kernel architecture also introduces important changes (in particular to mechanisms such as interprocess communication) to support the new platform security model introduced from Symbian OS v9. (Strictly speaking, the security model is present in Symbian OS v8 but implements a null policy. The full security model, which depends on the new kernel architecture, is present from Symbian OS v9.)

The Client–Server Model

In Symbian OS, all system resources are managed by servers. The kernel itself is a server whose task is to manage the lowest level machine resources, CPU cycles and memory.

From the kernel up, this pattern is ubiquitous. For example, the display is a resource managed by the Window Server; display fonts and bitmaps are managed by the Font and Bitmap Server; the data communications hardware is managed by the Serial Server; the telephony stack and associated hardware by the Telephony Server; and so on all the way to the user-interface level, where the generic Uikon server (as specialized by the production GUI running on the final system) manages the GUI abstractions on behalf of application clients.

Threads and Processes

The client–server model interacts with the process and threading model in Symbian OS. While this is in keeping with a full object-oriented approach, which objectifies machine resources in order to make them the fundamental objects in the system, it can also cause confusion.

In Symbian OS, threads and processes are defined in [Sales 2005, Chapter 3] as follows:

- threads are the units of execution which the kernel scheduler schedules and runs

- processes are collections of at least one but possibly multiple threads which share the same memory address space (that is, an address mapping of virtual to physical memory).

Processes in other words are units of memory protection. In particular each process has its own heap, which is shared by all threads within the process. (Each thread has its own stack.)

A process is created as an instantiation of an executable image file (of type EXE in Symbian OS) and contains one thread. Creation of additional threads is under programmer control. Other executable code (for example, dynamically loaded code from a DLL file) is normally loaded into a dynamic-code segment attached to an existing process. Loading a DLL thus attaches dynamic code to the process context of the executing thread that invokes it.

Each server typically runs in its own process,[10] and its clients run in their own separate processes. Clients communicate with the server across the process boundary using the standard client–server conventions for interprocess communication (IPC).[11]

As Peter Jackson comments, Symbian OS falls somewhere between conventional operating system models in its thread and process model.

Peter Jackson:

Most of the threads versus processes issues are to do with overhead. In some operating systems, processes are fairly lightweight, so it's very easy to spawn another process to do some function and return data into a common pool somewhere. Where the process model is more heavyweight and the overhead of spawning another one is too great, then you invent threads and you let them inherit the rest of the process, so the thread is basically just a way of scheduling CPU activity. In Symbian OS, you can use whichever mechanism is appropriate to the requirements.

Server-Side and Client-Side Operations

Typically a server is built as an EXE executable that implements the server-side classes and a client-side DLL that implements the client-side interface to the server. When a client (either an application or another

---

[10] There are some exceptions for reasons of raw speed.

[11] [Sales 2005] defines the Symbian OS client–server model as inter-thread communication (ITC), which is strictly more accurate than referring to interprocess communication (IPC). However, arguably the significance of client–server communications is the crossing of the process boundary.

system service) requests the service, the client-side DLL is attached to the calling process and the server-side executable is loaded into a new dedicated process (if it is not already running).

Servers are thus protected from their clients, so that a misbehaving client cannot cause the server to fail. (The server and client memory spaces are quite separate.) A server has responsibility for cleaning up after a misbehaving client, to ensure that resource handles are not orphaned if the client fails.

At the heart of the client–server pattern therefore is the IPC mechanism and protocol, based on message passing, which allows the client in its process, running the client-side DLL, to communicate via a session with the server process. The base classes from which servers and their client-side interfaces are derived encapsulate the IPC mechanisms.

The general principles are as follows:[12]

- The client-side implementation, running in the client process, manages all the communications across the process boundary (in the typical case) with the server-side implementation running in the server process.

- The calling client connects to the client-side implementation and creates a session, implemented as a communications channel and protocol created by the kernel on behalf of the server and client.

- Client sessions are typically created by calling `Connect()` and are closed using `Close()` methods, in the client-side API. The client-side calls invoke the standard client–server protocol methods, for example `RSessionBase::CreateSession()` and `RProcess::Create()`. On a running server, this results in the client session being created; if the server is not already running, it causes the server to be started and the session to be created.

- The client typically invokes subsessions that encapsulate the detailed requests of the server-defined protocol. (In effect, each client–server message can be thought of as creating a subsession.)

- Typically, client-side implementations derive from `RSessionBase`, used to create sessions and send messages.

- Typically, the server side derives from `CServer`.

Servers are fundamental to the design of Symbian OS, and are (as the mantra has it) the essential mechanism for serializing access to shared resources, including physical hardware, so that they can be shared by multiple clients.

---

[12] The best description is [Stichbury 2005, Chapter 12].

Andrew Thoelke:

It's not so much that there is a server layer in the operating system as a hierarchy. It's very much a hierarchy and there are a lot of shared services. Some of them are shared by quite a few components and some of them really support just a very small part of the system, and of course those shared services may build on top of one or more client–server systems already.

Client–server is a deep pattern that is used as a structuring principle throughout the system.

Asynchronous Services

Another deep pattern in the system is the design of services to be asynchronous.

System responsiveness in a multitasking system (the impression that applications respond instantly and switch instantly) depends on asynchronous behavior; applications don't wait to finish processing one action before they are able to handle another.

The alternatives are blocking, or polling, or a combination of both. In a blocking request (the classic Unix pattern), the calling program makes a system call and waits for the call to return before continuing its processing. Polling executes a tight loop in which the caller checks to see if the event it wants is available and handles it when it is. (Polling is used by MS-DOS, for example, to fetch keystrokes from the keyboard.)

Blocking is unsatisfactory because it blocks others from accessing the system call which is being waited on, while it is waiting. Polling is unsatisfactory because code which is functionally idle, waiting for an event, is in reality not idle at all, but continuously executing its tight polling loop.

Blocking reduces responsiveness. Polling wastes clock cycles, which on a small system translates directly to power consumption and battery life.

Charles Davies:

Asynchronous services was driven by battery life. We were totally focused on that. For example on one of the Psion devices, we stopped the processor clock when it was idle. I don't know if that was innovative at the time. We certainly didn't copy it from anybody else, but we had a static processor. Usually in an idle process, the operating system is doing an idle loop. But we didn't do that, we stopped the clock on the processor and we turned the screen off, and that was fundamental to the design.

Typically, client–server interactions are asynchronous.

The Plug-in Framework Model

A final high-level design pattern, the plug-in framework model is used pervasively in Symbian OS, at all levels of the system from the UI Framework at the top to the lowest levels of hardware abstraction at the bottom.

A framework (as its name suggests) is an enclosing structure. A plug-in is an independent component that fits into the framework. The framework has no dependency on the plug-in, which implements an interface defined by the framework; the plug-in has a direct, but dynamic, dependency on the framework.

Frameworks are one of the earliest design patterns (going back to the time before design patterns were called design patterns, in fact) [Johnson 1998]. While, in principle, nothing limits them to object-oriented design, they lend themselves so naturally to object-oriented style that the two are strongly identified. A key principle of good design (again, not limited to object-oriented design but closely identified with it) is the separation of interface from implementation. On a small scale, this is what designing with classes achieves: a class abstracts an interface and its expected behavior and encapsulates its implementation. Frameworks provide a mechanism for this kind of abstraction and encapsulation at a higher level. As is often said, frameworks enable a complete design to be abstracted and reused.[13] Frameworks are therefore a profound and powerful way of constructing an object-oriented system.

In detail, a framework in Symbian OS defines an external interface to some part of the system (a complete and bounded logical or functional part) and an internal plug-in interface to which implementers of the framework functionality (the plug-ins) conform. In effect, the framework is a layer between a calling client and an implementation. In the extreme case, a 'thin' framework does little more than translate between the two interfaces and provide the mechanism for the framework to find and load its plug-ins. A 'thicker' framework may do much more, providing plug-in interfaces which are highly abstracted from the external visible client interface. Symbian OS contains frameworks at both extremes and most points in between.

Because in Symbian OS a framework exposes an external interface to a complete, logical piece of the system, most frameworks are also implemented as servers.

As well as providing interface abstraction and separation from implementation, and flexibility through decoupling, frameworks also provide a natural model for functional extension. This approach is used for example by the telephony-server framework to provide an open-ended design. The core framework supports generic telephony functionality based around a small number of generic concepts. Framework extensions implement

---

[13] A framework is 'reusable design' as [Johnson 1998] puts it.

the specialized behaviors which differentiate landline from mobile tele-phony, data from voice, circuit- from packet-switched, GSM from CDMA, and so on.

As well as this 'horizontal' extension of the range of functionality of the framework, such a plug-in also defines the interfaces which are implemented 'vertically' by further plug-ins that provide the actual services.

Because the plug-in framework model is pervasive, Symbian OS pro-vides a plug-in interface framework. (Available since Symbian OS v7.0s but universally enforced since Symbian OS v8.0 as part of the phased introduction of Platform Security.) The plug-in framework (also known as ECom) standardizes the mechanisms and protocols that allow frameworks to locate and load the plug-ins which provide their implementations, and for plug-ins to register their presence and availability in the system as implementation modules.

Clearly, plug-ins pose a potential security threat because they provide a mechanism for untrusted (that is, externally supplied) code to be loaded into the processes of some system components (although the microkernel architecture keeps them well away from the kernel). The plug-in framework therefore enforces the security model on plug-ins before they are loaded [Heath 2006].

Another area in which plug-ins pose potential risks to the system is in performance. Potentially, a badly designed or poorly implemented plug-in can damage the performance of the framework that loads it. The plug-in model can also make it hard to understand the dynamic behavior of the operating system and, in particular, can make system-level debugging tricky, since the system can become (from the perspective of the debugger) highly indeterministic, unpredictable and unreproduceable.

However, enabling a pervasive model of run-time rather than static loading can boost system performance. Plug-ins are loaded on request; if they are not requested, they are not loaded, saving loading time and system resources (including RAM, on systems that do not provide execute-in-place).

An interesting example of just how pervasive the plug-in framework pattern is in Symbian OS is the original implementation of applications as plug-ins to the application and UI Framework rather than as more con-ventional executables. (This architecture changes somewhat in Symbian OS v9, where applications are implemented as EXEs rather than DLLs, while retaining other characteristics of plug-ins.)

In implementation terms, an ECom plug-in is implemented as a poly-morphic DLL and a resource (RSC) file. The DLL entry point is a factory function that instantiates the plug-in object. All system plug-ins are stored into well-known locations, as required by the security model.

The plug-in framework provides a standard and universal mechanism for binding implementations (plug-ins) to interfaces (frameworks) at run

time, together with the mechanisms for packaging multiple interface implementations into a single DLL (that is, loading multiple implementations
at once, to improve performance), plug-in registration and implementation versioning, discovery and loading including boot-time discovery
optimizations to avoid run-time overhead, and cleanup after unloading
plug-ins. (A plug-in instance cannot destroy itself, because its destructor
code would be part of the code being removed from memory.) The framework also provides security-policy definition and policing mechanisms.

The plug-in framework is implemented as a server, in effect a broker
between frameworks and conforming plug-ins, managing those plug-ins
as a resource to its framework clients.

Microkernel Architecture

Symbian OS has a microkernel architecture, which sets it apart from
operating systems such as Microsoft Windows and Linux.[14] In Symbian
OS, core services that would be inside the kernel in a monolithic operating system are moved outside. The pervasive use of the client–server
architecture, and the protection of system code from clients which follows from it, guarantees both the robustness and high availability of these
services. The goal is a robust system that is also responsive and extensible;
experience suggests that the design achieves it.

Andrew Thoelke:

The actual client–server architecture, the division into processes across the
operating system and the boundary of the kernel, means that the actual
privileged mode software is much smaller than in desktop operating systems.
It's very nearly theoretical microkernel, but not completely truly microkernel
because device drivers all run kernel side, and a true microkernel would say
that device drivers should run user side, and who knows maybe we'll get there
in a few years time. But all file system services, all higher level comms services
including networking, and the windowing software for example, all run user
side.

If anything the new EKA2 kernel architecture goes beyond the microkernel design and encapsulates the most fundamental kernel primitives
within a true real-time nanokernel, supporting an extended kernel that
implements the remaining Symbian OS kernel abstractions, but is equally

---

[14] There are microkernel implementations of Unix, based on the Mach microkernel.
Mac OS X is an example; it is built as a Berkeley Unix variant with a Mach microkernel
and proprietary user interface layer. Other microkernel designs include QNX, which is an
operating system similar to Unix, but not Unix; Chorus, which is not just a microkernel but
also object-oriented and which, like Mach, is capable of hosting Unix; and iTron, which is
an important mobile-phone operating system in Japan.

capable of supporting 'personality' layers to mimic the interface of any other operating system. But the essential elegance of the Symbian OS kernel design goes right back to its earliest days.

Martin Tasker:

The Symbian model is that you're either a user thread or a kernel thread, and if you're a user thread then either you're an application thread, which has a session with the window server and interacts with the user, or you're a server thread which has no interaction with the user. And if you're a server thread, well then you sit around waiting for client requests to happen and when they do you service them, and in fact the kernel has a server and it does just that. There are a couple of kernel calls which are handled by something known as fast execs, which don't involve the kernel server. But the design philosophy of the kernel is to make those things very short and sweet and to put most of the work into the server. I think that's a cool architecture. Some of it goes down to Colly Myers's explainability requirement, that it takes more than an average programmer to implement any of this stuff, but any average programmer should be able to use it.

The lineage of course can be traced back to the precursor Psion systems.

Andrew Thoelke:

It owes its design very much to the heritage of Series 3. Colly Myers took that same OS structure, that you've got a small amount of protected mode software that can do everything, and that even all the file system and file services actually operate in a separate process from that and have less privileges, and that you have a very tightly integrated client–server architecture that actually binds everything together. That is definitely quite different to what you see in a lot of other systems.

Notwithstanding the move to the EKA2 kernel architecture, at a high level the lineage is still visibly present.

Martin Tasker:

The change from EKA1 to EKA2 is a hugely significant change. But at the system-design level, you know that change hasn't actually radically altered the system design at all. It's still either application processes or server processes, and that design was actually pioneered all the way back in SIBO, and it hasn't changed much since then, and the reason is: it's a proven design.

3.6 The Application Perspective

Symbian OS has been designed above all to be an application platform (although it might be argued that that has begun to change, and that in the latest devices it has become primarily an engine for driving fast, mobile data communications). Applications have always been an essential part of the system. The early operating system shipped with a complete set of productivity and communications applications targeting connected PDAs. Although Symbian OS no longer supplies a GUI and user-ready applications but only common application engines, Symbian OS phones now ship with more built-in applications than ever before, supplied either with the licensee GUI or as extras provided by the phone vendor or network operator.

> **Charles Davies:**
>
> Symbian started off as an operating system plus an application suite. We never designed it as an operating system independently of the suite of applications.

Just as importantly, both S60 and UIQ are also explicitly pitched as open platforms for third-party applications and provide extensive support for developers including freely available SDKs, support forums and tools.

From the beginning the approach to applications has been graphics-based. Like much else, the approach has evolved and, in particular, it has evolved as Symbian's user interface strategy has evolved. However, the principles of application structure have been essentially mature since the first release of S60 and UIQ in 2002.

Uikon is the topmost layer of Symbian OS. It provides the framework support on which a production user interface is built. The three currently available custom user interfaces are S60, UIQ and MOAP, but there is no engineering reason why any licensee should not build its own bespoke user interface, which indeed is precisely the origin of S60 and MOAP. Uikon abstracts application and control base classes in the Application Architecture and Control Environment class hierarchies to create generic GUI application classes (that is, classes free of a look and feel policy) which are customized by the custom user interface. The custom user interface abstracts the Uikon policy-free base classes to provide the policy-rich classes that applications derive from.

Uikon thus integrates the underlying support of the Application Architecture and the Control Environment to create a framework from which (as abstracted by the custom user interface), applications derive. Uikon is a framework and applications behave recognizably as plug-ins. Uikon is implemented as a server.

The Structure of an Application

Every application is built from three basic classes:[15]

- an application class, derived from the Application Architecture (`CApa-Application`)
- a document class, derived from the Application Architecture (`CEik-Document`)
- an application user interface class, derived from the Control Environment (`CCoeAppUiBase`).

These classes provide the fundamental application behavior. However, two important parts of the application are missing from this structure: the application view, which is the screen canvas the application uses to display its content, and the application data model and data interface implementations, which encapsulate the application 'engine'.

The classic application structure expects that the data model (the data-oriented application functionality) exists independently of the GUI implementation of the application and is, therefore, independent of any user interface classes. It is hooked into the user interface by a member pointer (`iModel`) in the document class. The classes specific to the user interface then interact with it purely through the APIs it exposes.[16]

Charles Davies:

We always had that structuring of applications, the idea of separating the UI from the application engine. That was an early design principle and it was the design guidance for application writers. We knew about Model–View–Controller, and we thought of an application engine as a model, and our design guidance was to keep the application logic separate from the UI. Not because we anticipated at that time multiple UI flavors, but because we recognized something more fundamental in terms of writing an application. That you might write an application and decide to improve the design of the UI, where the refinement of the UI was just pragmatic, the basic functional application logic stayed the same. So if you could separate those two things, that was good, and that led to the terminology of application engines.

---

[15] This is the 'classic' application structure, with roots in the Eikon applications of Psion Series 5. Both UIQ and S60 extend the design patterns for applications. See [Edwards 2004, p. 184] for discussions of the 'dialog-based' and 'view-switching' S60 application structure. UIQ applications also extend the basic pattern with custom view classes.

[16] This is in fact a very powerful design principle, implying, for example, that the data model can run without a direct user interface at all. Engines designed this way are independently testable and intrinsically highly portable between different user interfaces. The principle runs deep in the Symbian ethos, as witnessed by the presence of engines independent of the user interface in the operating system itself.

In Symbian OS, a control is a drawable screen region (in other words, the owner of screen real estate). The Application view class is derived directly from the Control Environment control base classes.

On small devices, where screen real estate is scarce, desktop-style windowing is not appropriate. A more natural approach for small displays is to switch whole-screen views, for example switching between a list-style view of contact names and a record-style view of the details of a single contact. Applications therefore typically define a hierarchy of views, with the main application view at the root.

Because Symbian OS is multitasking, multiple applications can be running at once, even though only one (the foreground application) will be presenting its view on the display. Both S60 and UIQ support switching directly between views in different applications, including launching the view of a new application inside the context of the current one (for example selecting a phone number from within a Contact entry and immediately switching to the phone application and dialing the number).

Symbian's application structure makes much of the detail of the application user interface programmable solely via resource files. Resource files are compiled separately as part of the application build process and linked into the built application, providing a natural mechanism for language localization (all text strings used within an application can be isolated in resource files and recompiled to a new language without having to recompile the application). Resource files are also compressed.

Charles Davies:

We lived in tougher times as far as Moore's law was concerned in those days. Resource files were around in contemporary GUI systems at that time. But from the beginning we did Huffman compression on resource files, and we were careful about the amount of information we put in them.

Uikon

The most striking fact about Symbian OS at the user interface level is its support for a replaceable user interface, and indeed the fact that it ships without a native user interface at all. (User-interface-dependent components are shipped only with a TechView test user interface.)

While it seems fair to say that Symbian did not get its user interface strategy right first time (in particular, the Device Family Reference Design (DFRD) strategy looks, with hindsight, to have been naïve), nonetheless the operating system has been able to support multiple licensees, each having a distinct user-interface philosophy, occupying different positions in the market and spanning diverse geographical locations. Those differences are encapsulated in the differences between the user interfaces that have evolved for Symbian OS.

S60 builds on the classic Nokia user interface to provide a simple, key-driven but graphically rich and arresting user interface. In contrast, UIQ is firmly pen-based and targets high-end phones with rich PDA-like functionality including pen-based handwriting recognition. MOAP aims squarely at its solely Japanese market, providing a graphically busy user interface featuring Kanji as well as Roman text and animated cartoon-style icons.

File System or 'Object Soup' Storage Model

FAT is the 'quick and dirty' file system that MS-DOS made famous. When work on EPOC started, the Apple Newton was a leading example of a different way to approach consumer computing (different, for example, from the MS-DOS-based Hewlett Packard machines which were the leading competitor for Psion's Series 3). Instead of a conventional file system the Newton employed an 'object soup' storage model.[17]

On any useful system, data requires a lifetime beyond that of the immediate context in which it is created, whether that means storing system settings, saving the memo you have just written to a file, or storing the contact details you have just updated.

Charles Davies:

We had a normal file system on the Series 3. When we went to C++, we talked a lot about persistent models of object-oriented programming, and we went for stream storage. We narrowly rejected SQL in favor of stream storage. I remember the design ideas around at the time, and it was done in the interests of efficiency. Different applications were having to save the same system objects and we were having to duplicate that code. So for something like page margins, which was a system structure, if that object knew how to serialize itself, that would solve the problem. You do that by having serialization within the object, so objects that might reasonably want to be persisted could persist themselves. And that was in the air, I mean Newton had its soup at that time which I think was object-oriented, and there was a belief at that time that object-oriented databases were it, and that objects ought to be seen as something that existed beyond the lifetimes of processes.

Objects, in other words, can be viewed as more than just the run-time realizations of object-oriented code constructs. However, in terms of the standards of the day, approaches based on something other than a file system were certainly the exception. The big challenge in maintaining data is that of data format and compatibility, ensuring that the data remains accessible. Any device which aims to be interoperable

---

[17] 'Object soup' is described in [Hildebrand 1994].

(in any sense) with other devices faces a similar challenge. In both cases, the design is immediately constrained in how far it can deviate from the data-format conventions of the day. For EPOC at that time, compatibility with desktop PCs was an essential requirement. For Symbian OS now, the requirement is more generalized to compatibility with other devices of all kinds. Probably the most important test case for both is readability of removable media file systems. (All other cases in which a Symbian OS device interoperates with another device can be managed by supporting communications protocols and standard data formats, which are independent of the underlying storage implementation.)

While external compatibility does not determine internal data formats, the need to support FAT on removable cards probably tipped the balance towards an internal FAT filing system. One (possibly apocryphal) story has it that the decision to go with FAT was a Monday morning *fait accompli* after Colly Myers had spent a weekend implementing it.

Peter Jackson:

There were periods when we explored all sorts of quite radical ideas but in the end we always came back to something fairly conservative, because if you take risks in more than one dimension at a time it doesn't work. So I spent quite a lot of time at one stage investigating an object-oriented filing system. But one day I think Colly Myers had a sudden realization and he just said, 'Let's do FAT', and he was probably right.

But FAT is not the whole story. In fact, Symbian OS layers a true object-oriented persistence model on top of the underlying FAT file system. As well as a POSIX-style interface to FAT, the operating system also provides an alternative streaming model.

It is an interesting fact that data formats, whether those of MS-Word or Lotus 1-2-3 or MS-Excel, have proved to be powerful weapons in the marketplace, in some cases almost more so than the applications which originated them. (The Lotus 1-2-3 data format lives on long after the demise of the program and, indeed, of the company.) Data in this sense is more important than the applications or even the operating systems with which it is created.

Peter Jackson:

The layout of the file is an example of a binary interface and, as software evolves, typically those layouts change, sometimes in quite an unstructured or unexpected way, because people don't think of them as being a binary interface that you have to protect. So the alternative way of looking at things is to say you don't think about that, you ignore the layout of the file. What you do is you look at the APIs, and you program all your file manipulation stuff to use the same engines that originated the data in the first place.

In effect, this is the approach that Symbian adopted. But it has a cost.

Charles Davies:

We went for an architecture in which applications lost control of their persistent data formats, and in retrospect I think that was a mistake, because data lasts longer than applications. The persistence model is based on the in-memory aggregation in the heap of whatever data structure you're working with. For example, if it's a Contacts entry, then it consists of elements and you stream the elements. One problem is that if you try to debug it and you're looking at a file dump, its unfathomable. It's binary, it's compressed, so it's very efficient in the sense that when you invent a class it knows how to stream itself, so it's a sort of self-organizing persistence model, but the data dump is unfathomable. The second problem is that when you change your classes it changes how they serialize. So it works. But if you add a member function which needs to be persisted, then you change the data format. You lose data independence, and that stops complementers from working with your formats too. So we sacrificed data independence. And because that data has to carry forward for different versions of the operating system, you get stuck with that data format and you end up with a data migration problem. So I think that was a mistake. It would have been worth it to define data-independent formats. In my view that's what XML has proved, the XML movement has shown that data sticks longer than code.

In some ways, implementing a persistence model on top of a FAT system leads to the worst of both worlds, on the one hand missing out on the benefits of MS-DOS-style data independence, and on the other missing out on Newton-style simplicity.

Peter Jackson:

If you implement your permanent store structure in terms of a database design then you have all the advantages of being able to use database schema idioms to talk about what you're doing, and it turns out that those idioms now are fairly stable and universal. So I think there are examples where we have pruned away the databaseness of an application because we thought our customers didn't really want a database – but that may be a bad thing if one day our customers decide they want more than just flat data.

Store and DBMS

The native persistence model is provided by Store, which defines Stream and Store abstractions. Together they provide a simple and fully object-oriented mechanism for persistence:

- A Stream is an abstract interface that defines `Externalize()` and `Internalize()` methods for converting to and from internal and external data representations, including encrypted formats.

- A Store is an abstract interface that defines `Store()` and `Restore()` methods for persisting structured collections of streams, which represent whole documents. Store also defines a dictionary interface which allows streams to be located inside a store.

Symbian OS also includes DBMS, a generic relational database API layered on top of Store, as well as implementations including a lightweight, single-client version (for example, for use by a single application that wants a database-style data model which will not be shared with others). Databases are stored physically as files (single client databases may also be stored in streams).

Database queries are supported either through an SQL subset or a native API. Since the introduction of platform security, the DBMS implementation supports an access-policy mechanism to protect database contents.

3.7 Symbian OS Idioms

C++ is the native language of Symbian OS. Symbian's native APIs therefore are C++ APIs (although API bindings exist for other languages: OPL, Java and, most recently, Python). C++ is a complex, large and powerful language. The way C++ is used in Symbian OS is often criticized for being non-standard. For example, the Standard Template Library (STL) is not supported, the Standard Library implementation is incomplete, and POSIX semantics are only partly supported. Since Symbian OS competes with systems which do support standard C++, there is also little doubt that the operating system will evolve towards supporting more standard C++. But, like it or not, true native programming in C++ on Symbian OS requires understanding and using its native C++ idioms.

Among some developers inside the company the view has been unashamedly one of, 'Those who can, will; those who can't should use Java, Python, or even OPL'.[18] While that may not make for mass market appeal for Symbian C++ itself, the fact is that programming on

---

[18] For example, see the remarks by David Wood in Chapter 18.

any platform requires specialist expertise as well as general expertise, and, in that, Symbian OS is no different. The skill level required is commensurate with the programming problem. It is far from easy to write software for consumer devices on which software failures, glitches, freezes and crashes – things people put up with regularly on their PCs – are simply not an option. Mobility, footprint, battery power, the different user expectations, screen size, key size and all the other specifics of their small form factors make mobile devices not at all like desktop ones; phones, cameras, music players and other consumer devices are different.

Symbian OS idioms are not casual idiosyncrasies; they are deliberate constraints on the C++ language devised to constrain developer choices, consequences of the market the operating system targets, and of the embedded-systems nature of ROM-based devices. Strictly speaking, they are less architectural than implementational but, in terms of the overall design, they are important and they have an important place in the history of the evolution of the system. Understanding them is essential to understanding what is different about Symbian OS, and what is different about mobile devices. There are some large-scale differences.

- Lack of a native user interface means that the development experience is significantly different for device creation developers using the TechView test user interface than for developers later in the product lifecycle using S60, UIQ or MOAP.

- The build system is designed for embedded-style cross-compilation, which is a different experience from desktop development.

- Idioms have evolved to support the use of re-entrant, ROM-based DLLs, for example disallowing global static data.

- Other optimizations for memory-constrained, ROM-based systems result in some specific DLL idioms (link by ordinal not name, for example).

There are what might be described as language-motivated idioms:

- descriptors
- leaving functions
- the cleanup stack
- two-phase construction.

And there are some design-choice idioms:

- active objects and the process and threading model
- UIDs

- static libraries and object-oriented encapsulation
- resource files to isolate locale-specific data, for example, text strings.

Active Objects

Active objects are an abstraction of asynchronous requests and are designed to provide a transparent and simple multitasking model.

An active object is an event handler which implements the abstract interface defined by the `CActive` class and consists of request and cancellation methods, which request (or cancel) the service the object should handle, and a `Run()` method which implements the actual event handling. When the requested service completes and there is a result to be handled, a local active scheduler invokes the active object's `Run()` method to handle the completed event.

An active scheduler is created by the UI Framework for each application. All active objects invoked by an application (but only that application's active objects) share a single thread, in which they are not pre-empted (i.e. they are scheduled in priority order by the scheduler).

Active objects are a pervasive Symbian idiom and provide a non-pre-emptive multitasking alternative to explicitly creating multithreaded programs (although that option remains available to developers), as a solution to the problem of managing multiple paths of execution within a program, in the context of an event-based, reactive application model. From the perspective of a GUI application developer they offer a much easier solution than multithreading, in effect handing off the awkward details to the system.

Charles Davies:

Our model for events was very much asynchronous events and signals and requests. So what we had first of all, and it's what other systems have too, is that you make one or more requests for events, and events include timers and serial events and all kind of events that can come out of anywhere, not just user-originated events. So you just set off a large number of events and then you wait for any one of them to come through. So things need to be able to respond to events from multiple sources. Now Windows had a way of handling this. There's a Windows API, though it's not very elegant. The problem is, it's tied to the GUI programming model. In Windows you have to run up the whole GUI to get the event model going, and we thought that was a real weakness in mobile devices. We thought that servers needed this as well, that servers sit there waiting for events from multiple sources, events like 'my client has died', which comes from a different source than the message channel saying 'here's the next request from the client'.

The event-driven model is essentially a state-machine model. But, except within niche areas such as communications programming, these were not widely used patterns, especially for applications programming. And except for those familiar with Windows at the time, or with other GUI systems such as Amiga and Macintosh, the event-driven application model was not widely or well understood.

Charles Davies:

When I was interviewing people I used an example of a terminal emulation program. Here is a program that indisputably gets events not just from the user. The normal, naïve way of writing an interactive application at that time would be to wait for a keypress, see what keypress it was, and respond to it; was it a function key, was it any other key? You'd have some horrible case statement responding to a keypress. So I would ask, 'How would you write an application where you don't know whether your next input is coming through the serial port or from the keypress?' And if they had a good answer to it they got hired, and if they didn't, they didn't.

Well we started off programming it the way that anybody would program it, you make asynchronous requests on whatever event sources you want to respond to. There are many pitfalls in doing that, for example if you don't consume that event in the right way. You end up with an event loop that's quite messy, and it's pages long, and people were making mistakes. Every event loop was buggy, and horrible bugs too, so we said 'Let's make it modular.'

Martin Tasker had the benefit of a background of programming IBM mainframes:

Martin Tasker:

I've written plenty of event-handling loops, in communications programs or command handlers where by definition you don't know what's going to happen next. Every time I wrote one of these loops I remember thinking, 'Have I got this right?' Dry running through every possibility, you used to have to tell people coming on to the team, 'No, if you handle your loop that way you're either going to double-handle some event or fail to handle some event, or you're not going to handle event number 2 if event number 2 happens while you're handling event number 1, or you're not actually going to handle event number 2 until event number 3 comes along...' These are all mistakes that everybody makes when they're writing event-handling programs. Over the lifetime of a program you tend to add in more and more events, or you remove them, and you change things around. And in those circumstances, when you're modifying existing code, it's tremendously difficult to get event-handling loops right.

Active objects were devised explicitly to solve such problems, by creating an easy-to-understand and easy-to-use mechanism for firing off event handlers asynchronously, deliberately breaking the dependencies between events which are implied by the big, single-block switch statement which is the typical implementation. More generically, active objects enable multitasking within applications without the use of explicit multithreading.

Charles Davies:

We could have done it with threads and created a multithreaded UI, which by the way is what Java does. But the bad thing about threads is that you can pre-empt at any time, and then you've got to protect the data, because you have no idea when you're processing one thread what state the data is in. The solution was active objects, for any program that responded to events from multiple sources. So it came about because people were getting it wrong, because the old way was so complicated. So what are active objects? They're really non-pre-emptive multitasking within an application. And that is a very strong pattern. But it is also something that throws people, because it wasn't copied. It was invented here, and it's widely used, and it has been useful, but it is a particular strength of Symbian OS.

Active objects are used widely throughout the operating system, as well as providing a ready-made mechanism for developers creating native Symbian OS applications.

Martin Tasker:

Colly Myers was right, active objects are a fantastic solution. For people who know they are dealing with event-handling programs, they are an absolute joy. And the whole single-threaded nature of an application process is also great for programmers. In an event-handling system, active objects are a natural way of handling things, and they are easier for programmers to work with than pretty much all of the alternatives.

Cleanup, Leaving and Two-Phase Construction

The native Symbian OS error-recovery model evolved explicitly to handle the kinds of errors that should be expected on resource-constrained and mobile devices: low-memory situations, low-power situations, sudden loss of power, loss of connectivity or intermittent connectivity, and even the sudden loss of a file system, for example when a removable media card is physically removed from the device without unmounting. These are all likely or even daily occurrences in the mobile phone context, causing errors from which the system must recover gracefully. In contrast, for a large system these may be rare enough occurrences for system failure with an 'unrecoverable error' message to be acceptable.

The Symbian OS model is proven, playing a large part in the unrivaled robustness of the system, and going back to the earliest days of the operating system, and indeed to Psion systems before it.

Charles Davies:

We had `Enter()` and `Leave()` in the 16-bit system, which was Kernighan and Ritchie inspired. When we went to C++, the standards for exception handling were still being written, so they certainly weren't available in compilers. So we carried forward `Leave()` and `Enter()` rather than adopting native C++ exception handling, because at that time it consisted of `longjump()` and `setjump()`. It was very unstructured, and we didn't like that. We liked `Enter()` and `Leave()`, and we stuck with it.

In Symbian OS, `Leave()` is a system function (provided by the User Library) which provides error propagation within a program. Typically, `Leave()` is used to guard any calls which can fail (for conditions such as out of memory, no network coverage and disk full). The system unwinds the call stack until it finds a prior `Leave()` call wrapped by a TRAP macro, at which point the TRAP is executed and the failure is handled by the program in which it occurred.[19]

Functions which may fail because of a leave, whether because they directly invoke the action which might fail or do so indirectly by calling some other function that does, are described as 'leaving' functions. By convention, leaving functions are named with a trailing 'L', which makes it easy for programmers to see where they are invoked and trap appropriately.

The second leg of the error-handling strategy uses the 'cleanup stack' to store pointers to heap-allocated objects whose destructors will fail to be called if the normal path of program execution is derailed by a leave.[20] As well as unwinding the call stack to handle the leave, the cleanup stack is also unwound and destructors are called on any pushed objects.

The third leg of the strategy is 'two-phase construction', which guarantees that C++ construction of an object will always succeed, by moving any leaving calls out of the C++ constructor into a secondary constructor. (It is important that construction succeeds, since only then can the object's destructor be called; if the destructor cannot be called, memory may have been leaked [Stroustrup 1993, p. 311].) Again, a number of system functions are available to regularize the pattern and take care of underlying details for developers. (In its earliest implementation, two-phase construction was matched by two-phase destruction. The eventual consensus was that this was an idiom too far.)

---

[19] See [Stichbury 2005, p. 14] for a detailed explanation.

[20] See the discussion in [Harrison 2003, p. 150]. This is the authoritative programmers' guide.

Charles Davies:

We had an ethic that said that memory leakage was something the programmer was expected to manage. So something like the Window Server, which might be running for a year at a time, needed to make sure that if an exception was called it didn't leak memory. The cleanup stack was an invention to make it easier for people to do that. You'd have an event loop, and at the high end of the event loop you'd push things on the stack that needed to be unwound, whether they were files that needed to be closed or objects that needed to be destroyed. That was a pragmatic thing, you know. 'Let's provide something that encourages well-written applications from the point of view of memory leakage.'

Cleanup is pervasive in the system ([Harrison 2003, p. 135]), permeating every line of code a developer writes, or reads, in Symbian OS, with its highly visible trailing 'L' naming convention, its `Leave()` methods and TRAPs, and its cleanup stack push and pop calls.

For new developers, it is both highly visible and immediately unfamiliar, which leads to an immediate impression that the code is both strange and difficult. However, the conventions are not intrinsically difficult, even if the discipline may be. The purpose is equally straightforward: to manage run-time resource failures. On a small device, memory may rapidly get filled up by the user (whether by loading a massive image, downloading too many MP3s, or simply taking more pictures or video clips than the device has room for). Other resources, whether USB cable connections, infrared links, phone network signals, or removable media cards, can simply disappear without warning at any time. Mostly these hazards simply do not exist on desktop systems. On phones, they are the norm.

Martin Tasker:

I think the cleanup stack was a brilliant solution to the problem that we were faced with at the time.

Descriptors

Descriptors are the Symbian OS idiom for safe strings. ('Safe' means both type safe and memory safe and compares with C++ native C-style strings, which are neither[21]) Descriptors were invented (by Colly Myers) because there was no suitable C++ library class, or none that was readily available.

---

[21] Nor are Java or Microsoft Foundation Class strings for that matter, according to [Stichbury 2005, p. 55].

In principle, descriptors simply wrap character-style data and include length encoding and overrun checking. (Descriptors are not terminated by NULL; they encode their length in bytes into their header, and refuse to overrun their length.) As well as this basic behavior they also provide supporting methods for searching, matching, comparison and sorting.

Descriptors support two 'widths', that is, 8-bit or 16-bit characters, based on C++ #define (typedef) and originally designed to enable a complete system build to be switched, more or less with a single definition, between ASCII-based and Unicode-based character text support.

More interestingly, descriptors also support modifiable and unmodifiable variants and stack- and heap-based variants. The content of unmodifiable (constant) descriptors cannot be altered, although it can be replaced, whereas that of modifiable descriptors can be altered, up to the size with which the descriptor was constructed.[22]

Another important distinction is between buffer and pointer descriptor classes. Buffer descriptors actually contain data, whereas pointer descriptors point to data stored elsewhere (typically either in a buffer or a literal). A pointer descriptor, in other words, does not contain its own data. A final distinction is between stack-based and heap-based buffer descriptors. Stack-based descriptors are relatively transient and should be used for small strings because they are created directly on the stack (a typical use is to create a file name, for example. Heap-based descriptors, on the other hand, are intended to have longer duration and are likely to be shared through the run-time life of a program (see Table 3.1).[23]

Table 3.1 Descriptor classes.

	Constant	**Modifiable**
Pointer	TPtrC	TPtr
Buffer (stack-based)	TBufC	TBuf
Heap-based	HBufC	

See [Harrison 2003, p. 123] for a fuller explanation of the descriptor classes.

---

[22] Although modifiable, once allocated there is no further memory allocation for a descriptor, so its physical length cannot be extended. For example, to append new content to a descriptor requires that there is already room within the descriptor for the data to be appended.

[23] [Stitchbury 2005] contains a good overview.

Descriptors differ from simple literals, which are defined as constants using the_LIT macro, in that they are dynamic (literals are created at compile time, descriptors are not). A typical use of a pointer descriptor is to point to a literal.

Martin Tasker:

The 8-bit/16-bit aspect was ASCII versus Unicode, though, in retrospect we should have been braver about adopting Unicode straight away. But bear in mind that the ARM 3 instruction set we were then using didn't have any 16-bit instructions or, more accurately, it didn't have any instructions to manipulate 16-bit data types, so it was not efficient to use Unicode at that time. But maybe we should have had more foresight and courage, because it turned out to be a distraction. But as a kind of memory buffer, I think they were reasonably distinctive.

Given the state of the art at the time, Peter Jackson believes that the distinction between 8-bit and 16-bit was understandable but that a more naturally object-oriented approach would have been preferable.

Peter Jackson:

I think it would have been more elegant to have a descriptor that knew internally what kind of descriptor it was, whether it was the 8-bit or 16-bit variant. I never liked the fact that some of these things were done by macros.

Descriptors are not only type safe, they are memory safe, making memory overflow ('out-of-bounds' behavior) impossible. Descriptor methods will panic if an out-of-bounds attempt is detected (see Figure 3.1).

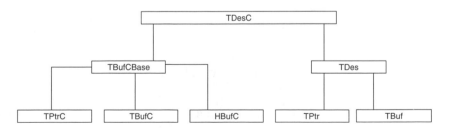

Figure 3.1 Descriptor class hierarchy

Charles Davies:

Descriptors were Colly Myers's thing, definitely, and the idea was rather like the cleanup stack, to stop people doing memory overwrites. That's a big protection against worms and other attacks, deliberate and malicious overwriting of the heap, although at the time that wasn't the driving reason to do it. We did it to stop programmers making mistakes.

C and T and Other Classes

As well as the use of the trailing 'L' (for 'leaving') and 'C' (for 'constant') to flag properties of methods, Symbian OS also uses some similarly straightforward class-naming conventions to flag fundamental properties of classes.

Martin Tasker:

If you look at the C and T types, they offer a very, very simple guide to the programmer as to how to use these types. They are as simple as Java's objects and built-ins. We don't do garbage collection because C++ doesn't do garbage collection, so we have to cope with that. We have to do it manually, but otherwise I think our conventions are as simple as Java.

The most important naming conventions are summarized as follows: [24]

- T classes are simple types which require no destructor and behave like C++ built-in types.

- C classes derive from CBase and should always be explicitly constructed, thus ensuring that they are always allocated on the heap. CBase classes also therefore require explicit destruction. CBase provides a basic level of additional support, including a virtual destructor, allowing CBase-derived objects to be deleted through the CBase pointer and performing cleanup stack housekeeping. CBase also overloads operator new to zero-initialize an object when it is first allocated on the heap. All member data of derived classes is therefore guaranteed to be zero on initialization.

- R classes indicate resource classes, typically a client session handle for a server session. Since an R class typically contains only a handle, it does not require either construction or destruction. R classes therefore may safely be either automatics or class members.

---

[24] [Stichbury 2005, Chapter 1] provides a comprehensive discussion.

- M classes are 'mixin' classes (abstract interface classes), the only form in which multiple inheritance is supported in Symbian OS.

- Descriptors are immediately recognizable as either `TPtr` pointer descriptors, or `TBuf` (stack-based) or `HBufC` (heap-based) buffer descriptors.

Manifest Constants

Symbian OS uses manifest constants – implemented as `typedefs`, that is, system-defined types – instead of the native types supported by a standard C++ compiler on standard hardware. This is partly, of course, because the cross-development model means that the eventual intended target platform is not the same as the development platform, hence the 'native' types of the platform on which the code is compiled may differ from those of the platform on which it is intended to run. The use of type definitions also has its roots in designing to support both ASCII and Unicode builds, which is now superfluous since Symbian OS has been all-Unicode since before v6.

Supporting emulator builds (that is, running Symbian OS programs on PC as well as ARM, and not just developing on PC) creates the additional complexity of requiring not one supported compiler but two (or more); originally Microsoft compilers were specified for emulator builds and GCC for ARM. More recently Metrowerks and Borland compilers have been supported and, in Symbian OS v9, ARM's RVCT replaces GCC as the 'official' ARM target compiler (although GCCE is still supported to ensure a low-cost development option). Recent initiatives such as Eclipse, for example, or the adoption of the standard ARM EABI are likely to continue to change the story of the development tools.[25] Again, using manifest constants provides the necessary level of decoupling of code from compiler dependencies.

The key classes are summarized as follows:[26]

- `TInt` and `TUint` are the generic types for signed/unsigned integer values; `TInt8`, `TInt16`, `TInt32`, and `TUint8`, `TUint16`, `TUint32` are also provided; in general, the least specific types are preferred, that is, `TInt` and `TUint`

- `TInt64` is a 64-bit integer type intended for platforms without a native 64-bit type

---

[25] Symbian, like Psion before it, has always assumed that mainstream development is done under Microsoft Windows, although this is not the only solution that works. There are a number of independent open-source solutions for developers wanting to work on Linux or Mac OS X.

[26] Again, [Stichbury 2005, Chapter 1] provides a comprehensive discussion.

- `TReal`, `TReal32` and `TReal64` are single- and double-precision floating-point types; again the least specific type, `TReal`, is preferred

- `TText8` and `TText16` are 8-bit and 16-bit unsigned types for characters

- `TBool` is a 32-bit unsigned Boolean type

- `TAny*` is used instead of void*.

Unique Identifiers

Unique identifiers (UIDs, implemented as signed 32-bit values) are centrally controlled in Symbian OS. One common usage of them is to identify applications and other binary and data types. UIDs, for example, are used in Symbian OS to associate data types with programs and plug-in types with frameworks. UIDs are also used as feature IDs and package IDs (for SIS files).

> **Charles Davies:**
>
> The idea was that if you had polymorphic DLLs, dynamic libraries in other words, then there are situations where the DLL is a plug-in, and it all goes very wrong if the caller doesn't get the interface it's expecting from the DLL, so we needed to characterize the interface. And we came up with the idea of using a UID to do that.

UIDs are used in a three-tier construction to build `TUidType` objects:

- UID1 – a system level identifier that distinguishes EXE from DLL types

- UID2 – a specifier for library types that distinguishes between shared library DLLs and various types of polymorphic DLL (for example FEPs and other types of plug-in)

- UID3 – the individual component ID, also used by default as the secure identifier (SID) required by platform security.[27]

UID3 is used, for example, by developers to uniquely identify their applications, and can then be used by the streams, stores and files created by that application to identify themselves. UID3 is assigned through Symbian's UID allocation database, from which third-party developers can request blocks of UIDs for use in their applications.

Platform Security introduces two new types of UID, the SID (Secure ID), which by default is identical to UID3, and VID (Vendor ID).

---

[27] See the discussion in [Sales 2005, p. 328].

3.8   Platform Security from Symbian OS v9

Platform Security is the system-wide security model introduced in Symbian OS v9. Providing an open, third-party programmable platform has been an important principle in the development of Symbian OS. However, openness brings with it the risk of misbehaving software (whether accidentally or deliberately misbehaving) finding its way onto users' devices. The security model is designed to protect users from that risk, while still preserving the openness of the platform.

Architecturally, Platform Security is a set of pervasive changes at all levels of the system, based on a simple conceptual model,[28] which is deliberately as lightweight as possible, and supported by the Symbian Signed certificate signing program, which provides a means for creating a formal link between an application and its origin, as well as providing a review mechanism to promote best practice in designing and writing Symbian OS applications.

Will Palmer is one of the system architects who is currently responsible for the Platform Security project.

> **Will Palmer:**
>
> There are three principles to Platform Security. The first principle is the unit of trust, the idea of the process being the unit of trust. Since memory is already protected per-process on the processor, that fits quite nicely, and it also has the advantage of being a 'least-privilege' approach, based on the smallest element in the operating system. The second principle is the idea of capabilities, which are in effect authorization tokens. So to be able to access a potential resource, a process needs to possess a particular capability that allows it to do so. And the third principle is data caging, which is about read and write protection of files, which protects the integrity of data as well as protecting data from prying eyes.

The essential principles are:

- processes as the unit of trust,[29] which turns trust into another process-granular system resource

- capabilities as the tokens of trust, which are required to perform actions

---

[28] According to [Heath 2006, p. 18], the model conforms to the eight design principles of [Saltzer and Schroeder 1975], which include economy, openness, least privilege and psychological acceptability.

[29] This is an elegant extension of the kernel's process model, in which the process is the unit of ownership of all system resources (for example, memory protection is per process).

- data caging, which protects data from prying eyes (by policing read access) or interference (by policing write access) or both.

The direct consequence of defining the process as the unit of trust is that all threads in a process share the same level of trust (which is natural, since they have access to the same resources).

The goal is to protect device users from the kinds of intentionally rogue software, or 'malware', that plague the PC world. Symbian OS for a long time avoided some of the worst threats from malware because it was typically deployed in ROM-based devices, in which the system itself cannot be corrupted (for example, it is impossible to install trapdoors or trojans in system files) because system code is stored in unwriteable ROM memory. By design, Symbian OS also protected against some of the more trivial security holes found on other systems. Descriptors, for example, make buffer overrun attacks much harder. Similarly, Symbian's microkernel architecture helps to increase security and robustness; since the trusted kernel is deliberately the smallest possible subset of system functions, there is little privileged code to exploit, and the smaller codebase is easier to review and validate.

The nature of mobile devices, especially phones, also makes them different from desktop systems. The physical access model is different (personal devices are less likely to be shared) and the network access models are different (connections are transient).

On the other hand, phones also present new opportunities for malware. If a phone, or user, can be spoofed into making a call, real money is at stake. (Premium-rate-phone-number scams are an example.) From a network perspective, the cost of network disruption is immediately commercially quantifiable in a way that Internet attacks are not.

These differences all require appropriately designed security mechanisms.

Will Palmer:

When the capability model was designed there were a set of constraints about what it had to deliver: it had to be robust; it had to be simple; and it shouldn't get in the way of the operation of a phone so, for example, you couldn't use hundreds of extra clock cycles on it, because on a small device you have performance and power constraints. Also it had to be appropriate for an open operating system: people have to be able to install additional software on their phones and it has to be simple and easy to understand.

Data caging, for example, was chosen for its simplicity and economy (in terms of clock cycles and power). Another important consideration was that mechanisms which users are quite comfortable with on desktop computers – logging on, for example – would be quite inappropriate on a phone.

Will Palmer:

Authorization based on the process–capability model is simple to understand and it fits the phone case much better than an authentication system. So in an authentication system you log on and your password authenticates you to the system, and once authenticated you can do anything permitted by your authentication level. But a phone is different: it's a single-user environment; it's in your pocket; it belongs to you. Although things are getting more complex now because of requirements coming in for administrative rights. For example, the network operator might want to change settings on the phone.

The capability mechanism is used to protect both 'system' and 'user' (i.e., application-owned) resources. Will Palmer sums up the difference neatly.

Will Palmer:

It's not that some types of capabilities are more powerful than others, they just protect different things. System capabilities protect the integrity of stakeholders and of the device, whereas user capabilities protect the user's privacy and money.

Protected APIs are tagged at method-level with the capability required to exercise them and access any underlying resources (data files, for example). The capabilities of a method are part of its interface. To use protected APIs therefore, developers must request an appropriate set of capabilities, which is done through the Symbian Signed program.

A 'signed' application is granted a set of capabilities. Application capabilities are verified by servers when protected APIs are called by applications. Unsigned software is flagged to the user at installation time as being unsigned (and therefore untrusted). Thus, while unsigned applications can assign any user capabilities to any binaries as they see fit, the user is alerted at installation time and given the option to approve the application or not. Unsigned applications cannot use system capabilities, in other words they cannot use APIs which affect the behavior of the device. Data security is provided on a per-application basis by the data-caging model.

4

Introduction to Object Orientation

4.1 Background

Symbian OS is a full-blown, from-the-ground-up, object-oriented system. In context, the decision to 'go object-oriented' was a natural step. Object-oriented ideas had been increasingly adopted in Psion's preceding operating systems, from the first Organiser products to the 16-bit SIBO operating system for the Psion Series 3. However, the decision to apply object-oriented design to the whole system, and not just to the higher user interface and application-level layers, was none the less radical for that. In particular, the decision to adopt C++ as the implementation language for the operating system was a bold one. The earlier systems (once they had evolved beyond assembler) had been written in a home-grown object-flavored dialect of C.[1] Adopting C++, which was still far from the mainstream, was, with hindsight, far-sighted though not without risk.

In 1994, when the project to create what eventually became Symbian OS started up, C++ was still a new and evolving language. C++ compiler implementations for the PC were still being pioneered by small companies such as Zortech and Watcom (the 'industrial' C++ market was still based on Unix). Microsoft had only just entered the market.[2] The language standard was still some years away. Standardized tools were even further away.[3]

The immediate consequences were twofold. First, cross-platform development was difficult (compiling on Intel for eventual ARM targets) because the low-level language bindings were not consistent across hardware

---

[1] See also Chapters 2 and 17.
[2] See for example the Wikipedia article **http://en.wikipedia.org/wiki/Visual_C_Plus_Plus** for a history of Microsoft's C++ releases. VC1.5 was the big release.
[3] Tools standardization (enabling compiler and linker interoperability across vendors, for example) depends on agreeing the low-level application binary interface (ABI). The standardized ABI for ARM processors is only now emerging into the tools mainstream.

architectures. Secondly, some language features were missing, immature, or just unsuitable for the project's purposes. While C++ was explicitly intended as a systems language, and to some extent also inherited C's low-level–high-level mantle and its long history of optimized compiler internals, some features of the language were far from optimal for small, low-memory footprint, low-power devices.[4] By and large, the language made no claim to be particularly suitable for small systems of any kind. Its roots were in big, middleware systems running on big hardware (e.g., millions of lines of code phone switches).

There were some significant consequences for the evolution of Symbian OS; many of its hallmark idioms were invented because the C++ language as it stood could not meet requirements (type-safe strings, structured exception handling, and so on) that Psion's designers considered essential for the class of device they were targeting. Subsequently, as Symbian OS has itself begun the move into the mainstream, these legacies of early language immaturity and Psion's early adoption of C++ have presented obstacles to a new generation of developers who have grown up with a standard language. Inevitably, there is pressure on Symbian OS to do better at supporting the standard language.

But it is fair to say that this problem is related to the success of Symbian OS. The pressure comes from its exposure to a much broader range of developers than in the past. It seems inconceivable, or at least unlikely, that Symbian OS would now be poised on the edge of mass-market adoption had its architects not innovated far beyond the homegrown tools and language idioms of its predecessors. The choice of C++ was a prescient one, accurately predicting what turned out to be a language juggernaut, sweeping all before it (at least until the rise of Java). There were also benefits from adopting an object-oriented methodology across the whole of the operating system.

4.2   The Big Attraction

Of all the perceived benefits of the object-oriented approach to software creation, reuse is probably the most compelling. Software is expensive. Software is unreliable. Software is complex. These are the three truisms of software development and reusability meets them all head on, or at any rate purports to.

First of all, software is expensive because it is complex. Software projects overrun because the problem at hand always turns out to be more complex than was at first thought and things prove to be harder than they looked in the plan. But if software projects can be started from a baseline of existing, already proven code or finished components, or at

---

[4] For example, the overhead of vtables.

least proven design, the scope for misunderstood complexity might just be reduced, and this seems to be what reuse promises. The more artifacts there are to be reused, the less the complexity, and therefore the lower the cost.

Secondly, software is expensive because it is unreliable. It is unreliable because it contains defects and it contains defects because it is complex. Reuse seems to hold promise here too because reuse improves quality by reusing proven parts. It also improves quality by reducing the complexity which causes defects in the first place.

Reuse does indeed look like the key to conquering software complexity, and this is very much how it has been sold. Object orientation claims to deliver reuse and reuse is the big attraction. In the words of [Gabriel 1996], reuse was 'the hook that grabbed the mainstream world and pulled it toward object-oriented programming'. Since effort costs money, reusing effort must save money. And since effort is error-prone, reusing effort must reduce errors.

Of course, reuse is not the privileged domain of object orientation. The earliest innovations in what were not yet called operating systems[5] were as much about code reuse as about multiplexing processors and peripherals. The same is true of the early language standardization drives, from Fortran to COBOL to C and beyond.

There are other aspects of reuse too. Reuse also occurs at project level, as every programmer quickly learns and as [Gabriel 1996] points out. Today's new problem can be understood as a variation on last week's problem, and therefore last week's solution can be adapted to become this week's solution too.

Languages, however, have the advantage of working at several levels, from the individual to the team, from program level to project level. But all languages are not equal. The clever observation that heralds the discovery of full blown object orientation is that reusing data structures counts as much as reusing algorithms. Object orientation makes this a language feature and supports it with language constructs, not just code libraries and link-time tools.

Other benefits also arise from reuse. Object-oriented analysis is a good way of modeling real-world problems. For example, object-oriented language pioneers have claimed 'real world apprehension' and 'stability of design' as two benefits which follow from the directness of the correspondence between an object model and a real-world problem domain [Madsen et al. 1993, p. 2]. The object approach to modeling also provides its own natural model for program organization (code is naturally granular at the object level; code can be divided between interface definitions and implementations, and so on). It probably turns

---

[5] Possibly the earliest example was the Supervisor program of the Manchester University Atlas computer in the late 1950s (see [Hansen 2001]).

out, too, that this way of organizing a program makes it easier to extend than more traditional organizations.

We are now some years on from object orientation's initial promise.[6] Object orientation is the industry's dominant programming methodology and software is still expensive, software projects are still delivering late (when they deliver at all: abortive projects remain at an astonishing 30% across the industry) and software is still unreliable (i.e., it cannot be guaranteed to perform its intended function without error).[7]

It would hardly be fair to blame object orientation for this, although it is tempting to ask what became of the vision of reusable components, of a black-box component industry and a free market in ready-made, reusable software parts.[8] Either the vision fizzled out or our gaze moved on. If the market ever materialized, it failed to thrive.

There are still no magical solutions [Gabriel 1996]. The truth is that simple promises rarely deliver. Reality is always more complex and more interesting than that. Object orientation, meanwhile, has enjoyed an astonishing rise and, perhaps for other reasons, remains in the ascendancy, even if the search for the 'New New Thing'[9] in reuse has moved on.

Interestingly, following what seems to be an inevitable evolutionary trajectory, the focus has shifted, or turned back, to the next level of abstraction beyond languages and beyond the meta-languages of patterns, to projects, project organization and other 'soft' or 'human' aspects of programming, with methodologies such as *extreme programming* and *agile programming* dominating the quest.

4.3   The Origins of Object Orientation

Object orientation is an approach to design and programming rather than a fixed methodology.[10] This makes it a rather loose label. At root, it is a way of thinking, a programming style, a particular approach to modeling

---

[6] It is ten years since Richard Gabriel's book was published and he was using the past tense even then.

[7] The annual CHAOS report from the Standish Group includes IT project resolution statistics. The 1994 report claimed that 31% of software projects are cancelled, with a further 16% either over budget, late or reduced in the scope of their features or functions compared to the initial specification. In 2004, the numbers were respectively 29% and 18% (see **www.standishgroup.com**).

[8] These were the radical slogans which accompanied the announcement of the 'software crisis' and which were aimed at overturning the crisis, see [Assmann 2003, p. 6].

[9] This phrase is attributed to Netscape's Jim Clark, see [Lewis 1999].

[10] For a discussion of terminology and many interesting insights into object orientation, including the object-oriented conceptual framework, see [Madsen *et al.* 1993, p. 9]. In general, I try to follow the BETA language terminology: 'object orientation' is an outlook or perspective; 'object-oriented' is an attribute of specific tools or techniques (e.g., language implementations or analysis techniques).

the world in software. Object orientation as a programming style is distinct from any particular object-oriented language implementation.

In the first place, object orientation grew up around the need for a descriptive language for use in simulating discrete physical systems. In particular, it emerged from the work of Dahl and Nygaard at the Norwegian Computing Centre through the early and mid-1960s, which resulted in the Simula languages.[11] These ideas were in turn picked up in the early 1970s by Alan Kay's research group at Xerox PARC in California and drove the development of Smalltalk, which was initially an experiment in devising a language to teach programming concepts to children [Kamin and Samuel 1990].

Both Simula and Smalltalk (but Simula in particular) served as explicit influences for Bjarne Stroustrup, working at Bell Labs in the early 1980s and looking for a way of introducing what had become known in the literature as abstract data types into a C-style language, to try to overcome problems in writing very large systems. The specific context was large projects at AT&T, including telephone switch software (which typically were programs containing millions of lines of code). Coincidentally, an independent effort to harness the plain syntax and underlying efficiency of C to an object model was being pursued by Brad Cox and led to the appearance of Objective-C more or less simultaneously with C++.[12]

Just as both C++ and Objective-C set out with an explicit goal of creating a better C, so later twists in the story of object orientation have seen Java claiming a place as a better C++, and C# claiming in turn to be a better Java. James Gosling's group at Sun started work on what became Java in 1990, addressing the perceived shortcomings of C++ in the particular context of small, consumer devices such as set-top boxes. Java certainly achieves greater simplicity, greater language uniformity and a purer object model than C++, as well as wider goals of platform independence, language safety, and tamper resistance.

The work at Microsoft to create a better Java began in the late-1990s, as part of the Java-like managed code model for the .NET internet services framework. The result is C#, a rather small increment to Java in language terms, and a rather larger increment to C++, but one which so far is only available on the Microsoft platform. (Albeit that makes it a large marketplace.)

As well as this relatively linear evolutionary mainstream, a whole host of object-oriented languages have sprung up through several decades of research. Some have been shortlived, some have persisted, and almost all have contributed something of interest to the wider object-oriented research effort. From Beta to Sather to Eiffel to Dylan to Self to Python to Ruby, all have had some following, if only within the research community,

---

[11] See the discussion and timeline by Sklenar at *http://staff.um.edu.mt/jskl1/talk.html*.
[12] Stroustrup tells the history in [Stroustrup 1994, p. 175].

and one or two have found a more permanent niche. Many other already-established non-object-oriented languages have adopted object-oriented extensions. Smalltalk style, for example, caught on in the Lisp community in the 1980s with Common Lisp Object System (CLOS), which became a model for similar extensions to languages such as Pascal, as well as more esoteric ones such as Prolog and ML. Similarly, the true inheritors of the Pascal mantle are the Modula languages, of which Modula-3 is an object-oriented language, and Oberon which again is object-oriented (and, interestingly, is not class-centric).[13]

It is hard to think of a major programming language which has not been touched, in some way or another, by object-oriented ideas.

4.4   The Key Ideas of Object Orientation

The goal of the original Simula language was to reconcile natural models of description (of complex real-world behavior) with computation (specification of algorithms which could compute such complex behavior or compute with it), to support programmed simulations. From that starting point, the key ideas of object orientation emerged.

While traditional computing languages cut the world into algorithms and data structures, object-oriented languages instead cut the world into objects, each of which encapsulates both algorithms (behavior) and data (state). Running an object-oriented program becomes more like running a physical model of the world.[14] This different approach captures a number of insights, in particular that the real world is more naturally understood as discrete and not continuous (or at any rate that we can benefit from modeling it that way) and that, in the real world, behavior comes packaged with context (context-free behavior is of formal interest only).

A few high-level principles provide the basic modeling tools of object orientation:

- *Abstraction* hides detail by finding the commonalities between things, so that difference becomes variation

- *Data hiding* hides data inside objects as state

- *Interfaces*, or behavior hiding, expresses public behavior in public protocols and hides private behavior.

Different object-oriented languages vary in the ways they support these principles, but a small number of mechanisms are almost universal (at

---

[13] See the official page at ***www.oberon.ethz.ch***.

[14] 'A program execution is regarded as a physical model, simulating the behavior of either a real or imaginary part of the world.' [Madsen *et al.* 1993, p. 16].

any rate in the mainstream object-oriented languages, including Smalltalk, C++ and Java):

- *Encapsulation* supports data hiding; in class-based mainstream object-oriented languages, classes are the units of encapsulation

- *Inheritance* provides the mechanism for structuring relationships within object-oriented programs and for supporting code-sharing and reuse

- *Polymorphism* (sometimes referred to as dynamic binding), the headline characteristic of object-oriented languages, is the result of abstraction and the basis for reuse; the mechanisms that enable objects to display multiple behaviors are superclass (generalization) and subclass (specialization).

While a lot of theory has evolved around object orientation, object-oriented ideas are intended to be intuitive. As Coad and Yourdon put it, quoted in [Madsen *et al.* 1993], 'Object-oriented analysis is based upon concepts that we first learned in kindergarten: objects and attributes, classes and members, wholes and parts'.

Object orientation emerged very naturally in the context of computer simulations of physical processes. The purpose of simulating a process is to understand it; but, in order to simulate it, it must be modeled and modeling requires understanding. To break the regression, think of modeling as a way of transforming one kind of understanding into another kind (information in this respect is like energy or matter: it resists lossless compression). A model reduces a problem in a systematic way to recognizable objects, parts, and the relationships between them, allowing a deeper understanding to emerge from the complex dynamics which arise in the running system from the interactions between objects. A good model represents an object in a way which reveals more information about the object than was available without the model.

Arguably, all programming is based on the principle of abstraction[15] (all problem decomposition is abstraction by one means or another), but every language lends itself to a particular programming style (the one it makes easiest). Each language provides a different conceptual toolkit and encourages and enables different design and implementation techniques. Abstraction, inheritance and polymorphism are the essential characteristics of object-oriented languages.

'Abstraction', as [Koenig and Moo 1997, p. 9] rather neatly puts it, 'is selective ignorance'. Inheritance and polymorphism are what make abstraction in object-oriented languages different from abstraction in other programming languages.

Inheritance builds on the 'is-a' relationship as a way of capturing similarities between types. Objects in a program are defined into a

---

[15] There is an interesting discussion of abstraction in [Koenig and Moo 1997, p. 75].

hierarchy of increasingly specialized types, each of which inherits the more general properties of its parents, while adding specialist properties which can in turn be inherited by child classes that provide further specialization. For example, in a financial application, *current account* and *savings account* specialize the properties and behavior of a generic *bank account*. A *current account* 'is-a' generic *bank account* that has been specialized; so is a *savings account.*

Polymorphism (the ability to take multiple forms) enables objects to respond either as specialized types or as the types from which they inherit, allowing the programmer 'to ignore the differences between similar objects at some times, and to exploit these differences at other times' [Koenig and Moo 1997, p. 35]. Thus in the financial application example, a *current account* can be treated either as a *current account* or as a generic *bank account.*

Encapsulation

Object-oriented languages are strongly influenced by the idea of abstract data types (ADTs). The central idea of an ADT is that it defines a data structure and the operations which may be performed on it [Bishop 1986, p. 4].[16] To use an ADT it is enough to have access to the (public) operations it supports, without requiring any knowledge of its internal structure, and especially without requiring any knowledge of its implementation (that is, the internal data it contains and how it implements the operations it supports).

ADTs are a powerful idea and mark a big step forward in enabling programmers to create their own, user-defined, complex types, having something like equal status with the built-in types of a language. ADTs really belong to the 'data abstraction' revolution (the revolution before the object-oriented revolution), which spawned the Modula-2 language and culminated in the definition of the Ada language.[17] Ada brought ADTs into the mainstream, but C++ is the language that has taken Ada's ideas and made them successful.[18]

Support for ADTs, that is encapsulation, does not itself define a language as object-oriented (Ada is not object-oriented). However, it is a central idea of object-oriented languages. Encapsulation is the most basic pattern an object-oriented system can use. It is also a key programming

---

[16] For a different view, see [Madsen *et al.* 1993, p. 278] and [Craig 2000, p. 17].

[17] See [Bishop 1986] for a discussion.

[18] For an insight into why, the aside in [Stroustrup 1994, p. 192] about the relative sizes of the Grady Booch component library is illuminating: 125 000 lines of uncommented Ada to 10 000 lines of C++. Ada wasn't much liked by anyone (see the note in [Kamin and Samuel 1990, p. 248] of Tony Hoare's Turing Award lecture remarks). 'What attracted me to C++ had more to do with data abstraction than with object-oriented programming. C++ let me define the characteristics of my data structures and then treat these data structures as "black boxes" when it came time to use them.' [Koenig and Moo 1997, p. 12].

insight, an important step away from a focus solely on algorithm and implementation. In class-based object-oriented languages, encapsulation of objects is provided automatically by the machinery of class definition.[19] In the case of C++, encapsulation of user-defined data types through the mechanism of class definition is probably the key concept of the language.

Classes define objects whose instances are created at run time. Objects hold values and an object's methods provide the means of access to its values, whether to set, update, retrieve or perform more complex operations upon them. An object's methods define the interface that the object exposes or, in Smalltalk terminology, the protocol that it understands. (Terminology varies between languages: Java has interfaces and methods; Smalltalk has protocols and methods; and C++ has interfaces and what are interchangeably called either methods or functions.)

Object-oriented languages also allow objects to be extended to create new objects. In class-based, object-oriented languages, inheritance provides the extension mechanism. (But prototype languages, for example, use a copy-and-modify 'cloning' mechanism to create new objects from old.)

In C++, there is no requirement to follow the logical separation of interface from implementation with a physical separation of code. In contrast, Java formalizes the separation by separating the class declaration from the class definition (implementation). The interface provided by a class for manipulation of instantiated objects of the class is declared in an interface file, with only one class per file.

Inheritance

Inheritance is the mechanism in class-based languages that allows new classes to be defined from existing ones. Not all object-oriented languages are class-based (e.g., there are actor- and prototype-based object-oriented languages[20]), but most are. Therefore while, strictly speaking, inheritance is not universal in object orientation, it is certainly typical.

Inheritance is a parent–child relationship between types, usually called subclassing in Smalltalk and Java (a class is subclassed from a superclass) and derivation in C++ (a class is derived from a base class). Whereas an abstract data type is a black box ([Stroustrup 1994, p72]) which can't be varied or adapted except by redefining it, inheritance provides a mechanism that allows flexible variation of abstract data types, in order to express both the differences and similarities between general types (such as *BigCat*) and their specializations (*Lion* and *Tiger*).

---

[19] [Beaudouin-Lafon 1994, p. 15] says, 'a class is simultaneously a type and a module', where type implies interface and module implies implementation.

[20] Actor languages with an object-oriented flavor include ABCL and Obliq; Self is probably the best known prototype language and is thoroughly object-oriented [Craig 2000].

The key differences in the way that languages approach inheritance are in whether multiple inheritance is supported or not, and in whether the inheritance hierarchy is singly rooted or not. Smalltalk and Java are singly rooted, meaning that there is a single privileged root class from which all other classes ultimately derive and which defines (and implements) a universal set of common class behavior. In both languages, all classes are subclasses of an `Object` class; Eiffel is similar, with all classes derived from the `ANY` class, either implicitly or explicitly. In C++, on the other hand, there is no universal base class: the inheritance hierarchy may have multiple roots. C++ also allows multiple inheritance, so that classes are unconstrained in the number of parent classes from which they may derive. Similarly, Eiffel allows multiple inheritance. Smalltalk allows only single inheritance, that is, a class may only have one parent, while Java allows multiple inheritance of interfaces, but only single inheritance of implementation.

Inheritance is not just additive. It does not just consist of adding new definitions in child classes; it also enables the redefinition in child classes of the existing behavior of parent classes. Typically this is known as overriding, the child overriding the behavior of the parent with its own specialized behavior.

Object-oriented languages typically distinguish between abstract behavior, which defines an interface to an object but which does not provide an implementation, and concrete behavior, which both defines and implements an interface. Abstract behavior is provided by defining abstract methods (in C++, virtual methods). Abstract methods emphasize the point that inheritance relationships are defined by methods, but not their implementations. Classes can also be defined as abstract. Abstract (pure virtual, in C++) classes cannot have instances. In C++, abstract classes provide the mechanism for polymorphism. Child classes are required to implement the abstract methods of a parent.

Inheritance is explicitly a mechanism of class-based languages. Non-class-based object-oriented languages, for example prototype languages, provide equivalent mechanisms based on the idea of cloning new objects from template objects ('prototypes'), to create 'pseudo-classes' of similar objects, rather than true classes, but the purpose is essentially the same [Appel 1998, p. 310].

Polymorphism

Intuitively, the operations that can be performed on a value depend on the type of the value. Adding numbers makes sense and concatenating strings makes sense, but adding strings or concatenating numbers do not make sense, or not in any generally agreed way.

Different programming languages treat the notion of type in different ways. At one extreme, the functional programming world favors complete type-inference systems that amount to full logics (i.e., languages) in their own right and are completely independent of any physical machine representations of values. At the other extreme, procedural languages such as C, as well as older languages such as Fortran, have type systems which have evolved naturally, and informally, from the physical representation of values in machine memory (bits, bytes, words, long-words, double-words, and so on).

Object-oriented languages fall somewhere between these extremes. Every object in an object-oriented program is really an instance of a fully encapsulated, and possibly user-defined, type. In a class-based language, class definition is the same as type definition. The inheritance relationships between objects are type relationships.

Polymorphism simply means 'having many forms' [Craig 2000, p. 4]. In an object-oriented context, it is often alternatively described as 'dynamic typing'. Polymorphism exploits a simple principle of substitutability: two objects related by inheritance (or an equivalent mechanism) share a common subset of methods. However, the implementation of those methods may differ.

Methods can be invoked on a child object based simply on what we know about its parent. We know that a set of methods is supported, whatever their implementation and whether or not we know what other specializations have been added. Sometimes we only know the parent class of an object and not which specialization we are dealing with. (For example, we may know that we have an event, but not what type of event we have, or that we are dealing with a document, without knowing what kind of document). We therefore know what common methods are supported by the object, whether or not we know what their behavior is, or what other methods are supported. Often we may not even care about the details, for example if we simply want to tell an object to print itself.

At other times, we may explicitly want to use the specialized behavior of the derived object. Polymorphism is the ability of the object to switch between these different behaviors, to appear in the run-time context of a program variously as an instance of the parent object or as the derived object; in other words the ability of an object to behave differently at different points of the program execution.[21]

How polymorphism is implemented varies between languages. For example, Smalltalk uses universal run-time type checking to provide the underlying support for run-time polymorphism. C++, on the other hand, employs static type checking, but allows a 'virtual' dispatch mechanism to support constrained run-time polymorphism.[22]

---

[21] See [Koenig and Moo 1997, p. 77] for a printing example.

[22] Polymorphism is also frequently referred to as 'dynamic binding'. [Bar-David 1993, p. 87] gives a slightly different slant to his definition of dynamic binding as 'the ability of an object to bind – dynamically at run time – a message to an action (or code fragment, if you will). The idea is that, in a given system, many different objects may respond to the same

A weaker notion of polymorphism is usually qualified as parametric polymorphism. It refers to functions which can be applied to arguments of a different type. This is not polymorphism in the same sense as dynamic typing, because the implication is that such functions execute identical code [Appel 1992, p. 7] whatever the argument type; in other words, overriding of implementation is not allowed. A simple example is the language operator (i.e., the built-in function) denoted, in the C language, by &; it creates a pointer to its argument, irrespective of the argument type [Aho *et al.* 1986, p. 364]. Functional languages such as ML and Scheme support parametric polymorphism systematically, while conventional procedural languages such as C and Pascal do not (although they may support occasional instances, such as the & operator in C). Object-oriented languages typically support polymorphism in its stronger sense.

Different languages adopt different strategies for type checking. The primary distinction is between static and dynamic type checking. Static type checking means that types are checked at compile time: if the compiler encounters static type errors, it rejects the program. Dynamic type checking occurs at run time, that is, during program execution: if the program encounters dynamic type errors, it halts the program or flags a run-time error in some other way. A different way of stating the distinction between them is to say that static typing concerns the type of the declaration (for example, a C++ reference to a variable or a C pointer to a variable), while dynamic typing concerns the type of the value (for example, a Smalltalk object) and the difference emphasizes the different underlying programming philosophies.

Statically typed languages include Pascal, C, C++, Ada and the functional languages ML, Scheme and Haskell. Statically typed languages are regarded as strongly typed if the type system enables static analysis to be sufficient to determine that execution of a program will be type correct [Aho *et al.* 1986, p. 343], although it is not required that the compiler necessarily be able to assign a type to every expression. Such expressions require run-time evaluation. Strongly typed languages include Pascal, C, Ada, Java, Simula, Modula-3 and C++ (except for the single case of a dynamically typed method).

Dynamically typed languages are those in which all expressions are typed and checked at run time. For example, Smalltalk and Eiffel use 'dynamic method lookup' [Appel 1992, p. 7]. (Smalltalk is sometimes described as untyped, like Lisp, but it makes more sense to say that the type information has been moved where it belongs, into the object as part of the object's encapsulation).

---

message – say "print" (i.e., display yourself to a display device); they just respond differently'. Alternatively, see [Ambler 2004]: 'Different objects can respond to the same message in different ways, enabling objects to interact with one another without knowing their exact type'.

Most languages that perform static analysis (such as Pascal, C, Ada, C++ and Java) require type declarations in programs for all declared types, whether data, operations (i.e. procedures, functions or methods, depending on the language's terminology) or user-defined types. (ML and Haskell are exceptions that use static type inference).

C++ is something of a hybrid. While it mostly checks types statically, it explicitly enables a mechanism for dynamic typing for polymorphic objects, as well as a limited form of type analysis (it is really mangled name matching) for objects loaded at run time, such as precompiled libraries.

Dynamic typing in C++ is enabled by addressing an object through a pointer or reference (although not every pointer or reference implies polymorphism of the object on the other end[23]). A C++ pointer addresses an object of the type of the pointer or of a type publicly derived from it. The type is resolved at run time, in principle at each point of execution in the running program. In C++ (and in Java), this allows the use of a parent class reference to address a local variable, a class instance variable or a method parameter instantiated by an object of a child class. In this case, it is the real type of the object which determines which methods are called, in cases where methods are overridden in a class hierarchy.[24] This enables a program to invoke a method on an object with a single call that is 'right first time', regardless of where in the class hierarchy the object is defined and regardless of the actual behavior of the method. (A `calculateBonus()` method in a payroll system, for example, performs the correct calculation, depending on the real type of the object, not on the type of the pointer.) The alternative, if polymorphism were not available, would require testing for all possible types of the object to isolate the particular case in every case every time, which is laborious and error prone, as well as verbose.

Java is statically typed but all Java methods are bound at run time.[25] All Java objects are typed and their types are known at compile time. The compiler performs static type checking. Dynamic binding means that polymorphism is always available, that is, all methods are like virtual methods in C++ and can be overridden. In other words, every subclass can override methods in its superclass [Niemeyer 2002, p. 11].

Both the static and dynamic approaches have their adherents. The really significant difference between them is that each lends itself to a certain style of programming.

The most common arguments in favor of statically typed languages are run-time efficiency (all types are computed at compile time, so

---

[23] See the discussion in [Lippman 1996, p. 21].

[24] See the discussion in [Warren et al. 1999, p. 33–34].

[25] Run time polymorphism, that is, dynamic typing, applies in C++ only through virtual functions [Koenig and Moo 1997, p. 35]. A virtual function counts here as a pointer, i.e. a pointer to a function in some class, its base class or a class derived from it.

there is no run-time overhead) and program safety. Thus, says [Appel 1992], programs run faster and errors are caught early. In statically typed languages, many programming errors are trivial type errors due to programmer oversight, which can be caught and corrected at compile time. In dynamically typed languages, these may arguably become run-time errors. (Arguably, because adherents of dynamically typed languages would probably claim that the rigidity and inflexibility of the type system caused the errors in the first place.)

Type declarations probably do improve code readability and make programmer intentions clearer. On the other hand, dynamically typed languages such as Smalltalk and Python allow greater expressivity and explicitly license a more exploratory programming style, as well as avoiding some of the binary compatibility problems of applications and libraries written in statically typed languages.

4.5   The Languages of Object Orientation

Smalltalk remains the canonical object-oriented language, but almost certainly more object-oriented code has been written in C++ and quite possibly in Java too. These three languages constitute the object-oriented mainstream. Python, a newer language more specialized for scripting and rapid development, may well be on its way to joining them in the mainstream; if it can oust Perl from its position as the universal language of the Web, it will certainly succeed. C# is another, newer language which has set its sights on conquering the Java world as part of Microsoft's .NET services effort. However, it currently remains a niche language.

The differences between these languages and the other object-oriented languages which come and go, are in large part about style (and history). However, in the differences between Smalltalk and C++ in particular, there are insights into more interesting, and deeper, differences about what matters most in programming, for example the trade-off between flexibility and correctness or, perhaps more precisely, what is the best route to correctness and to well-behaved programs which are also capable of evolving to serve the evolving needs of their users. Differences of language style reflect different intuitions about programming style (that is, not just about the style of programs, but also about the different styles of programming practice, the actual activity of designing and writing programs).

The key language differences can be fairly easily summarized:

- single versus multiple inheritance

- a single root class versus ad hoc class hierarchies

- dynamic versus static type checking and method binding.

Some other differences seem to have been relegated to questions of academic interest only by the success of the mainstream languages:

- encapsulation versus delegation

- classes versus prototypes.

Languages which seemed to hold promise for a more concrete and intuitive approach to exploratory programming (for example, Self or Squeak, both Smalltalk derivatives) seem to have been rapidly sidelined.

One seemingly arcane research topic which has migrated in the other direction, from the fringe to the language mainstream, is *reflection* or introspection. Both Java and C# now support reflection, as does Objective-C; run-time program objects are reflective (introspective) and are able to consider themselves as data to be manipulated. Smalltalk also uses reflection, in particular as the mechanism which enables objects to examine themselves to discover their own types.

Java supports reflection for similar reasons, but with a different mechanism, providing a set of reflective classes that allow users to examine objects to obtain information about their interfaces [Craig 2000, p. 197] and to serialize objects. (In Smalltalk, reflection is a meta-property of all class objects.)

Reflection is a rather esoteric property of a few languages (Smalltalk, Self, Java and C#), but it should be seen as part of the search to define more flexible languages, with more natural support for distributed and parallel programming, and part of a longer tradition of languages which include meta-level operations enabling a program to represent itself and describe its own behavior. Smalltalk, like Lisp, can manipulate its own run-time structures [Craig 2000, p. 184].

Other areas of object-oriented research focus less on language techniques than on run-time issues, such as just-in-time compilation techniques (for Java and C#, as well as Python, which are all interpreted languages). It seems unlikely that the familiar object-oriented languages will evolve very radically. The more likely areas of change will be the drive towards binary-object encapsulation for distributed programming (in the style of CORBA), which perhaps suggests an eventual convergence between object-oriented techniques and more declarative programming language styles, under the influence of the success of XML. (Declarative programming supports greater semantic transparency.)

Meanwhile, with C++ and Java, and perhaps Python, as the dominant languages, the programming mainstream now seems very squarely object-oriented.

Smalltalk

Smalltalk dates back to 1972 when the research project from which it originates began, although it came of age with the Smalltalk-80 release.

It drew its inspiration from Simula and was developed by Alan Kay's research group at Xerox PARC [Beaudouin-Lafon 1994, p. 57]. In many ways, Smalltalk is the canonical object-oriented language and it was certainly the first to achieve critical mass. It was launched into the spotlight in 1984, when *Byte* magazine devoted an entire edition to it.

Smalltalk gathered significant commercial momentum. However, since its peak in the late 1980s and early 1990s, it has largely been in decline. It has been decisively beaten (in terms of the programming mainstream) by C++ and Java. Its most interesting legacy has been its promise of a very different way of creating large programs, a more evolutionary and exploratory approach than is encouraged by the 'specification first', top-down style of C++.

Smalltalk is a dynamically typed, class-based, message-passing, pure object-oriented language:

- Everything is an object and every object is an instance of a class.

- Every class is a subclass of another class.

- All object interaction and control is based on exchanges of messages.

Conceptually at least, Smalltalk is remarkably clean and uniform, applying the object approach consistently and deeply. In particular, Smalltalk has a single root class, called `Object`, from which all objects ultimately inherit. `Object` itself inherits from the class named `Class`, which inherits from itself (to satisfy the rule that all classes are subclasses of another class).

In Smalltalk, a class whose instances are themselves classes is called a meta-class. Thus `Class` is an abstract superclass for all meta-classes and every class is automatically made an instance of its own meta-class. This mechanism is used to introduce the notion of *meta-class* methods ('class methods'), which all subclasses inherit and which define the canonical shared class behavior. For example, class methods typically support creation and initialization of instances and initialization of class variables.

The Smalltalk system at run time consists only of objects. All interactions between objects take the form of messages. The message interface of an object is known as its protocol and message selection determines what operations the receiving object should carry out. Each operation is described by a method. There is one method for each selector in the interface of the class.

All objects are run-time instantiations of classes. Classes are defined by class descriptions that specify the class methods (i.e. the meta-class methods), instance methods and any instance variables [Goldberg and Robson 1989, p. 79]. Method specifications consist of a message pattern (equivalent to a function prototype in C++) which specifies the message selector and argument names, and an implementation [Goldberg and Robson 1989, p. 49]. A protocol description for each class lists the messages understood by instances of the class.

The message-passing model is uniformly applied as the single control mechanism for objects. Objects respond to messages and issue messages, and there is no other control mechanism in the system. For example, a new object is created by sending a message to the required class, which is itself an object (because `Class` is itself an object) and can therefore receive messages. The class object creates the new class instance. Message expressions specify a receiver (the intended target object), a selector (the message name) and any arguments [Goldberg and Robson 1989, p. 25].

Inheritance is used as the mechanism which enables sharing between classes. In other object-oriented languages, classes are definitional constructs that define instances and it is these instances which are objects (i.e., an object instantiates a class but a class is not itself an object). This is not the case in Smalltalk, in which everything is an object, including numbers, characters, Booleans, arrays, control blocks, and even methods and classes. Smalltalk has a rich hierarchy of ready-made classes (230 classes in Smalltalk-80 with 4500 methods) [Mevel and Gueguen, p. 5].

The object purity of Smalltalk extends all the way down to what in other languages would be the purely syntactic level of control structures. This makes its syntax idiosyncratic compared with other languages. Probably the most unfamiliar aspect of Smalltalk syntax for anyone with a background in procedural languages is the absence of familiar control constructs such as `if-then-else`. Instead, control blocks act as switches. For example, compare a conventional C-style `if-then-else` with a Smalltalk conditional block, using a Boolean object and `ifTrue:` and `ifFalse:` messages. Certainly it can appear radically unfamiliar for anyone coming from a more conventional programming background.

A final idiosyncrasy (although it may seem more natural to newer generations of programmers brought up on IDEs rather than the command-line) is that Smalltalk cannot be invoked as a simple language interpreter or compiler, but is instead part of a complete graphical programming environment. Smalltalk programs do not compile into conventional executables and libraries, with conventional linkage models, but instead dynamically update the running image of the complete live environment. The Smalltalk system can thus be modified at run time (unlike a conventional compiled executable). The language (and its associated tools) are thus embedded in a live, interactive environment, which is consistent with the origins of the language and its goals (a teaching language for novice programmers, based on a 'physical world' metaphor). Snapshots of the environment can be created as persistent images.

An irony is that where Smalltalk aims for simplicity, the language (and the associated 'object theory') turns out to be surprisingly complex. While Smalltalk failed to gain much hold as a teaching environment, it found a number of commercial niches (it remains popular for financial modeling applications) and it has retained its place as an 'extreme' language for (far from novice) object purists. A great deal of advanced object-oriented programming practice and theory, from patterns to the philosophy of reflection to extreme programming praxis, have originated in the Smalltalk world.

An interesting Smalltalk spin-off is the Self language, designed by Randall Smith and David Ungar, which originated at Xerox as a vehicle for exploratory programming and an experiment in an object-oriented language not based on classes. Instead of classes, Self is based on the notion of prototypes. New objects are derived by cloning and modifying existing prototype objects. Self takes the idea of a language embedded in an environment modifiable at run time to an (interesting) extreme. By removing both the theory and the machinery that comes with classes (inheritance, polymorphism and so on), it removes almost all of the complexity, while still retaining the power of object-based abstraction. Self espouses as a central principle that an object is completely defined by its behavior.[26] The corollary is that programs are not sensitive to the internal representations chosen by objects or, indeed, any other hidden properties of programs.

C++

C++ originated from a networking-support research project at Bell Labs in 1979, as an attempt by Bjarne Stroustrup to improve C by adding class concepts derived from Simula to support powerful but type-safe abstract data type (ADT) facilities. Indeed those origins are made transparent by its first incarnation as 'C with classes'.

The central concept of C++ is that of class [Koenig and Moo 1997]. Classes enable user-defined types that encapsulate behavior and data. Originally, C++ classes began as elaborations of C structs. While structs allow structured data to be defined and managed to create user-defined complex data types, classes extend the idea to include method definitions as well as data. (C++ retains the notion that a simple class that defines no methods is synonymous with a struct.)

In its first implementations, C-with-classes and later C++ were implemented as pre-processors, which translated C++ into plain C and then invoked the standard C compiler. (Again, the history is in the name: the first C++ implementation was named Cpre) [Stroustrup 1994, p. 27]. In a general sense, C++ thus includes the C language but C++ is not a pure C superset (unlike Objective-C, for example).

The goal of C++ is to enable the same level of type safety as is enjoyed by built-in language types to be enjoyed by user-defined data types. C++

---

[26] See the *Self Programmers Reference Manual*, p. 55 at ***http://research.sun.com/self/language.html***.

also provides improved type safety for built-in types compared with C, for example with language constructs designed to support immutable values (the `const` construction and the 'reference' operator). Its secondary goal is to do so without compromising either the efficiency of C or C's ability to operate (when necessary) close to the machine level.

Compared with Smalltalk, its goals make C++ inherently a hybrid language, sacrificing purity in favor of pragmatism. C++ is often said to be not an object-oriented language at all, but a language which can be used to program in a number of different styles. The more use that is made of advanced language features, the closer the style becomes to object orientation. However, there are advanced features which have little to do with object orientation (as understood in the purer sense of Smalltalk at any rate), for example the templating support for parametric (also known as generic) programming styles.

In summary, C++ is a strongly statically typed language with support for classes.

- Objects are optional, but when used they are based on classes, which, at one extreme, may be C-like structs and, at the other, may define pure virtual (polymorphic) objects or may fall somewhere between.

- Objects are created from classes by constructor methods and are deleted by destructor methods, which may be defined (for complex classes) or default to standard methods if not defined.

- Objects can control access to their data and methods by declaring them private, shared or public.

- Values can be made immutable by declaring them `const` and all objects can be passed by value, by reference or by pointer.

- Separation of interface from implementation is encouraged but not enforced (for example, methods may be declared inline and their implementation specified at the point of definition, within a class definition). Definition and implementation can be mixed and there is no requirement for separate interface definition files. Multiple definitions and implementation specifications can be provided within a single file. C-style `#include` preprocessor directives are used to manage definition inclusion.

- There is no notion of a root class and, therefore, no definite class hierarchy. Multiple hierarchies can be created within a single program and multiple inheritance is allowed (so that a single class may inherit from multiple parents). Advanced object-oriented features such as reflection are not supported.

Run-time polymorphism is the exception in C++ rather than the rule and, for all other cases, type checking is performed at compile time.

Run-time polymorphism is enabled only for classes which are defined as pure virtual, in which case method dispatch is completed at run time through a 'vtable' (a virtual method dispatch table). Virtual methods use run-time binding and are not determined at compile time.

C++ retains a conventional C-style execution and linkage model. There is no automatic garbage collection in C++. Memory management is the responsibility of the programmer, making the language flexible and powerful but also dangerous (carelessness leads to memory leaks).

Java

Just as C++ began as an exercise to improve C, so Java began as an exercise to improve C++ and, in particular, to simplify it, straighten out inconsistencies and make it less dangerous (for example, proof against memory leaks) as well as more secure (in the sense of tamper-proof, the origin of the Java 'sandbox' application model) and, therefore, more suitable for a wider range of devices (in particular, for smaller, consumer-oriented systems). Perhaps even more importantly, from the beginning the Java implementation model aimed at maximum platform neutrality and a write-once–run-anywhere model.

Java language programs are thus compiled into an interpreted inter-mediate language which is executed by a Java virtual machine (VM) running on the target hardware. Any Java code runs on any Java VM, thus providing abstraction from physical hardware. In other words, Java pro-vides a software environment for code execution rather than a hardware environment.

In this sense, Java is like the pure object-oriented model of Smalltalk, which similarly provides a software execution environment based on a VM. Unlike Smalltalk, Java programs are separable from the execution environment and its linkage model is more akin to a conventional executable and library-linkage model.

The VM approach also allows Java to meet its goals of robustness and security. The VM controls access to the resources of the native environment, thus enabling a garbage-collected execution environment (so that memory management is the responsibility of the environment, not the program), as well as a security sandbox, isolating Java programs from the native environment (malicious software can at worst only attack other code executing on the VM and has no access to the VM itself, nor to the underlying system).

Java programs pay a price for the execution model, in the overhead of interpreting Java intermediate byte code. However, Java VM technol-ogy exploiting sophisticated compilation techniques has eroded the raw speed differences between executing Java byte code on a VM and exe-cuting native processor instructions, to the point where execution speed differences are almost insignificant.

Java has been less successful at reducing latency of program startup, however, which requires the complete Java environment to be initialized. Java has also struggled to slim down its substantial platform memory footprint. For desktop PCs and 'single-function' consumer devices such as set-top boxes, this is less of an issue than it is, for example, on mobile phones, where Java competes for resources with native code. Pure Java solutions such as JavaOS, which replaces the native operating system with a lightweight Java system sufficient only to host the VM, have not been successful to date, although the Jazelle project has challenged conventional solutions by providing a Java solution in dedicated hardware. Jazelle remains a contender in the mobile phone space.

From a language perspective, Java makes an interesting contrast with C++. It succeeds in its goals of providing an object-oriented language that is simpler and purer than C++, while avoiding the syntactic eccentricities of Smalltalk; it remains syntactically quite conventional and close to its C++ origins.

Like C++, Java is strongly statically typed. Unlike C++ and like Smalltalk, it is a purely class-based language, with an `Object` root class.

- Native number, character and string types are defined by the language; all other types (including all user-defined types) are objects.

- Every object is an instance of a class and every class is a subclass of another class, except for the root class.

- Objects are created from classes by constructor methods and are deleted by destructor methods, which may be defined (for complex classes) or default to standard methods if not defined.

- Objects can control access to their data and method members by declaring them private, shared or public.

- Values can be made immutable by declaring them `const`, and all objects can be passed by value, by reference or by pointer.

- Separation of interface from implementation is enforced. Every class consists of an interface definition and an implementation specification in separate files, with only one class per file.

- Unlike C++, all objects are run-time polymorphic (all methods employ late binding).

- Garbage collection is automatic. All program resources are cleaned up and recovered by the VM when a program completes (or is terminated).

Java's success has been striking and, in many ways, it is a model language. However, compared with C++, it is relatively inflexible and constrained (by design) and its deliberate isolation from the underlying device makes it generally unsuitable as a system-level language.

Microsoft has made its own attempt at improving Java and providing a managed-code solution of its own (for the .NET services platform, which competes with Java) in the form of C#. As a language, C# contains some interesting features, including a reflection model. However, the history of C#, which first emerged as a set of unilateral Java extensions, makes it somewhat unconvincing as a genuine language advance.

Other Languages: Objective-C, Eiffel and Modula-3

Objective-C was written by Brad Cox in the early 1980s. It has a visible Smalltalk influence, for example in some of its syntax, and in its adoption of run-time typing (in contrast to C, C++ and Java). Also unlike C++, it is a true superset of ANSI C, that is, it is a pure extension of C that leaves the core of C unrefined.

It was adopted for the NeXTStep, which employed a Mac-based flavor of Unix, and from there it was inherited by Mac OS X, in which it remains highly visible. (For native application development, the object hierarchy remains based on Objective-C, complete with the NeXTStep, i.e. NS, class-naming convention.) Objective-C was also an explicit influence and, indeed, the inspiration and model, for the Psion in-house object-flavored C that preceded the adoption of C++ for what became Symbian OS.

Eiffel emerged at around the same time as Objective-C, that is, after Smalltalk but before C++ had become dominant. Eiffel was designed as a commercial, pure object-oriented language intended to compete with Smalltalk, with a more conventional syntax. It included a comprehensive and pure object-oriented class library, including ready-to-go container, collection and iterator classes, well in advance of anything comparable in the C++ world. (The C++ Standard Template Library emerged well after the C++ language.)

In the Pascal lineage, Modula-3 evolved by way of Modula-2, adding object-oriented features and garbage collection.[27] Both Eiffel and Modula-3 are influenced by Simula, but while Simula and C++ allow a choice between static and dynamic binding, with dynamic binding provided via virtual methods, Eiffel and Modula-3 offer a pure polymorphic model with universal dynamic typing and run-time binding, for which run-time efficiency is the trade-off.

In other respects, both languages share similarities with C++. Classes in these languages are elaborations of the concept of a record, a description of a list of fields together with the methods that operate on them (just as C++ classes are elaborations of the concept of a C struct; structs and records are, in essence, synonymous). Again like C++, both Eiffel and Modula-3 allow multiple inheritance.

---

[27] According to **www.m3.org**, the language was first defined in 1989.

Part 2
The Layered Architecture View

5

The Symbian OS Layered Model

5.1 Introduction

This book explains the architecture of Symbian OS using the system model (see the fold-out and Figure 5.1), which represents the operating system as a series of logical layers with the Application Services and UI Framework layers at the top, the Kernel Services and Hardware Interface layer at the bottom, sandwiching a 'middleware' layer of extended OS Services.

In a finished product, for example a phone, Symbian OS provides the software core on top of which a third-party-supplied 'variant' user interface (UI) provides the custom GUI with which end-users interact and which directly supports applications. Typically, the variant user interface, including the custom applications supplied by the phone manufacturer, is considerably bigger than Symbian OS itself.

Beneath the operating system, a relatively small amount of custom, device-specific code (consisting of device drivers and so on), insulates Symbian OS from the actual device hardware.

5.2 Basic Concepts

The remainder of this chapter summarizes the key concepts of the system model and then describes the operating system layer by layer, starting at the top with the UI Framework and working down to the Kernel Services and Hardware Interface layer.

The basic approach taken by the model is to decompose the operating system into layers, and to further decompose the layers as necessary into blocks and sub-blocks before finally arriving at collections of individual components. Layers are the highest level abstraction in the model; components are the lowest level abstraction, the fundamental

Symbian OS

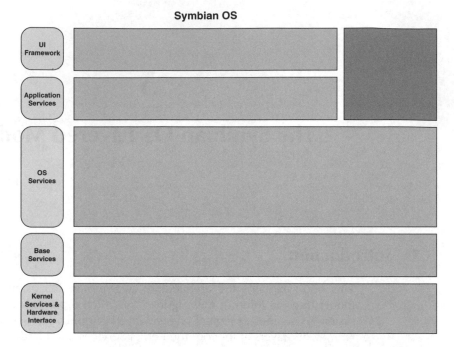

Figure 5.1   Symbian OS layered system model

units of the model; blocks and sub-blocks decompose layers by functionality – roughly speaking, by broad technology area. The key concepts
used by the system model therefore are layers, blocks and sub-blocks,
component collections and components.

Components provide the essential mapping from the logical model to
the concrete system. While layers, blocks and sub-blocks are essentially
logical concepts, components are physically realized in software, typically consisting of multiple files in the operating system delivery (e.g.
source code files including test code; built executables including libraries;
data and configuration files; build files; and documentation). However,
from the perspective of the model, components are treated atomically
and constitute the smallest units of architectural interest.

The complete component set shown in any particular version of the
model represents the superset of all components delivered by that release
of the operating system and intended to run on any Symbian OS device,
whether a phone or some other product, a development board or other
test hardware, or an emulator build of the system running on a host
operating system (such as Microsoft Windows).

Test components and tools are considered outside the scope of the
Symbian OS model, although they form an essential part of the model
of the complete delivery of the operating system as shipped by Symbian

to customers. (They are shown in a full product model as the Symbian Toolkit.)

Because the model reflects the concrete system, a new version of the model is published for each release of Symbian OS. The model has also evolved in its own right since the first versions were published for Symbian OS v7, in particular to bring it closer to the concrete system.

Layers

The model adopts a conventional software architecture interpretation of layers [Buschmann *et al*. 1996]: each layer abstracts the functionality of the layer beneath and provides services to the layer above.

Within each layer, components are either grouped directly into collections according to functionality (and to some extent also according to collaborations and shared dependencies); or are grouped into collections within blocks and possibly sub-blocks, which are broadly based on technologies.

The goal of the model is to impose manageable granularity onto the operating-system architecture, to make it easier to understand and to navigate. Hence, layers are useful approximations of structure, not precise specifications of architectural relationships. There is no concrete mechanism that instantiates layers in the existing system (i.e. there is no make file or equivalent).

However, the broad principles of the layering hold good: although there are some exceptions, dependencies in general flow downwards from higher layers to lower layers; and dependencies in the reverse direction are considered to be anomalies. In general, services are abstracted through the layers, with higher layers abstracting the services of lower layers, although for reasons of efficiency there is no requirement that layers only access the services of the layer immediately below them; thus the functionality of lower layers is accessible to all layers above.

One reason for showing the system as layered is to show how system functionality is increasingly abstracted away from hardware (at the bottom) and towards users (at the top); successive groups of tasks are increasingly abstracted from more basic tasks. A widely accepted principle for creating a layered model of a system is the 'inverted pyramid of reuse', characterized by the slogan 'Keep the base layer slim' [Buschmann *et al*. 1996, p. 39].[1]

Layers in the system model are defined with the following guidelines in mind:

---

[1] 'Layers' is a well known architectural pattern, the best known example probably being the OSI Seven-Layer Model. The Layers pattern is described and discussed in [Buschmann *et al*. 1996].

- all the services provided by a layer are at a similar level of abstraction

- a layer is relatively logically cohesive and relatively self-contained (both inexact terms, used with commonsense meaning)[2]

- a layer provides services to higher layers ('upwards')

- a layer delegates tasks to lower layers ('downwards')

- dependencies flow consistently from higher layers to lower layers (but dependencies are allowed sideways within layers)

- requests travel downwards

- notifications travel upwards

- higher layers abstract the services of lower layers away from machine-centric services towards user-visible functionality

- a layer provides services as far as possible via well-defined external interfaces, which can be separated from the internal interfaces available within the layer

- a layer could be a delivery unit (although, in the current system, no layer is delivered independently).

Blocks

A block or sub-block in the system model (see Figure 5.2) roughly corresponds to a 'technology domain'.

Blocks are used as a pragmatic way of partitioning layers into meaningful divisions according to commonsense criteria, with sub-blocks providing finer grained divisions for convenience. There is no concrete mechanism that instantiates blocks or sub-blocks in the existing system (i.e. there is no make file or equivalent).

Blocks in the system model are defined with the following guidelines in mind:

- a block is relatively logically cohesive and relatively self-contained

- a block is relatively cohesive and relatively decoupled (measured in terms of the coupling of the component collections it contains)

- a block provides services to blocks in the same layer ('sideways') or to blocks or component collections in higher layers ('upwards')

- a block delegates tasks downwards or sideways

---

[2] Cohesion and coupling are standard concepts used to analyze software complexity. See, for example, [Henderson-Sellers 1996] and the influential papers by Lionel Briand and others at the Fraunhofer Institute, such as [Briand *et al.* 1997].

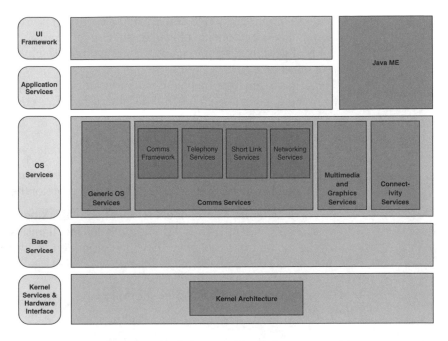

Figure 5.2 Block decomposition in the system model

- a block 'consolidates' the sum of services provided by the component collections it contains into a technology domain

- a block is not a delivery unit – it makes sense to partially deliver, update or remove a block.

Components and Component Collections

Components are the basic entities of the model and the smallest units of architectural interest. Importantly, components have a concrete interpretation in the source system, corresponding approximately to parts of the source tree controlled by a single high-level build file (an MRP file in the Symbian build and delivery idiom but, more generally speaking, a high-level make file).

Components are also the basic units of optionality in the system, the level at which common, optional and replaceable functionality is defined and at which it may be (respectively) included, removed or re-implemented by the respective licensees of Symbian OS.[3]

Component collections group individual components into coherent sets of collaborating components. In principle, a component collection

---

[3] For a more detailed discussion, refer to Appendix A.

delivers a complete, discrete and identifiable subset of system functionality. In practice, component collections are derived from a 'commonsense' analysis of existing system functionality, as well as the physical organization of the source tree.

There is no concrete mechanism that instantiates component collections in the existing system (i.e. there is no make file or equivalent).

Component definition follows these principles:

- a component is the smallest architectural unit of the system

- a component is understood as a set of implementation units that are built together to provide a discrete, reusable piece of the system

- in concrete terms, a component is identified with a single MRP file that ensures alignment with build and delivery mechanisms (in versions of the model up to Symbian OS v8, a component is identified with a high-level `bld.inf` file rather than an MRP file.)

Components should also obey the following guidelines and display these properties:

- a component is relatively cohesive (in essence it has been designed as a discrete part of the system)

- a component is a reusable unit of the system

- a component is the smallest unit of architectural significance and the finest grained unit of description, management and distribution of the system

- a component is implemented by at least one and possibly many collaborating sub-units

- no part of any component is shared by other components

- all interfaces defined at higher levels of the model are implemented by components.

In all, the system model for Symbian OS v9.3, the latest version of the operating system at the time of writing, defines approximately 250 individual components.[4] However, there is still a significant degree of idealization in the component definitions and, in many cases, the detailed mapping from the model to the system as built and delivered is approximate. In other words, the model serves as a useful logical

---

[4] Appendix A documents 258 components, for example, and does not include Toolkit components.

description, but cannot necessarily be unambiguously followed down to file level. (Improving alignment is an ongoing task.)

Component collections are defined with the following guidelines in mind:

- a component collection is relatively cohesive and relatively decoupled (in terms of the coupling of the components it collects)

- a component collection provides services to other collections within its block or layer ('sideways') or to blocks or component collections in higher layers ('upwards')

- a component collection delegates tasks downwards or sideways

- a component collection groups logically related functionality

- a component collection exposes the interfaces provided by the components it collects

- no component collection is shared between blocks or layers

- no component is shared between component collections

- a component collection is not a delivery unit – its individual components may be delivered, updated, or removed singly.

5.3   Layer-by-Layer Summary of the Symbian OS v9.3 Model

A high-level view of the system model for Symbian OS v9.3 is included in this book as a fold-out diagram.

All releases of the operating system from Symbian OS v7.0 to Symbian OS v9.3 share the same layer decomposition.

- UI Framework layer: The topmost layer of Symbian OS provides the frameworks and libraries for constructing a user interface, including the basic class hierarchies for user interface controls, and other frameworks and utilities, including concrete widget classes used by interface components.

- Application Services layer: This layer provides support independent of the user interface for applications on Symbian OS. These services divide into three broad groupings:

  o system-level services, such as basic application frameworks, used by all applications

○ services providing technology-specific logic, such as messaging and multimedia protocols, that are used by multiple classes of application

○ services that support specific individual applications, such as personal information management (PIM) and office applications.

Also included are a number of application engines that are used and extended by a licensee.

- Java ME: In effect, Java spans the UI Framework and Application Services layers, abstracting (as well as implementing) elements of both for Java applications. Symbian's Java implementation is based on Java ME MIDP 2.0 and CLDC 1.1. Java support has been included in Symbian OS from the beginning, but the early Java system was based on Personal Java and JavaPhone. A standard system based on Java ME first appeared in Symbian OS v7.0s. Since Symbian OS v8, the Java VM has been a port of Sun's CLDC HI.

- OS Services layer: The 'middleware' layer of Symbian OS provides the servers, frameworks and libraries that extend the bare system into a complete operating system. The services are divided into four major blocks that provide all technology-specific but application-independent services:

  ○ generic operating system services

  ○ communications services

  ○ multimedia and graphics services

  ○ connectivity services.

- Base Services layer: The foundational layer of Symbian OS provides the lowest level of user-side services, depending only on the operating system kernel (and related components), which it extends into a useable (but minimal) system. In particular, no services higher than those in the Base Services layer are required for a minimal base port to new hardware (in other words, a minimal base port requires only the two lowest layers of the system).

- Kernel Services and Hardware Interface layer: The lowest layer of Symbian OS contains the operating system kernel itself and supporting components that abstract the interfaces to the underlying hardware, including logical and physical device drivers and variant support that implements pre-packaged support for the reference hardware platforms. Releases up to Symbian OS v8 use the original Symbian OS kernel, Kernel Architecture 1 (EKA1 kernel). In Symbian OS v8.1b and from Symbian OS v9, all systems are based on the new Kernel Architecture 2 (EKA2) real-time kernel.

5.4   What the Model Does Not Show

The System Model shows a static view of the system, in effect a source view based on architectural relationships and abstracted from the details of what code appears in which files. It is not, therefore, a source tree or repository view.

The model also reflects only static (i.e. build-time) dependencies. It does not model processes, the memory contexts that are created on a device when the operating system runs, the threads that run within those memory contexts, or the services that those threads provide.

5.5   History

The system model was first published internally in 2004 (and therefore somewhat after the fact), as a description of Symbian OS v7.0. It was almost immediately updated for Symbian OS v7.0s. That model was first published for a wider audience in [Harrison 2004].

Since then, a revision of the model has been published for each release of the operating system. Since Symbian OS v9, the model has been used in the broader design and specification processes that are part of all operating system releases, providing a design base for each release and supplying the build definition to the software build system.

6

The UI Framework Layer

6.1 Introduction

The UI Framework layer is the topmost layer of Symbian OS (see Figure 6.1) and the immediate interface to the 'variant' user interface supplied by the manufacturer on a phone.

Symbian OS is delivered to licensees in a 'headless' configuration, with a minimal test user interface which is neither complete nor of production quality. (Known as TechView, it is considered to be a test and validation

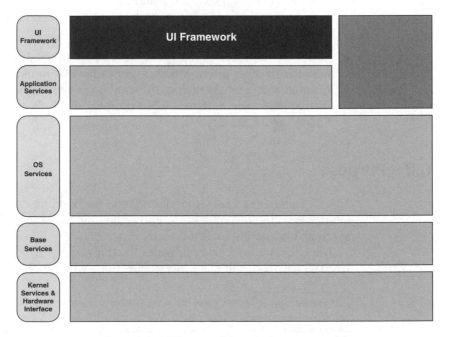

Figure 6.1 UI Framework layer in the system model

tool and is not, therefore, part of the operating system proper, although in the past it has been exposed to developers through 'preview' SDKs.)

Mobile phone manufacturers who license Symbian OS either replace the test user interface with a production quality user interface of their own, or license a suitable variant user interface. Typically in the latter case, the user interface is pre-integrated and pre-tested with Symbian OS, to simplify the task of bringing a device to market.

Currently two user interfaces are available for licensing: S60 (from Nokia) and UIQ (from UIQ Technology AB). Another important user interface is the MOAP user interface developed in Japan by DoCoMo's FOMA consortium of handset vendors, and used by consortium vendors on FOMA phones.

- **S60** is developed and licensed by Nokia. It ships on Nokia phones based on Symbian OS. Lenovo, LG and Samsung, among others, license and ship S60 phones based on Symbian OS. Licensees have also included Panasonic, Sendo and Siemens.

- **UIQ** is developed and licensed by UIQ Technology AB (until recently a fully-owned, Swedish-based Symbian subsidiary, now acquired by Sony Ericsson). Sony Ericsson, Motorola and Arima license and ship UIQ phones based on Symbian OS.

- **MOAP** is developed by the FOMA consortium in Japan as part of the DoCoMo common software platform for 3G FOMA handsets. FOMA members, including Fujitsu, Mitsubishi, Sony Ericsson and Sharp, ship MOAP phones based on Symbian OS.

- **Series 80 and Series 90** were developed by Nokia but are not licensed to other phone vendors. Series 80 was found on the Nokia Communicator family of devices based on Symbian OS. Series 90 can still be found on the Nokia 7710 phone, but has been merged with S60 for future devices.[1]

6.2 Purpose

The UI Framework layer is the foundation for building customized user interfaces on top of Symbian OS and is the immediate interface between Symbian OS and the variant UI layer.

The UI Framework layer provides the frameworks which custom user interfaces extend, the class hierarchies from which controls specific to

---

[1] Interestingly, Series 90 began life as the Hildon user interface, developed by London-based Mobile Innovation, now part of Macromedia. Hildon has since been ported as a widget set to the GNU GTK+ user interface toolkit, in which form it appears on Nokia's Linux-based 770 Internet Tablet.

the user interface are derived, and additional supporting components used primarily by user interfaces. It provides some specialist generic frameworks, for example animation, which are used by user interfaces but which are also available directly to applications.

The basic graphical and behavioral user interface abstractions are encapsulated in UI Framework layer components such as the control hierarchy, window interactions and graphics contexts, which determine basic application behavior.

The UI Framework layer is also used by the Java implementation, although Java also makes quite heavy direct use of some lower-level graphics frameworks. (Any such dependencies are transparent to Java applications, which see only Java APIs.)

6.3   Design Goals

The UI Framework layer is intended to enable user interface differentiation without fragmentation. This requires balancing the sometimes conflicting goals of providing a common, consistent functional and behavioral core to all user interface variants in order to provide a consistent development target for application writers while also providing the greatest possible flexibility and customizability to enable maximum user-interface differentiation for phone vendors

The design goals are that:

- the system should be a platform

- the different user interfaces should be distinct platforms.

6.4   Overview

Conceptually, the UI Framework layer has become thinner (functionality has migrated upwards) as the user interfaces built on top of it have become larger and richer; rich user-interface functionality is overwhelmingly in the user-interface variants.

However, important core user-interface functionality is retained in the framework base classes, from which the user interface variants derive. The framework approach means that many of the key user-interface design decisions (the basic user-interface architecture and broad division of responsibilities, managing input methods, the way user interfaces are customized, the basic control hierarchies, and some of the basic GUI application architecture and behavior) are encapsulated in the frameworks, which 'plumb in' the underlying operating system support for event handling (including all input–output events), window management and drawing, font and graphics support, and so on.

- The Uikon framework provides abstracted (i.e. high level or generic), customizable control of the overall GUI look and feel and encapsulates the main classes used to create applications. The underlying implementation of the generic application architecture is provided by lower-level frameworks, such as the Application Services layer.

- The Control Environment hierarchy (widely known as CONE) provides generic screen controls ('widgets') that are free of a look and feel and policies.

- FEP Base, the front-end processor framework, provides input-event capture (by key, pen or voice) and support for language preprocessing engines, for example, for handwriting recognition and exotic script input.

Supporting components provide additional graphical and other utilities (font, color and drawing support for user interfaces, including graphics effects such as fading and animation), as well as some useful frameworks that are used by both user interfaces and applications:

- The UI Graphic Utilities and Graphics Effects components contain common, general-purpose utilities used by user interfaces, for example drawing window borders and fading effects.

- The Animation and BMP Animation components provide frameworks for window animation and bitmap-based and sprite-based animation including animated clocks and animated user-interface elements.

- The Grid framework is a legacy framework specifically supporting cell-like (spreadsheet-style) layout.

Additional support for user-interface customization is included as part of the toolkit delivery (outside the scope of this book), which provides components that customizers may choose to re-implement as part of the variant user interface.

6.5 Architecture

Uikon and Control Environment (CONE) are the two most significant components in the UI Framework layer from an architectural point of view, since they determine the overall user interface architecture. Both also provide essential application support.

For most purposes, applications do not use Uikon directly, but instead use a Uikon-derived custom framework specific to the user interface (for example, Avkon in S60 and Qikon in UIQ). However, there are exceptions in which applications directly use Uikon; for example, applications directly use the many useful static methods of the user interface Environment object of class `CEikonEnv`.

The Control Environment is used both directly and indirectly by applications. Frequently the main application view is derived directly (from `CCoeControl` or `MCoeView`), bringing all the flexibility of the generic user interface control framework directly to it. Indirectly, applications use the Control Environment through the custom framework or the custom control set of the user interface variant.

Uikon

Uikon can be thought of as the common core on top of which are built the variant user interfaces that actually appear on phones.

Uikon provides a framework for creating user interfaces including the base classes which interface to lower-level system services such as application launching; key mapping and command handling; alarms and notifications; and graphics services.

Uikon supplies the base classes from which user interface variants derive essential application classes (Application, Document and AppUI) and encapsulates the relationships between them.

Uikon supplies the factory classes used by the user interface variant to create the hierarchy of custom concrete user interface control classes, including list boxes, scroll bars, buttons, dialogs and popups. (Basic menus are not controls but windows.)

Uikon loads a static library implementation (interface defined by the UI Look and Feel component) of the core library look-and-feel (LAF) component, which is supplied by the user interface variant. The UI Look and Feel component defines a standard set of methods which the variant user interface implements to define the concrete behavior of user interface elements, for example, layout and behavior of windows; choice of fonts and bitmaps; default location of resource files; system font and text rendering defaults; and the look and feel of toolbar, dialog, button and button container classes. The Uikon Error Resolver Plug-in is a small component that is used by the user interface variant to map system error codes to localized strings. Strictly speaking it is not a plug-in, but a resource file which is built as a dummy DLL.

Uikon provides a server stub which is run to launch other servers expected by the framework or by applications (the alarm server, notifier server, and server-side support for user-interface status panes) and to load implementations specific to a user interface variant for password and alarm notifications. (The Notifier is run inside the Uikon server thread to ensure that memory is always preallocated for those notification dialogs which must never fail, for example the 'Out of memory' dialog itself.)

Uikon provides servers to manage backup and shutdown (used to close running applications when the user starts a backup, and to handle shutdown when the user switches off the device). In earlier releases of the operating system (up to Symbian OS v7.0) Uikon also supplied a core library of concrete controls and dialogs, EikCoeControl; these are now supplied in the customization toolkit, and may be selectively re-implemented by the user interface variant or discarded.

The Control Environment (CONE)

Controls in Symbian OS are window-using, possibly nested, rectangular screen areas that accept user input and other events. (Windows do not necessarily own any controls; menus and sprites, for example, do not.)

Events (such as redraw events, standard events and foreground – 'focus' – events) are supplied by the Window Server to the Control Environment framework.[2] Of these, key, pointer and draw events are routed by the Control Environment to controls. Additional events may be generated by controls themselves, including change of focus events between controls. In effect, controls bring together:

- screen and window behavior as controlled by drawing, redrawing and other events

- graphics states, for example, color, font, brush and other settable attributes

- user-input handling (the Window Server serializes system events, such as key presses and pen taps, and delivers them to the currently active control of the foreground application).

The Control Environment defines the base classes that encapsulate these basic behaviors and the relationship between controls and their environment and define abstract controls. Applications can derive their own types of controls directly or use derived classes provided by Uikon and the user interface variant. The Control Environment, in effect, is the abstract middle layer between the low-level windowing functionality provided by the Window Server and the concrete user-interface classes provided by Uikon and libraries specific to the user interface variant.

- The CCoeControl class is the base class for derived controls.

- The CCoeEnv class encapsulates the application session with the Window Server, as well as providing utilities to manipulate the graphics state and for other system interactions (for example, it creates an application session with the File Server). Every application owns a singleton object of this class derived from CEikonEnv (which is

---

[2] Note that Window Server focus events are not the same as 'focus events' as understood by controls.

implemented as an active object responsible for routing input-event messages from the Window Server to the application framework `AppUi` class). Typically, the object is accessed from the application framework classes through the derived `CEikonEnv` class. From an application control, the object's methods are accessed through the control's `iCoeEnv` member.

The Control Environment also defines the user interface base class `CCoeAppUi`, providing the application user interface framework (brokered to applications via Uikon and the user interface variant) that manages input events. Key events are managed in the context of the stack of application controls (assigning a key event to the appropriate control).

Front-End Processor Framework

The Front-End Processor (FEP) Framework provides the abstractions that implement user-input capture and preprocessing, for example for handwriting recognition or multitap input systems, in order to capture, process and map user input events onto standard key events.

The FEP Framework provides the base classes for creating FEPs and defines the plug-in interface. The FEP Framework extends Control Environment base classes and is implemented as a DLL that is statically linked to by code which wants to derive from it. The Control Environment manages the creation, ownership and destruction of FEPs. FEPs are also available to Java and OPL applications.

FEP implementations are based on the `CCoeFep` class, which owns a high-priority, invisible control loaded by the Control Environment. Controls are organized as a priority queue. Since FEPs have high priority they receive keyboard events before (nearly all) other controls. The FEP captures and preprocesses sequences of input events which are then returned to the control stack as new events for consumption by lower-priority controls.

Only one FEP instance is allowed per application, since it must run within the application process and thread (in order to access the control stack). A FEP can exist on top of an application without the application being aware of it.

Animation

The animation framework is used to create bitmap-based and sprite-based animations. Animations are created as framework plug-in DLLs (with the extension ANI), which are recognized and loaded directly by the Window Server. While bitmap-based animations are rectangular and restricted to a single window (hence they are also known as 'window' animations), sprites can have irregular shapes and can overlap windows.

Because animations are run inside the Window Server thread, they run with higher priority than would otherwise be possible for any application thread, solving possible problems of slow running due to the high latency of redrawing.

Animations have been used since the early days of the Symbian OS and the framework still contains visible legacy of this, for example in the choice of timing periods.[3]

6.6   A Short History of the UI Architecture[4]

As early as 1997, when the Nokia Communicator project was already underway in Symbian, proposals were made for separating Eikon's look and feel (LAF) from its basic functional machinery. In the end, the Communicator project, like other early licensee projects, settled for adaptation (branching the codeline). However, it was clear that this could only be a short-term solution and that a principled approach was required to support the numbers of licensees and devices which were envisaged.

Reference Designs

As part of the Symbian OS v6 release project, therefore, the earlier look-and-feel separation proposals were revived. The result was Uikon. Its goal was to create a modular, streamlined and extensible user interface framework that would support multiple user interface styles whose look and feel could be customized from a common base. This approach became a central part of the DFRD strategy, which proposed to create reference designs for a generic product matrix that would be licensed to customers as the basis for real products.

Recognizing that each licensee had a distinct product philosophy, the reference designs in effect defined a set of distinct products. Reference designs specified the basic use cases and device style (classic phone or PDA; pen or keyboard input) and physical form factor (tablet or clamshell, as well as screen size, resolution and orientation), and were intended to be followed up with reference implementations including a reference user interface based on custom extensions to Uikon.

Uikon Architecture Evolution

The Uikon architecture consists of a common functional core (Uikon), a standard but non-core supporting library (EikStd), a graphical utility library (EGUL), and a LAF customization framework (UikLaf).

---

[3] See Douglas Feather's Window Server chapter in [Sales 2005].
[4] See also Chapter 16.

Early on, the implementation of common dialogs and controls was split between a core set and an optional set, with printing, file browsing, infrared beaming, and other similar functionality classed as optional.

- Core modules were intended for use unchanged.
- Standard modules were based on the Eikon baseline but were evolved in collaboration with the DFRD teams (Crystal, Quartz and Pearl).
- DFRD-specific libraries were created by DFRD teams.

Initially, 'mixin' classes were used to enable control implementations to reside in LAF-specific custom classes. Invoking `Set()` functions in the mixin classes loaded the custom library dynamically and allowed the core libraries to 'set' the custom concrete implementations.

Largely for performance reasons, this evolved into a stub library model in which the core links statically against a stub library which then loads and initializes the concrete custom library (or libraries, since there may be several). The advantage was that only one copy of the custom DLL was now loaded and one-off initialization was also faster than on-demand initialization. As well as providing a custom library, each variant user interface also implements a LAF module DLL that supplies the specific look-and-feel implementation for the Uikon core, to achieve a consistent look and feel across core, standard and custom libraries. The custom library replaces the Uikon internal library, `UikLafGT`.[5]

In its current architecture, Uikon principally provides application base classes for use by a variant user interface implementation. In early Symbian OS releases, it also provided a core set of controls (such as window borders) and dialogs (standard information and query dialogs).

Additional (optional) standard controls and dialogs, which are directly modified by customizers to form part of a variant user interface, are supplied in the UI Toolkit (part of the larger Symbian Toolkit delivery) and are not described here. Each variant user interface also defines its own custom controls, which vary between user interfaces.

6.7 Component Collections

The UI Framework layer contains two collections of components, as shown in Figure 6.2.

Figure 6.2 Component collections in the UI Framework layer

---

[5] 'GT' is a legacy Symbian internal name that originally stood for Generic Technology.

UI Application Framework Collection

The UI Application Framework collects the main frameworks related to user interfaces (see Figure 6.3).

It provides generic user-interface framework components that support user-interface customization. Additionally, it provides support directly to applications.

Table 6.1 UI Application Framework Components

Component Name	Development Name
Uikon	UIKON
Control Environment (CONE)	CONE
FEP Base	FEPBASE
UI Look and Feel	UIKLAFGT
Uikon Error Resolver Plug-in	ERRORRESGT

- The Uikon component provides a concrete framework for user-interface and application creation. Applications, typically, should not derive directly from Uikon classes. Instead, they should derive from equivalent classes provided by the variant user interface, because these provide the appropriate look and feel and other device-specific behavior. However, applications implement virtual methods inherited from Uikon and call inherited methods.

- The Control Environment (CONE) provides a control hierarchy and environment. It provides policy-free abstract controls (interactive screen elements) and control context, as the basis for interaction between the user and the application. It includes the application interface to user and keyboard events and View Server encapsulation. Derived concrete controls are provided by the variant user interface. All applications also use CONE (i.e. CCoeEnv and CCoeControl) directly within the application framework context.

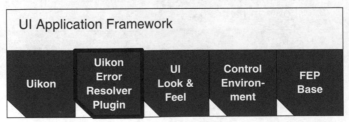

Figure 6.3   UI Application Framework collection

- The FEP Base component provides base classes for creating FEPs. FEPs selectively intercept and preprocess user input events, which are returned to the system as simplified events for handling by applications, to enable keyboard mapping, multitap keyboard input, handwriting recognition, voice recognition and other input preprocessing.

- The UI Look and Feel component defines the look-and-feel properties of the user interface. It defines standard methods (i.e. an API) for which user interface customizers provide an implementation in the UikLaf library of a variant user interface. The role of the user interface LAF component is to provide other parts of the application framework with a way of requesting look-and-feel information from a variant user interface, including the layout and behavior of windows; which bitmaps and fonts to use; and the location of various resource files.

- The Uikon Error Resolver Plug-in is a resource file that maps system-error numbers to helpful error-text strings, which a variant user interface extends and customizes. Errors are flagged when a user interface thread leaves normal execution inside the active scheduler of an application.

UI Support Collection

UI Support (see Figure 6.4) collects miscellaneous frameworks, utilities and libraries that are used by variant user interfaces and which, in some cases, may also be used directly by applications.

- The Graphics Effects component supports flicker-free animation of windows and window contents and composition of animation effects with other graphics objects, to enable GUI special effects (such as animated icons and 'exploding' menus) and moving and resizing windows (known as 'transition effects').

- The UI Graphics Utilities component consists of libraries used by user-interface framework components, the variant user interface and applications. They provide color, font, icon, text, drawing, and number conversion utilities. The utilities include those to query the relative

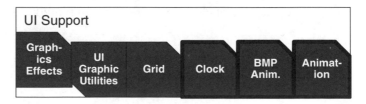

Figure 6.4   UI Support collection

Table 6.2 UI Support Components

Component Name	Development Name
Graphics Effects	`GFXTRANSEFFECT`
UI Graphics Utilities	`EGUL, NUMBERCONVERSION`
BMP Animation	`BMPANIM`
Animation	`ANIMATION`
Grid	`GRID`
Clock	`CLOCK`

positions of nested rectangles, to draw borders, to store color schemes and map logical to physical colors, to perform various font manipulations, to perform number conversions, to find pixel widths of text objects and to package icons as bitmap-plus-mask pairs.

- The Animation component supports window- and sprite-based frame-sequence animation. It enables animated effects to be included in the normal drawing of a window by a client or to be managed server side as a sprite. It also defines a plug-in interface enabling new animation types to be created and loaded as plug-ins directly into the Window Server. Hence, they run in its high-priority thread rather than in an application thread. Sprites can have irregular shapes and can overlap windows. Window animation is used, for example, to create fade effects.

- The BMP Animation component is a Window Server plug-in utility that enables bitmap-based frame-sequence animation. Bitmap-based animations are rectangular.

- The Grid component is a simple layout engine providing presentation, print preview and printing for complete spreadsheets and for spreadsheet cells, rows and columns. It is now considered a legacy component.

- The Clock component is a shared library for creating animation-based digital and analog clocks, used by user interfaces and applications.

7

The Application Services Layer

7.1 Introduction

The Application Services layer provides user-interface-independent support for applications on Symbian OS (see Figure 7.1). Broadly speaking, services whose clients and users are specifically intended to be applications or application engines (rather than system components and servers) can be found here. A number of essential application frameworks are also included. Note that the Java ME implementation also uses the frameworks and services found in the Application Services layer.

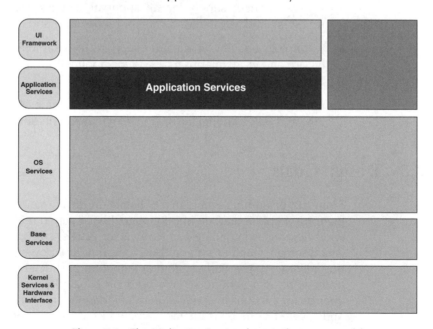

Figure 7.1 The Application Services layer in the system model

Services range from those used by all applications (basic application frameworks), to those providing technology-specific logic (for example, support for device management, messaging and multimedia protocols), to services targeting specific individual applications (PIM and office applications support).

Test or 'reference' user interfaces, where required, are supplied in the customization toolkit for licensees but are replaced in licensee products (including SDKs) and are not described here.

7.2   Purpose

The Application Services layer builds on the underlying services of the operating system to provide services intended primarily for use by applications and their engines, and includes some essential application frameworks which are used by all applications, either directly or as mediated by higher-level frameworks. The Application Services layer is also used by Java ME components.

The Application Services layer provides services used by all applications but mediated by the UI Framework layer and the variant user interface above it, for example, application installation and launching, view switching, and the basic application architecture relationships. It also provides:

- generic services supporting all application types, for example, text rendering and MIME-based content recognition and handling

- technology-specific application support; for example, Versit support (vCard and vCal); alarms for PIM-type applications; and Internet, web and multimedia session protocols

- application-specific services, for example, engines and plug-ins for PIM and office applications; device management; and provisioning.

7.3   Design Goals

From the beginning, Symbian OS has been designed as an application platform. In particular, an important goal has been to make it possible to write rich and compelling applications for pocket-sized, mobile devices (small screen, small ROM and RAM footprint, low power, connected but not 'tethered'). The early system architecture abstracted the application framework as a generic service used by all applications and supplied engines for the built-in application suite independent of the user interface, and layered both beneath the frameworks which supported the GUI-specific aspects of the user interface.

The basic separation of applications into user interfaces and engines and, in particular, the adoption of an MVC-like approach has a long history in Symbian OS. As the system has evolved, there has been an increasing distinction between engines and services. Services are understood as providing generic support for working with data models, for example, generic recognizers, translators and protocol handlers for typed data at the application level. Engines are understood more narrowly as the application-specific logic forming the part of an application implementation that is independent of the user interface. According to this definition, application services would be expected to expose Symbian OS interfaces but application engines would not.

Applying this definition to the system has the effect of moving functionality out of engines (which become narrower in scope and more specific to an application, user interface or vendor), while increasing the common functionality available to the wider set of applications on the phone. This is the direction in which the operating system has been evolving. In the latest operating system releases therefore, the Application Services layer supports application engines but does not include them (except for legacy engines).

There are good reasons for this evolution. Compared with its beginnings, Symbian OS now supports a wider range of devices in diverse markets and geographies. Increasingly, the APIs provided by the generic engines have been perceived by licensees as being too broad (providing too much functionality), while not delivering functionality required in specific markets, for example in Japan and the Far East. Supplying generic engines, with APIs big enough and comprehensive enough to support all application implementations, risks fragmenting the platform rather than unifying it, since licensees are more likely to choose to provide their own specialized (and small) engines than reuse bulky generic engines which nonetheless need extending. Generic engines, in other words, can prove to be a false economy, neither delivering the expected benefits in time to market nor avoiding platform fragmentation.

Providing rich services is a more effective, more generic, more granular, and more customizable way of increasing the capabilities of the platform while serving licensees better.

Some services and application engines may now be considered redundant. For example, Bluetooth profiles are more relevant to phones than WYSIWYG printing; and phones do not typically need a full spreadsheet or word-processor engine, as found in the Office Application engines components.

7.4 Overview

Applications have always been central to the vision of Symbian OS. The original design conception called for more than simply an operating system with an application suite; applications and application support were considered intrinsic to the operating system. The application architecture was embedded into the object-oriented design and specific application logic – shared data models and data persistence – was provided at the level of operating system services.

- The operating system has evolved to become a common software platform for diverse categories of device, not simply for a single device family as first envisaged.

- The device categories it targets have evolved from PDAs through PDAs with phones to phones, and continue to evolve more generally in the direction of connected, mobile, consumer devices, including phones but not limited to them.

- Open standards have become increasingly important. Efficient and deep integration of open standards for multiple technologies into the operating system platform has become one of its distinguishing features.

- Support for specific, shared data models has become less important, for example the office-style application engines are considered to be legacy functionality.

The Application Services layer includes support for important application-level standards:

- the Versit specification, specifically vCard and vCalendar

- data synchronization, device management and client provisioning, including on-device and 'over the air'

- email standards, including POP, IMAP and SMTP

- phone-messaging standards for GSM and CDMA including SMS, MMS and WAP messaging

- Internet document and data protocols including HTML, XML, WAP, HTTP and Synchronized Multimedia Integration Language (SMIL)

- application-session protocols, Real-time Transport Protocol (RTP) and Session Initiation Protocol (SIP).

Although many of the services based on these standards have been designed to support specific standard applications (messaging and phone applications, for example), they are also generally available to third-party developers creating new applications.

7.5   Legacy Application Engines

The Word, Sheet and Data engines should be considered legacy func-tionality.[1] While the functionality may continue to feature on specific devices, it should not be considered part of the generic operating system delivery.

Other services, for example printing, should also be considered legacy for different reasons. The original goals of the printing support in Symbian OS (to provide WYSIWYG document printing) have been overtaken by the nature of the content being printed (from photos and contact details to web pages, but rarely a full business document) and by newer protocols (such as the Bluetooth printing profile).

7.6   Architecture

A goal of the user-interface architecture in Symbian OS is to enable as much common functionality as possible on the system side and to make it available to as wide as possible a range of applications. This allows applications to be written with a minimum of new code and the maximum reuse of system-provided code. Applications gain in robustness and reliability because, as far as possible, the most complex code is written only once, on the system side where it is tested and validated, and is reused by application authors. While the strategy for delivering this goal has shifted from providing full application engines to providing comprehensive services, with engines moving up to the licensee layers, the goal remains the same. And while the classes that define the basic architecture of a Symbian OS application differ between variant user interfaces, they all derive from generic Symbian OS classes; Symbian OS implements the underlying generic behavior.

For the application writer, this is interesting. On the one hand, it is extremely powerful, because a little application code goes a long way. On the other hand, having so much richness in the system presents a steep learning curve to the application writer. The Application Services layer provides 'rich system' support for applications.

Application Framework

Model–View–Controller (MVC) is the classic object-oriented abstraction of a graphically based, data-centric, interactive user application. (MVC was originally part of Smalltalk-80 and, according to [Johnson 1998], was 'the first framework that was recognized as a framework'.)

---

[1] In practical terms, their public APIs are likely to be deprecated in some future release.

Symbian OS, from its first inception, applied an MVC-like model to applications. It is not quite pure MVC, because it elevates the application itself (as an abstraction for system-owned resources) into a first-class concept and because the variant user interfaces do not necessarily code the MVC classes directly. How they interpret MVC is strictly the business of the variant user interfaces.

Applications, documents, UIs and views

The first rule of object orientation (in C++ anyway) is, according to [Koenig and Moo 1997], to 'use classes for concepts'. There are four key concepts in the application model: Application, Document, AppUI and View. The Application Framework supplies the base classes for Application, Document and AppUI, and variant UIs supply appropriate custom specializations. The View class is typically derived directly from `CCoeControl`.

An application is built as a EXE that is recognized by the application architecture and launched in its own process.[2] The framework-defined entry-point function calls the factory function that creates the application instance. The application encapsulates the relationships between the application instance, its document, its document-owned user interface, and its view or views, as well as application-owned resources, for example the application icon and more abstract properties such as UIDs. Applications may have multiple views; every application must have at least one view (i.e. one window-owning or window-controlling control).

Strictly speaking, the application document abstracts a data model and not a file, although applications may be file-based. The document is responsible for storing and restoring the application's persistent data, whether to or from a file or a database. Documents can also be embedded, so that documents may contain other documents (including documents belonging to other applications). The application document is also responsible for creating the application user interface (although the framework takes ownership of the user interface and is responsible for destroying it). Just as the document exists to persist the data state of the application, the user interface exists to manipulate the data state.

A 'data model' in this context really means data plus the APIs defined to create and manipulate it (getters and setters, the 'data logic' defining the translations and other functions that can be applied to the data to return results of some kind). In Symbian OS, this is often loosely referred to as an application 'engine'; the engine is really a code implementation of the machine that transforms the data state, driven by the user interface: the document encapsulates the data model state.

---

[2] In releases before Symbian OS v9, applications were built as DLL plug-ins and shared process space; the changes are required by the system-wide security model.

'Engine' classes do not have any framework significance (and hence do not derive from a framework class) and they are not required, although they are a useful design pattern for encouraging separation of logic from data.

Since the document creates the user interface and every application needs a user interface, every application must have a document. Each application instance is associated with a single document.

The application view provides a view onto the state of the application data. Views are implemented using controls. On a typical Symbian OS device, desktop user interface idioms (such as multiple overlapping windows) are not appropriate, for a number of reasons: display size is hugely limited compared with a desktop device; handheld operation (and, in particular, one-handed operation) rules out mouse-style interaction, and so on.

Typically, the view metaphor is closer to a stack of sticky notes or a deck of cards. The top card conceals the other cards in the deck. Cards can be brought to the top, shuffled to the bottom of the deck or shuffled unseen within the deck. While applications can have multiple views, only one is visible at any time.

View switching

The View Server provides a framework for sharing application views by 'view switching'. Originally designed to support switching between flip-open and flip-closed modes on the Ericsson R380 (an ER5-based phone), it migrated into the Quartz user interface (which became UIQ) and was eventually adopted back into the operating system. Applications can register views with the server. A registered view owned by one application can then be used by any other application (or indeed by another view in the same application) that requests the view to be activated.

In UIQ, for example, the Contacts application can request activation of the New Message view from the Messaging application when a user taps on an email address in a contact detail. View switching provides a clever shortcut to passing data between applications.

Note that while the View Server manages view switching and owns the framework, it is not used directly by applications: instead, switching is enabled via the application user interface (which is a Control Environment wrapper). View Server uses the Window Server client API to effect view switching.

Support for Generic Applications

While applications are highly dependent on the frameworks supplied by the variant user interface, the underlying support for the application logic is largely provided by Symbian OS. This is an important part of the platform promise that Symbian makes to developers: application logic should in principle be reusable across the whole range of devices based on Symbian OS. As discussed above, in recent releases the emphasis within the operating system has shifted away from providing reusable application engines towards application services.

Legacy engines

The earliest versions of Symbian OS included a number of fully fledged applications, ranging from standard PIM and Office applications (Agenda, Data, Sheet, Word) to Time World (a time zone browsing and setting application), a Help system, and so on. While there was no Contacts application on the original Series 5, by the time of the later Psion devices (such as the Revo) it had joined the set of standard applications.

Increasingly, providing common services and standardizing APIs is seen as providing more value to licensees than providing ready-made, one-size-fits-all engines. However, the legacy engines still form part of the operating system.

Along with phone-specific functions (messaging and email as well as the phone application itself), PIM applications – most importantly, a phonebook and a simple calendar – are at the heart of what a modern phone provides to its users. Underlying these standard applications are a number of common services, including support for basic text handling, the vCard and vCalendar standards, alarms, backup and restore notifications, and file and date conversions.

Text handling (EText) and formatting (FORM)

Text Handling supports the storing of editable text and its formatting attributes, while Text Formatting provides text view and layout classes (CTextView, CTextLayout, MLayDoc) that control scrolling, selection, cursor management, margin setting, and other attributes of displayed text. Managing display attributes (layout and drawing) is thus distinguished from managing logical text attributes (including text content).

Text content is managed by the text-handling APIs, and consists of Unicode characters, including space characters and paragraph delimiters, as well as formatting attributes, including properties such as paragraph alignment, character fonts, and so on. (Formatting attributes are not the same as text formatting layout attributes.)

The text-handling APIs and the rich text model underlying them have a long history in Symbian OS. They have used Unicode since the ER5u release, the first release to be used in phones, in 1997.

The Text Formatting layout framework is used directly by applications (to lay out text in application user interfaces and documents) and by user interface and system components (to lay out text in dialogs, etc.); for example, text views are used by the Uikon Core API for editable text windows ('editors'), as well as directly by applications to format and display rich text.

vCard and vCalendar

vCard and vCalendar are standards that define formatting conventions for card (address detail) and calendar (diary appointment) entries. The standards allow entries to contain more than simply text (character, number, date and time) content. For example, they can include sounds (for example, alarms) and pictures.

The vCard and vCalendar component provides parsing APIs for vCard and vCalendar entries and enables conversion into Symbian OS native formats.

Alarm server

The Alarm Server manages a queue of system-wide, time-based alarms and provides APIs for applications to set, modify and query alarms. Note that the Alarm Server does not actually notify, sound or show alarms (the Alarm Alert Server performs those functions).

The Alarm Server is a conventional Symbian OS server managing a shared resource (the alarm queue). Clients create a session and connect to the server to use the APIs. The Alarm Server has a long history in Symbian OS.[3]

Backup and restore notification

The backup and restore notification mechanism provides an alert (based on Publish and Subscribe) to signal to PIM applications that backup or restore is in progress or has completed. Applications may need to refrain from writing data to file during backup or may need to re-read files after restore. Other applications should use Publish and Subscribe.

Chinese calendar converter

The Chinese Calendar Converter provides a simple API for converting between Gregorian and Chinese calendar dates.

File converter plug-ins

The File Converter Plug-in is a simple converter that translates HTML to Symbian OS rich text format. It is used, for example, to convert text to HTML email format.

---

[3] Until Symbian OS v7.0s, a single component (known cryptically as EALWL) combined both World Server and Alarm Server functions and served as the engine for the TimeWorld application, see [Tasker 2000, p. 108]. In Symbian OS v7.0s, they were separated and rewritten. The new version of the Alarm Server replaced the old EALWL-specific alarm types (e.g. clock alarms and agenda alarms) to make them more generic.

Printing support

Printing Support implements a framework for managing printers and print jobs, generating graphics input to raster devices and treating printing as a special case of drawing to a device context, much like drawing to a screen or any other display device.

It is intended to be used by applications printing directly to supported printer types and is therefore most suitable for 'old-fashioned' PDA-style applications on 'converged' devices, such as Communicator-style phones, and less appropriate for the more lightweight kinds of application likely to be found, for example, on a phone without a keyboard. For such applications, full WYSIWYG printing is unlikely to be as important as sending a picture to a printer using Bluetooth technology. The print framework can, therefore, be seen as part of the legacy functionality of Symbian OS, along with the Office-style applications it most naturally supports.

It presents a simple application-level interface to underlying printing support provided by the Multimedia and Graphics Services. The printing API, among other things, manages:

- listing and selection of available printers

- encapsulation and setting of the device and print job properties

- selection of a printer port (where required by the printer).

Data synchronization and device management and provisioning

The Open Mobile Alliance (OMA) sponsors data synchronization services based on SyncML, Client Provisioning for OTA device configuration, and Device Management standards. The Application Services layer includes specific support for OMA standards.

Support for Generic Technologies

Standards-based messaging and browsing have become essential functions for mobile phones. The Application Services layer provides extensible support for messaging standards including SMS, MMS and email; for Internet browser protocols; and for newer, session-based multimedia protocols. Supporting services include content recognition, including MIME-type recognition, for data originating from the network; and support for 'smart' messaging (messages containing network-originated configuration and settings data intended to be used by the system rather than read by the end user).

Messaging

Comprehensive support for messaging of all kinds, from email to text and multimedia messages, is an important feature of Symbian OS. Messaging support has been available from the first release. As the operating system has become more phone-centric, messaging has arguably become even more critical than it was originally, although (interestingly) the use cases are subtly different for phones and PDAs.

The Symbian OS messaging implementation provides a complete messaging infrastructure for use by a messaging application, whether from a licensee or other source. It is based around a message server, which manages access to a unified Message Store and performs generic messaging actions that are exposed through a client-side API. It also owns an extensible framework allowing generic actions to be specified for particular message types. The framework is open and is intended to support enterprise-level customization (for example, for bespoke, corporate messaging systems or services) as well as licensee extension and customization (for example, to adapt the generic functionality to a particular user interface idiom – S60 and UIQ messaging applications behave differently from an end-user perspective). The client-side API enables client applications to manipulate the message store, for example, to browse and navigate the message-store folder tree, and provides basic functions, such as edit, copy and move. The framework also supports scheduled sending of messages.

The underlying communications services of the operating system are used to enable message transport over any available network connection, whether phone, short link (Bluetooth or infrared), or cable (serial or USB).

Extensions are provided by Message Type Module (MTM) plug-ins to the framework and the operating system provides product-quality implementations for a standard set of message types, including email (SMTP, POP3 and IMAP4 HTML mail), SMS (on both GSM and WCDMA, that is on 2 and 2.5G, 3G and CDMA 2000 networks: the SMS protocols are specific to each type of network) and MMS.

BIO messaging

An important secondary server and framework is the Bearer-Independent Object (BIO) Messaging Framework, which extends generic messaging to provide a 'smart' messaging server, a message type framework and a watcher framework. Bearer-independence means that the message handling is independent of the type of transport over which the message was received; 'smart' messages are those which are intended for processing by the system, or directly by applications. BIO messaging supports application message types, such as encapsulated vCard and vCalendar data, and system services such as network-access setup messages. The

BIO messaging APIs allow application developers to create their own application-specific 'smart' message types.

The message server provides the underlying mechanisms used by dedicated messaging applications, or other mail or SMS client applications, as well as providing a 'Send As' API as an extension to the client-side API, which allows any application to encapsulate a document and send it as a message type (including Fax), over any available bearer. Any application can also receive messages, using the watcher service, and 'smart' objects.

Messaging support includes handling of MIME and other recognized data types (provided by the Content Handling components); handling of attachments; managing local and remote mail boxes; and editing message contents and properties. The watcher frameworks support alerts for message-related external events, for example a fax-line ringing or an SMS or email being received, and for 'system' messages to be identified and handled.

The basic design principle in the messaging system is to clearly separate generic message handling performed by the framework from the detail of manipulating and handling different message types, which is delegated to the MTM extensions.

The Message Store is considered to be a shared system resource, for which the client–server design ensures multiple simultaneous access by client applications.

The plug-in-framework design allows for a modular and extensible implementation. (However, the MTM model is complex: creating a new MTM is a challenging system-level programming project.) An MTM implements concrete support for three client-side APIs and one server-side API. The client-side implementation consists of a user interface for viewing and editing message contents (and service settings), concrete data, such as icons that clients should display, and the message creation and management functions. The server-side implementation supports manipulation of messages on remote services. Messaging clients link to the client-side MTM. The matching server-side MTM is loaded as needed by the messaging server.

The BIO messaging server and framework is itself implemented as an MTM. BIO messaging plug-ins derive from and implement the framework classes and are loaded by the BIO messaging MTM.

BIO Messaging responsibilities are divided between the MTM (which implements the server and framework), a BIO database (which maps port numbers, MIME types, etc. to BIO types in order to identify the type of incoming BIO messages), and plug-in parsers that parse and process the BIO message payload. Because BIO messages arrive over other message transports, for example as a WAP push or an SMS, watchers are used to receive and tag incoming BIO messages. Watchers that watch for specific message types are created by deriving from and implementing the watcher framework classes.

The scheduled send framework is implemented by the Server MTM and provides classes that define the scheduling parameters, allowing messages to be scheduled (sent later), rescheduled or deleted from the schedule. MTM implementations for different message types can choose whether or not to support message scheduling.

At Symbian OS v9, the supported message types include email (POP3, IMAP4 and SMTP), SMS and OBEX. MMS messaging, which was included in Symbian OS v8, may be provided as part of a licensee user interface implementation.

Content handling

An important aspect of supporting messaging, browsing, and other network-oriented applications is the provision of content recognition, parsing and access services for protected content (key, certificate or other DRM-protected downloads, for example).

Symbian OS provides standard application services that support:

- file and data recognition based on MIME types (MIME Recognition Framework), standard web types (Web Recognizers) and multimedia file types (MMF Recognizers)

- parsers and handlers to support SMIL (SMIL Parser) and 'smart' messages and content (BIO Messaging Parsers) and WAP 'push' messages (WAP Push Handlers)

- handling and providing access to DRM-protected content (Content Access Framework for DRM).

These services are used by applications either indirectly via the various application-level messaging, web and multimedia frameworks and services or directly through the Application Architecture recognizer interface. These services are also used by system components, for example, the messaging framework.

The Application Architecture provides a 'Recognize Data' interface which is implemented by plug-ins to the MIME Recognition Framework. This enables recognition of non-native document types in order to associate documents with applications. (Native document types are identified and associated with applications using UIDs). Associating documents with applications allows appropriate applications to be started (or offered to users) when a user performs an action to open a document, as well as allowing default documents to be located when applications are launched. Data types as well as documents can be recognized.

File and data recognizers are written as plug-ins to the MIME Recognizer Framework (from Symbian OS v9 they conform to the ECOM Plug-in Framework) and are scanned for and loaded during operating system startup.

Data recognizers are provided for common MIME types, URLs, web bookmarks, HTML and XML, and multimedia file types. The supported multimedia types depend on the licensee implementation of multimedia plug-ins for supported media types.

Applications can register with the Application Architecture as handlers for specified MIME/data types. The Application Architecture maintains a list of all recognizers in the system and their supported data types.

The WAP Push handlers are intended to support WAP browser applications. They are plug-ins to the WAP Push Framework and respond to WAP Service Initiation (SI) and Service Load (SL) signals to take ownership of incoming messages and validate, parse, and extract the message content. SI and SL messages signal actions to WAP browser applications (to display content or a URL), unlike other WAP Push message types (MMS and OTA), which are pure message carriers for messaging, not browsing, services. The Web Push handlers are intended to support WAP browser applications directly.

The Content Access Framework provides a generic mechanism to support DRM implementations, based on defined interfaces for brokering controlled content between content agents (DRM applications) and content-consuming applications (for example, media players).

The BIO Messaging parsers plug into the BIO Messaging Framework to enable parsing of specific BIO message types, including vCard business cards, email notifications, Nokia Smart Messages and Nokia and Ericsson OTA setup messages. (Note that BIO Messages use WAP messaging.)

The SMIL Parser is an XML parser that uses a 'mini-DOM'-like API to parse and validate XML against simple DTDs. SMIL is an XML language that defines presentation attributes for encoded text, images, video and audio. It is provided primarily to support handling of MMS messages with SMIL content. (Note the earlier remarks about MMS not necessarily being supported on all devices from Symbian OS v9.) The parser however is also available for direct use by applications and provides APIs to perform simple XML parsing (not limited to SMIL). Heavier duty, generic XML parsing is provided by Base Services components.

A SMIL parser was first introduced in Symbian OS v7.0. The current implementation, which is able to parse any XML document against a simple DTD, was introduced in Symbian OS v7.0s and the original parser was deprecated.

Internet, web and multimedia protocol support

A number of components provide infrastructure support for Internet and web applications including web and WAP browsers and WAP messaging. Newer protocols such as RTP and SIP have also been introduced in the latest releases of the operating system to support new categories of interactive streaming applications.

Basic Internet, web and WAP support consists of framework, utility, and application engine components providing application-level interfaces to Internet protocols (HTTP, Telnet) and WAP Push messaging. The HTTP Transport Framework provides a generalized client interface for applications wanting HTTP transport sessions over TCP/IP or WSP sessions (the WAP equivalent of HTTP).

The HTTP Transport Framework provides a complete supporting framework for HTTP and WSP applications, such as HTML or WAP browsers. For WAP browsing the underlying support of a full WAP stack is required; this is no longer part of the Symbian OS delivery and therefore depends on the licensee platform to provide a full WAP stack. The framework adopts a session model based on a core client API and request–response message exchange transactions with a remote URL.

The WAP Push components provide an interface between the WAP stack and the messaging infrastructure to support WAP as a messaging transport. The WAP Push components are used by the messaging services, to support receiving WAP push messages and BIO messages, and by other system components including, for example, Java. Note that simple client access to WAP push is provided by the WAP Message API of the WAP Stack.

Implementers of a WAP stack need to be aware of the dependency of the HTTP Transport Framework on it. In effect, the lower level of the framework serves as an adaptation layer to the WAP stack, implying that work is required to adapt it to a WAP stack implementation.

The Telnet and FTP engines are rather simple application-level services based on clients creating a client session to the Symbian Telnet or FTP daemon, through which the client can conduct a dialog with a specified host. (Note that FTP does not expose public APIs.)

More specialized Web browsing support is provided by the stand-alone Bookmark Support component, which provides access to a bookmark database and APIs for creating, reading and deleting bookmarks and creating a folder tree. The database uses the Central Repository to store all data. There is only one bookmark database.

A folder object contains an array of CBookmarkBase objects. A bookmark must contain a URI, authentication data, the last time it was visited and an indication if it is the home page. Applications can set item visibility to public or private.

HTTP transport framework

The HTTP Transport Framework is based around a Core API, which manages the client-session interface to a session based on either a WSP or HTTP protocol, for example for WAP or web browsing. In both cases, secure versions of the protocols are also supported. Within a session, message-based transactions are conducted with the remote

URL. A message is a generic abstraction that packages contents of any type.

As well as the Core API, clients can configure a session to use Filter Plug-ins that are loaded by the framework and used by applications to handle, process or modify message content. Default filters are provided for message authentication, redirection and validation.

Beneath the Core API, protocol handlers and transport handlers interface to the underlying transport. WSP and HTTP protocol handlers are supplied by default. WAP Stack, WAP WTLS, HTTP and HTTPS transports are available. WAP and WTLS use the WAP Stack interface directly; HTTP and HTTPS use the Socket Server to provide TCP/IP sockets or secure sockets, in all cases using an appropriate network interface.

Real-time transport protocol

The real-time transport protocol (RTP) is a network transport service that provides real-time guarantees on packet latency to support uses such as interactive audio and video, for example, web conferencing. TCP-based packet services have a (relatively) high potential latency. For many applications, heuristics (buffering, selective dropping and repeating of packets, etc.) can be used to maintain service quality at a satisfactory level, even for demanding applications such as streaming. However, two-way interactive services have effective real-time requirements which cannot be met simply by smoothing packet arrival latencies.

RTP implements reliable and real-time bound transport using UDP packets over IP. RTP services support payload-type identification, sequence numbering, time stamping, and delivery monitoring of packets.

From a system perspective, RTP is provided to support the Multimedia Framework introduced in Symbian OS v8. It is designed as a core software stack that implements RTP/RTPC packet creation and handling using the underlying network infrastructure, and an upper API used by applications, which link to it.

RTP is available to applications using a socket interface. From the user perspective, it is created and used in essentially the same way as any socket-based transport. Within a socket server session, an RTP subsession is opened.

RTP provides APIs to:

- create and manage RTP sessions
- register for and handle events
- manage and access RTP packets and reports
- create, send and receive packet streams
- manage, send and receive reports.

RTP was introduced in Symbian OS v9.

Session initiation protocol

Session Initiation Protocol (SIP) is a simple but powerful protocol enabling peer-to-peer, multiple-participant sessions to be created over a TCP/IP packet network. The protocol is reminiscent of HTTP (in its use of URLs to identify participants) and SMTP (plain text messages). SIP messages are used to set up and terminate sessions.

The SIP Framework integrates a plug-in implementation into the underlying network infrastructure, including RTP. The operating system provides only the framework; licensees supply the service implementation.

The SIP Framework was introduced in Symbian OS v9.2.

Other Application Services

Improved secure installation services provided by the App Installer and a System Starter that manages server startup at boot time were introduced in Symbian OS v9.1 to improve the supporting infrastructure for applications (although they do not expose APIs directly to applications).

The App Installer uses the Certificate and Key Management services (see Chapter 8) provided by lower layers of the system to manage certificate- and key-secured applications.

The System Starter, while it does not expose public APIs, is configurable by licensees. In the original design of Symbian OS, true reboots were assumed to be rare events. The operating system was designed to support devices that would run for months and even years at a time between reboots. Booting-up time was therefore an insignificant cost. However, the phone use case is very different. Phone users switch phones off frequently and expect a fast boot when they switch them on.

Symbian's server model – ubiquitous use of servers to manage all system resources, logical and physical – leads to multiple servers being started at device boot time with a cascade effect. (Any server can arbitrarily start many other servers; in the past, some have.)

The System Starter allows a start-up policy to be specified and enforced. This enables careful management of the start-up sequence, to enable a device to become maximally responsive in minimum time, even if loading of the full server set continues in the background. If a server run list is found, it is used to select which servers start and in which order. In this scenario, some servers are not started until they are first called by a client.

7.7   Component Collections

The Application Services layer contains the component collections shown in Figure 7.2.

- System level services:
  - Application Framework

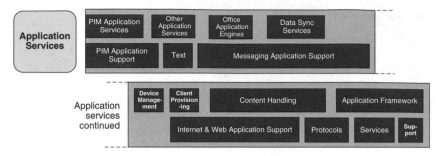

Figure 7.2   Component collections in the Application Services layer

  o   Application Launch Services

  o   Multimedia Protocols

- Application services and engines:

  o   Data Sync Services

  o   Device Management

  o   Client Provisioning

  o   PIM App Services

  o   Other Application Services

  o   Office Application Engines

- Lower-level application support:

  o   PIM Application Support

  o   Messaging Application Support

  o   Content Handling

  o   Internet and Web Application Support

  o   Printing Support

Application Framework Collection

- The Application Architecture component defines the key application responsibilities and interactions with data and the user interface.

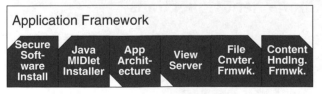

Figure 7.3   Application Framework components

Table 7.1 Application Framework Components

Component Name	Development Name
Application Architecture	APPARC
View Server	VIEWSRV
File Converter Framework	CONARC
Content Handling Framework	CONTENT_HANDLING
Secure Software Install	SECURESOFTWAREINSTALL
Java MIDlet Installer	JAVAMIDLETINSTALLER

It encapsulates the key application classes, which are abstracted via Uikon and, ultimately, by a vendor-specific variant user interface.

- The View Server component provides a mechanism for view sharing and view switching between applications. A running application can switch into and use a view belonging to another application.

- The File Converter Framework component supports creation of file converter plug-ins that enable applications to request file-to-file conversion based on the MIME types of the files. It is typically used to support conversion between Microsoft Office and Symbian OS proprietary formats.

- The Content Handling Framework component contains the File Converter and Content Handling frameworks, which are used to provide applications with common framework behavior independent of the user interface. The Content Handling Framework supports the finding, loading, processing and displaying of typed content by content handlers on behalf of applications.

- The Secure Software Install and Java MIDlet Installer components enable the installation of native applications and Java MIDlets. SIS files, based on versions of Symbian OS that do not include platform security, do not install on devices based on Symbian OS v9.

Application Launch Services Collection

This collection (see Figure 7.4) contains only one component, which enables policy-based startup of system servers at boot time. The server startup sequence is defined in a policy file, which can be customized by

Figure 7.4 Application Launch Services components

licensees to tune boot-up time and ensure that the device is responsive to the user as quickly as possible after switch on.

Table 7.2 Application Launch Services Components

Component Name	Development Name
System Starter	SYSSTART

Multimedia Protocols Collection

This collection (see Figure 7.5) provides support for the real-time transport protocol (RTP) and the session initiation protocol (SIP).

Table 7.3 Multimedia Protocols Components

Component Name	Development Name
RTP	RTP
SIP Framework	SIP_COM
SIP Connection Provider Plug-ins	SIPCPR, SIPDUMMYPRT, SIPSTATEMAC, SIPPARAMS, SIPSCPR

- The RTP component is a server- and user-side API providing socket-based access to RTP services. It provides an IP-based real-time network transport service.

- The SIP Framework and SIP Connection Provider Plug-ins provide support for SIP and integration into the networking infrastructure. It

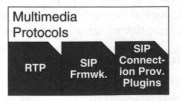

Figure 7.5 Multimedia Protocols components

does not provide the protocol implementation (which is provided as a plug-in by licensees). SIP is the main signaling protocol for 3GPP and is used by phone, multimedia and messaging applications.

Data Sync Services Collection

This collection (see Figure 7.6) provides support for data synchronization.

Table 7.4 Data Sync Services Components

Component Name	Development Name
Sync Initiation	`SYNCMLINITSERVER`
OMA SyncML Framework	`SYNCMLCLIENT`
OMA SyncML DM Interface	`SYNCMLDMCLIENT`
OMA Data Sync	`SYNCMLDSCLIENT`

SyncML is an open industry standard, primarily for data synchronization but extending to device management. SyncML has been adopted and standardized by the Open Mobile Alliance.

The SyncML protocol supports data providers (for example, application engines) and components requiring remote management (for configuring settings, for example) over various transports. It is implemented as a server, with a supporting plug-in framework, that supports synchronization and device management over HTTP, WSP and OBEX.

Client Provisioning Collection

This collection (see Figure 7.7) provides components for client provisioning.

Client Provisioning is an OMA standard for configuring application and network settings on mobile devices. It overlaps to some extent

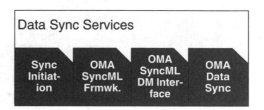

Figure 7.6 Data Sync Services components

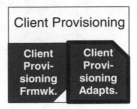

Figure 7.7   Client Provisioning components

Table 7.5 Client Provisioning Components

Component Name	Development Name
Client Provisioning Framework	`DEVPROV_CLIENTPROV_FRAMEWORK`
Client Provisioning Adaptors	`DEVPROV_CLIENTPROV_ADAPTERS`

with SyncML and competes with proprietary alternatives (Nokia Smart Messaging and Nokia and Ericsson OTA).

Client Provisioning supports components requiring either one-off settings configuration (network settings setup, for example) or continuous provisioning (for example, settings management plug-ins to applications or Symbian OS services) over various transports.

Device Management Collection

This collection (see Figure 7.8) provides a framework and plug-ins implementing OMA Device Management based on SyncML and supporting Remote Terminal Management and continued provisioning of devices by network operators.

PIM Application Services Collection

This collection (see Figure 7.9) provides specialized support specifically for the Agenda and Contacts applications.

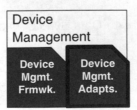

Figure 7.8   Device Management components

Table 7.6 Device Management Components

Component Name	Development Name
Device Management Framework	DEVPROV_DEVMAN_FRAMEWORK
Device Management Adaptors	DEVPROV_DEVMAN_ADAPTERS

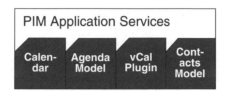

Figure 7.9 PIM Application Services components

Table 7.7 PIM Application Services Components

Component Name	Development Name
Contacts Model	CNTMODEL
Calendar	CALINTERIMAPI
Agenda Model	AGNMODEL
vCal Plug-in	AGNVERSIT

- The Contacts Model component is an application model providing a common contact or address book API implemented over an underlying database.

- The Calendar component is intended to replace the previous Agenda Model API. Calendar provides a cut-down API more suitable for a modern phone. The Agenda Model API is larger and has its origins in the needs of PDA users. Calendar partially supports the iCalendar standard. The vCal Plug-in is a library used by the Agenda Model to communicate with the vCard and vCal components.

Other Application Services Collection

This collection (see Figure 7.10) provides miscellaneous application support, originating from the Series 5 set of built-in applications, but extended more recently with the addition of the Timezone component.

Figure 7.10   Other Application Services components

Table 7.8 Other Application Services Components

Component Name	Development Name
Timezone	TZ, TIMEZONELOCALIZATION, TZLOCALIZATIONRSCFACTORY, TZCOMPILER, TZDB
World Server	WORLDSERVER
Help	HLPMODEL

- The Timezone component provides localization support, including a time-zone database, for Standard, Daylight, Short Standard and Short Daylight names for time zones. Localized names are stored in the resource file framework. Users can create cities and link them with time-zone information. Cities can also be grouped irrespective of time zone.

- The World Server component originated in the Time/World application of the original EPOC release. It is based on a world cities' database and server, and allows setting and easy switching between 'home' and 'away' locations and time zones, as well as time-zone browsing. It was deprecated in Symbian OS v8.1, in favor of the Timezone component.

- The Help component provides an engine implementation of a context-sensitive help system, providing read-only access to all help files on a Symbian OS device. Help files are essentially heavily compressed databases, each containing a series of topics relating to different applications or subjects.

Office Application Engines Collection

This collection (see Figure 7.11) provides legacy application-engine implementations of the original EPOC built-in applications: Data (database), Sheet (spreadsheet), and Word (word processor). Redundant on a modern phone, they are likely to be removed in a future operating system release.

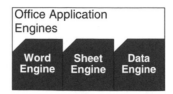

Figure 7.11   Office Application Engines components

Table 7.9 Office Application Engines Components

Component Name	Development Name
Data Engine	DAMODEL
Sheet Engine	SHENG
Word Engine	WPENG

PIM Application Support Collection

This collection (see Figure 7.12) provides services that may be useful to a variety of applications and application engines but which, typically, are quite closely tied to legacy applications.

Table 7.10 PIM Application Support Components

Component Name	Development Name
Alarm Server	ALARMSERVER
vCard and vCal	VERSIT
Chinese Calendar Converter	CALCON
File Converter Plug-ins	CHTMLTOCRTCONVERTER, CONVERT, RICHTEXTTOHTMLCONV
Backup Restore Notification	BACKUPRESTORENOTIFICATION

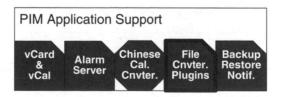

Figure 7.12   PIM Application Support components

- The Alarm Server component manages a queue of system-wide, time-based alarms, providing set, modify, query and notify APIs for client applications.

- The vCard and vCalendar components are parsers that convert between vCard or vCalendar entries and Symbian OS native formats.

- The Chinese Calendar Converter component provides a simple API for converting between Gregorian and Chinese calendar dates.

- The File Converter Plug-ins component supports conversions between HTML files and Symbian OS rich text objects stored in files, and between specific formats, for example Microsoft Excel, Microsoft Word and Microsoft font formats, and Symbian OS native rich text.

- The Backup Restore Notification component is used by legacy applications to notify of system-wide backup and restore operations. Publish and Subscribe provides a preferred alternative for new applications.

Messaging Application Support Collection

This collection (see Figure 7.13) provides Messaging and BIO Messaging frameworks and MTM plug-ins.

- The Message Store component provides a message server and framework, supporting standard message types (for example email and SMS).

- The BIO Messaging Framework component supports 'smart' message types (Bearer-Independent Objects), for example vCard or vCalendar messages and network setup messages.

- The BIO Watchers component provides a framework and service for notification of message arrival to applications.

- The Scheduled Send MTM component supports scheduled sending of any available message type and defines the scheduling parameters.

- The Email MTM components are plug-ins to the Message Store framework providing support for sending, receiving or editing POP3, IMAP4 (HTML mail) and SMTP email messages.

- The OBEX MTM components are plug-ins to the Message Store framework providing support for OBEX messages.

Figure 7.13 Messaging Application Support components

Table 7.11 Messaging Application Support Components

Component Name	Development Name
Message Store	MSG_FRAMEWORK
BIO Messaging Framework	MSG_BIOMSG
BIO Watchers	MSG_BIOWATCHERSCDMA
Scheduled Send MTM	MSG_SCHEDULEDSEND
POP3 MTM	MSG_EMAIL
IMAP4 MTM	IMAPSERVERMTM
SMTP MTM	SMTPSERVERMTM
OBEX MTMs	MSG_OBEXMTM
SMS MTM	MSG_SMS8.1
CDMA MTM	CDMASMSMTM
MMS Settings	MSG_MMS_SETTINGS
MMS MTM	MMS

- The SMS and MMS MTM components are plug-ins to the Message Store framework providing SMS message support for GSM/WCDMA and CDMA 2000 and the infrastructure support for MMS messages. From Symbian OS v9, licensees may provide the MMS MTM.

Content Handling Collection

This collection (see Figure 7.14) provides frameworks, handlers, parsers and recognizers for typed data and documents (including MIME and web types, SMIL and BIO messages) and DRM content.

Figure 7.14 Content Handling components

Table 7.12 Content Handling Components

Component Name	Development Name
SMIL Parser	GMXML
MIME Recognizer Framework	EMIME
WAP Push Handlers	WAPPUSHSUPPORT
Web Recognizers	RECOGNIZERS
Content Access Framework for DRM	CAF2 , CAF2CONFIG
Reference DRM Agent	DRMAGENT
MMF Recognizers	RECMMF
BIO Messaging Parsers	CBCP, ENP, GFP, IACP, WAPP

- The SMIL Parser component parses SMIL content based on a generic XML Parser and Composer with a 'mini-DOM' API able to perform syntax checking against simple DTDs. It replaces the SMIL Translator implementation of Symbian OS v7.0s.

- The MIME Recognizer Framework component supports for MIME data types.

- The WAP Push Handlers components are plug-ins to the WAP Push Framework implementing handlers including Several Interfaces, Single Logic (SISL).

- The Web Recognizers component supports URLs and web bookmarks and are implemented as plug-ins to the MIME Recognizer Framework.

- The Content Access Framework for DRM component provides generic APIs for brokering DRM-protected content between agents (DRM applications) and consumers (e.g. media players). It includes a reference DRM-agent implementation.

- The MMF Recognizers component provides support for multimedia data and document types.

- The BIO Messaging Parser components parse by BIO message type.

Text Rendering Collection

This collection (see Figure 7.15) enables not just applications but any components that want to display or manipulate text to use the Symbian OS text-handling and formatting APIs.

Table 7.13 Text Rendering Components

Component Name	Development Name
Text Formatting	FORM
Text Handling	ETEXT

- The Text Formatting component provides text view and layout classes to control scrolling, selection, cursor management, margin setting, and other attributes of displayed text. It supports the separation of display attributes (layout and drawing) from logical text attributes (styles). It is used, for example, by the Uikon Core API for editable text windows and, more generally, by applications to format rich text.

- The Text Handling component supports the storage of editable text and its formatting attributes, for example, paragraph alignment and character fonts. It is used with the Text Formatting text view APIs.

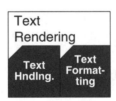

Figure 7.15 Text Rendering components

Internet and Web Application Support Collection

This collection (see Figure 7.16) provides Internet, web and WAP application support.

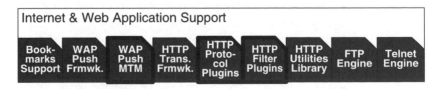

Figure 7.16 Internet and Web Application Support components

Table 7.14 Internet and Web Application Support Components

Component Name	Development Name
HTTP Transport Framework	`HTTP`
HTTP Protocol Plug-ins	`HTTP`
HTTP Filter Plug-ins	`HTTP`
HTTP Utilities Library	`INETPROTUTIL`
Bookmark Support	`BOOKMARK_SUPPORT`
Telnet Engine	`TELNET_E`
FTP Engine	`FTP`
WAP Push Framework	`WAPPUSH`
WAP Push MTM	`WAP-BROWSER`

- The HTTP Transport Framework component enables clients to establish a transport session for HTTP-like protocols, provides core APIs for transport sessions, transactions, and messages.

- The HTTP Protocol Plug-ins component provides dynamically loaded application and network protocol handlers, including TCP/IP, HTTP 1.1 and WSP 1.2.

- The HTTP Filter Plug-ins component provides dynamically loaded plug-ins to configure a transport session before use. It includes default HTTP and WSP filters that encapsulate responses to session events, for example, client authentication, message validation and message redirection.

- The HTTP Utilities Library component stores utility classes commonly used by Internet protocol parsing components. It contains implementations for URIs, a standardized time format, and simple text parsing utilities.

- The Bookmark Support component provides a bookmark database for web browsers.

- The Telnet Engine component provides a Symbian OS Telnet daemon and supports client sessions for communicating with a specified host.

- The FTP Engine Symbian OS FTP daemon, supports client sessions for communicating with a specified host. Does not expose public APIs.

- The WAP Push Framework component provides an interface between the WAP stack and the messaging infrastructure to support WAP as a messaging transport.

- The WAP Push MTM component provides a WAP stack implementation supporting messaging interfaces.

Printing Support Collection

This collection (see Figure 7.17) provides standard dialogs for setting up print jobs and controlling access by application clients. It is considered a legacy component for most devices.

Figure 7.17   Printing Support components

Table 7.15 Printing Support Components

Component Name	Development Name
Printing Services	PRINT

8

The OS Services Layer

8.1 Introduction

The OS Services layer (see Figure 8.1) provides the servers, frameworks and libraries that implement the core operating system support for graphics, communications, connectivity, and multimedia, as well as some generic system frameworks and libraries (Certificate and Key Management; the C Standard Library) and other system-level utilities (logging services). In effect, it is the layer that extends the minimal base layers of the system (the kernel and the low-level system libraries that implement

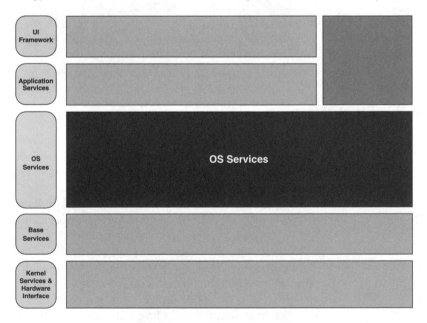

Figure 8.1 OS Services layer in the system model

the basic OS primitives and idioms) into an extensible, programmable, and useful operating system.

In terms of the number of components, it is by some margin the largest single layer of the system. To bring clear structure to it, the System Model organizes the layer into four major blocks by broad technology type (see Figure 8.2):

- Generic OS Services

- Comms Services

- Multimedia and Graphics Services

- Connectivity Services

These blocks are relatively self-contained. (Generic OS Services is used by the other blocks in the layer; Connectivity Services uses the transport technologies of Comms Services.)

This chapter describes the Generic, Multimedia and Graphics, and Connectivity Services blocks; the Comms Services block is described in Chapter 9.

Figure 8.2 Blocks of OS Services components

8.2 Purpose

Symbian OS is a microkernel operating system. The kernel is restricted to providing the minimum of essential services, specifically those required to implement process execution and memory access models. These are extended by the remaining (non-kernel) components of the base layers of the system, to support bringing up a bare system on hardware, providing access to peripherals and a file system, and to support a program execution model. Higher-level system services are built on top of this foundation.

In Symbian OS, the higher-level system services are located in the OS Services layer. These services provide the specialized system-level support required by other system components and by higher layers of the system, as well as by applications. Thus, for example, graphics support, communications support including networking and telephony, and the connectivity infrastructure are all provided as OS services.

Generic OS Services Block

This block provides a small number of generic services for use directly by applications, as well as some specific programming libraries intended for application and system use (including for use by the user interface and application support layers above).

- The logging and task-scheduling services are used by applications as well as by system components.

- The C Standard Library, providing a basic POSIX environment, is used by system components (for example, Java) and is also useful to those porting software from other platforms.

- There are libraries and frameworks supporting cryptographic and certificate-based security, including the key and certificate stores.

Multimedia and Graphics Services Block

This block provides all graphics services above the level of hardware drivers and provides the frameworks supporting multimedia services.

- It provides windowing, event handling, bitmap and vector graphics support including all font, drawing and bitmap functions, as well as low-level support for WYSIWYG printing.

- It defines a comprehensive set of multimedia APIs and provides a framework for implementation. It includes camera and broadcast tuner APIs, sound capture and recording APIs, still and moving image capture and recording APIs, display and play APIs, and conversion and manipulation APIs.

Figure 8.3   Generic OS Services block

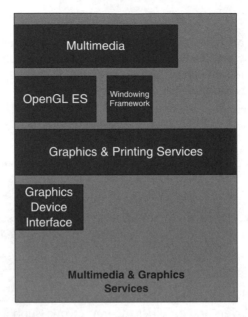

Figure 8.4   Multimedia and Graphics Services block

Connectivity Services Block

This block provides the device-side support for connectivity services, for example backup and restore, file transfer and browsing and application installation. (Data synchronization is provided in the Application Services layer, see Chapter 7.)

8.3   Design Goals

While the detail has changed considerably, most of the services located in the OS Services layer can be traced back to the original, early architecture of Symbian OS. In the earliest designs, the principal communications transport technology was serial, although networking support and, in particular, thorough support for standard Internet protocols had already been identified as an essential requirement, leading to the design of a networking infrastructure tightly bound to the communications services.

The first work on telephony-specific services, meanwhile, was well underway even before the first release of the OS, and relevant requirements were being evolved in collaboration with licensees. While at that time there were no specific multimedia services, the bitmap-based, windowing graphics system was central, and support for various audio formats was present from the beginning.

Figure 8.5 Connectivity Services block

Connectivity was also considered a vital service from the beginning, although it was a significantly simpler service based on Symbian's proprietary PLP protocol, a simple data transfer protocol over a physical (wired) serial port or emulated serial port over IrDA.

Since then, the rapid evolution of mobile telephony through successive technology generations, the ubiquity of the Internet and the increasing packetization of services, and the emergence of data exchange standards and protocols such as SyncML have all been powerful forces in shaping the evolution of the OS Services. The rapid convergence of multiple device functions with mobile phones has also had a dramatic impact on the kinds of services required from Symbian OS. Above all, multimedia technologies, which a decade ago were the province of top-end workstations, have migrated inexorably downwards onto smaller devices; simultaneously new categories of multimedia device have been invented (digital music players and digital cameras). New technologies, including digital broadcast TV for mobile devices and session-based multimedia protocols enabling two-way, real-time video and audio applications, continue to emerge and evolve.

In Symbian OS, providing support for all such services falls squarely in the realm of the OS Services layer.

8.4 Overview

All the core system servers, with the exception of the kernel server and the file server, are found in the OS Services layer.

- Generic OS Services:
  - The Task Scheduler provides a task-launching service for time-based and condition-based task triggers.
- Multimedia and Graphics Services:
  - The Window Server provides access to screen hardware and application and system events.
  - The Font and Bitmap Server provides font and drawing contexts for all bitmap-based devices.
- Connectivity Services:
  - The Software Install Server provides a secure software installation interface from remote clients.[1]
  - The Remote File Server provides a file system interface from remote clients.
  - The Secure Backup Socket Server provides a backup and restore interface from remote clients.

In addition, the essential communications servers appear in this layer (see Chapter 9):

- Comms Framework and Serial Comms
- Telephony
- Networking

Together these servers provide interfaces to system-level support for almost all the major services provided by the OS above the level of the kernel (persistent store and file system services are the notable exceptions). OS Services therefore really can be thought of as the essential infrastructure on top of which all application-level services are built.

It is also the location of Symbian OS support for many open standards including:

- OpenGL ES, FreeType, and graphics and audio file formats including GIF, BMP, WAV, MP3

---

[1] The Remote Software Install Server is *not* the same as the Secure Software Install Server; the former is used only by Connectivity Services components and manages software installation from a connected host device, typically a PC; the latter is the trusted computing base gatekeeper installation component, see Chapter 7.

- Cryptographic and key standards including RSA, DSA, DH, DES (not for use by end-users)
- ANSI C Standard Library, POSIX
- TCP/IP v4 and v6 networking
- 2G, 2.5G, and 3G telephony for GSM/UMTS and CDMA2000
- Serial RS232, USB, Bluetooth, Infrared/IrDA, OBEX
- Fax

From an application perspective, many of the provided services are sufficiently specialized that few applications use them directly, or even at all. Their functions are exposed to applications through higher-level frameworks and, for example, only specialized applications explicitly use telephony or networking, although any application may use the SendAs API.

However, all applications use the font services and perform event handling, window management, and drawing. Whether they know it or not, all applications depend on these core servers.

From another perspective, this layer brings out many aspects of the particular character of Symbian OS: multiple essential services are provided independently of the kernel; the client–server and framework and plug-in design patterns are ubiquitous; and object-oriented design idioms are widespread.

8.5   Architecture

The system model captures the broad division of responsibilities between components in the block structure of the layer. In general, each block is structured around one or more servers that collaborate to deliver a set of related services. Typically, servers also provide a plug-in framework, enabling extensible and flexible implementation of the underlying services. Frequently, the design includes multiple levels of frameworks through which services are implemented.

As an approximation, the key interfaces to each block are encapsulated in the principal servers and frameworks each block contains, although in many cases there are also additional utilities exposing library interfaces. In general, each block can also be thought of as layered to form a logical stack. The topmost layers of the stack expose the client interfaces (used by applications and system clients); the middle layers typically interface to other system services; and the lower layers expose framework interfaces (used by device implementers to create hardware adaptation plug-ins).

8.6   Generic OS Services Block

The Generic OS Services block provides a number of general-purpose utility-style services which are useful to applications (and other system

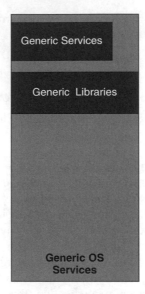

Figure 8.6   Generic OS Services block

components) and some specific frameworks and libraries that provide useful system services.

The frameworks and libraries include an implementation of the C Standard Library and framework support for secure certificates, keys and tokens. The more general-purpose utilities include logging and scheduling services and some legacy components.

Design Goals

For the most part, the Generic OS Services are system utilities or libraries which have (or had in the past) some specific association with particular applications but which have been seen as more generally useful for both system and application support and so have migrated downwards in the system as their services have been generalized.

The Event Logger and Task Manager were closely tied to the original PIM-style onboard applications, the File Logger to the telephony implementation (of which it was originally a component), the C Standard Library to Java (for which it was originally written to provide a minimal C wrapper for system calls), and so on. In all cases, these components have been part of the OS since its early releases.

The Cryptographic Token Framework and Certificate and Key Management components are relatively more recent, first appearing in Symbian OS v7. Their initial appearance in the platform represented the first steps toward providing complete and pervasive architectural support for secure network connections and secure browsing. The introduction of Platform Security, in Symbian OS v9, completes that process and takes it further, providing a complete architectural solution to the problems of security, privacy and trust.

ANSI C and POSIX Support

The C Standard Library first appeared early in the evolution of the OS and has remained largely unchanged through subsequent releases, providing a basic subset of the standard ANSI C library functions and POSIX system calls. It is designed to make it easier to port programs written in C or mixed C and C++ from other platforms to Symbian OS, although it does not claim to create a complete POSIX-like environment on Symbian OS. Instead, it supports the essential library functions, for example `malloc()`, `free()`, `printf()`, and so on, that almost any C program needs in order to run. All of `stdio.h` and `math.h` are supported.

The goal is to solve the most basic problems of porting and to enable basic C programs to run, allowing developers to focus on porting specific program logic and mapping to Symbian OS native idioms. In many areas, the underlying operating system semantics (of POSIX and Symbian OS) are quite different. For example, native process and thread semantics, file semantics, console behavior, and error and signal handling are very different in the two systems.

The C Standard Library implementation was originally written to support the first Java port to Symbian OS and included the bare minimum of the library needed by Java. (The porting problem was exacerbated by the original licensing conditions for Java, which limited source code availability; providing a minimal POSIX support layer was the simplest solution.) Since then it has been used by other system components, as well as by third-party code, especially for porting programs originally written for Unix. For example, recent ports of Python rely heavily on it.[2]

While Symbian OS v9 improves the support for 'standard C', there is still no support for accessing native idioms such as active objects from the standard C library. In other words, the library does not attempt to provide a complete C language interface to Symbian OS. Thus while POSIX can be seen as a valuable migration tool, for complex system ports its omissions are significant. It is likely that future releases of Symbian OS will include increasingly complete POSIX support as part of the support for a more standard C and C++ application platform.

Secure Certificates, Keys and Tokens

The Certificate and Key Management framework provides a complete framework for managing and storing security certificates and keys, and supports certificate storage and retrieval, certificate-chain building and

---

[2] Nokia provides a Python implementation for S60 through ***http://forum.nokia.com***. Tim O'Cock's Python for Symbian OS can be found at ***www.monkeyhouse.eclipse.co.uk***.

validation, and key operations including importing and exporting RSA, DSA and DH key pairs. (There is no support for generating keys.) The framework is not generally available to third-party applications, but is used by system clients (Application Installer, for example) and licensee applications (browsers and VPN client applications).

The Cryptographic Token Framework provides the additional support needed to manage certificate or key-protected hardware tokens (media cards such as SD or Memory Sticks, for example), and again is available to applications as well as system clients. It also provides an API for displaying security-related dialogs to the user (the implementation is supplied by the user interface).

Tokens are used to store secure keys. The framework provides an abstraction based on stored keys or certificates or PIN-style key authentication, as well as finder support to identify and enumerate secure media. Typical uses of secure tokens include DRM-protected content on physical media or their equivalent software 'emulations' (for example, DRM-protected games or films), as well as downloads (for example, of music).

Both frameworks make use of the unified key store and unified certificate store, abstractions allowing devices to have multiple, coexisting, key and certificate store implementations and providing a single point of access for clients, regardless of where an actual certificate or key resides (e.g. it might be on external media).

The key store is a repository of private PKI keys and provides APIs for storing and retrieving keys and for managing the store itself. The certificate store is a repository of root and user certificates and it provides APIs for storing and retrieving certificates and for managing the store itself. Root certificates typically belong to a certificate authority. User certificates belong to and are authenticated by the phone owner, and are always associated with a private key which is stored in the key store.

The security-related components within the Generic Libraries collection sit between the application-level services (such as the secure installers and the Content Access Framework, which is a generic framework for controlling content access in a way which is transparent to applications) and the low-level implementation of the cryptographic libraries (see Chapter 10).

Other Generic Services

The Task Scheduler provides a mechanism for performing time-based or condition-based tasks by scheduling the launch of an appropriate application when the task trigger is met. (This is not a notification service therefore, it is an application launcher.) From Symbian OS v9.1, conditions may include Publish and Subscribe variables becoming true. Typical uses include scheduled connections (connecting to email or message services, for example) and scheduled backup or data synchronization. Note that the Task Scheduler is a system server that always runs and which saves schedules to a permanent file store to ensure continuity across reboots.

Before Publish and Subscribe, the System Agent provided the means for storing and querying system state. From Symbian OS v9.1, most system-state values become Publish and Subscribe `RProperty` values to which clients can subscribe (given appropriate security-model capabilities). The System Agent retains only a few key services, for example, it defines and creates some default global system properties at startup and it maintains the Publish and Subscribe battery strength property.

The Event Logger provides an interface for logging and filtered querying of system events of interest to applications. Built-in and user-defined event types are supported. Typical uses are for creating call or message lists (a list of 'Recent Calls' in a phone application, for example). Events are expired when their lifetime is reached. However, the actual logging engine is optional and is supplied by the licensee (in the variant user interface on a particular device). If it is not present, calls to the logging APIs have no effect.

The File Logger, which provides a logging to file service, is deprecated in Symbian OS v9.1 and should be used only as a debugging tool. (It remains in the system for backwards compatibility purposes only.)

Component Collections

Components are organized into two small collections of servers, framework, and libraries. The common theme of the collections is general utility.

Generic Services Collection

This collection provides miscellaneous system services including some legacy components (retained for API compatibility).

- The Task Scheduler component is an application-launching server that supports creating, querying and editing of time- or condition-triggered tasks. From Symbian OS v9.1, clients should migrate to revised interfaces.

- The Event Logger component is only an interface (i.e. it is supplied only as a wrapper) supporting logging of events, for example, call

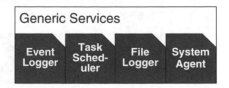

Figure 8.7 Generic Services components

Table 8.1 Generic Services Components

Component Name	Development Name
Event Logger	LOGENGONGOING
System Agent	SYSAGENT2
Task Scheduler	SCHSVR_ONGOING
File Logger	FLOGGER, COMMSDEBUGUTILITY

and message lists and retrieval, filtering and viewing by clients. The logging engine itself is assumed to be supplied by the variant user interface. If no engine is present, calls to the wrapper succeed but have no effect.

- The System Agent component is a legacy component that performed a number of useful functions for monitoring and reporting system state. From Symbian OS v9, the main System Agent functionality is taken over by the Publish and Subscribe service provided by the User Library (see the RProperty class). The System Agent retains a few key services only, for example, it defines and creates some default global system properties at start-up, and it maintains the Publish and Subscribe battery strength property.

- The File Logger component is a legacy utility for logging system or application messages to a log file. From Symbian OS v9 this is considered a debugging utility that is provided only for backwards compatibility.

Generic Libraries Collection

This collection provides system-level libraries for use by applications and system components.

- The Certificate and Key Management, Certificate Store and Key Store components provide a framework for certificate and key management that supports public key cryptography for RSA, DSA and DH key pairs (including storage and retrieval), assignment of trust status and certificate-chain construction, validation and revocation. Certificate

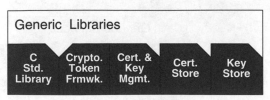

Figure 8.8 Generic Libraries components

Table 8.2 Generic Libraries Components

Component Name	Development Name
Certificate and Key Management	`CERTMAN`
Certificate Store	`CERTSTORE`
Key Store	`KEYSTORE`
Crypto Token Framework	`CRYPTOTOKENS, FILETOKENS`
C Standard Library	`STDLIB`

Store provides a single point of access for clients to certificates stored on the device. Key Store is a repository of private PKI keys that may be used to sign data, verify signatures, and so on, and provides APIs for storing and retrieving keys and for managing the store itself.

- The Cryptographic Token Framework supports the use of secure hardware tokens (i.e. encrypted media cards and file systems), for example DRM-protected games or films on SD cards or memory sticks, or their equivalent software emulations, for example, downloaded music tracks.

- The C Standard Library is a subset of the POSIX C library which maps C function calls in as simple a way as possible to native Symbian OS calls. It is a subset implementation and does not attempt to provide a complete POSIX environment on Symbian OS.

8.7   Multimedia and Graphics Services Block

Graphics has always been central to Symbian OS (see Figure 8.9), which was designed to support a sophisticated graphical user interface and sophisticated application graphics. In Symbian OS, there is no notion of character-based applications (except for test or development purposes); all applications are intrinsically graphical. Likewise, full-color support has always been an integral part of the OS design; even when running on 16-bit grayscale devices, 24-bit color modes were supported (which still remains beyond the capabilities of most phones).

Similarly, while the native font format is bitmapped (bitmap fonts are still preferred for small-screen devices, where pixel-perfect design is required to optimize for relatively small physical display size), support for FreeType vector fonts was introduced early on. Indeed sophisticated support for non-Roman fonts, including right-to-left and even bi-directional fonts, was always seen as central to the global aspirations of the OS.

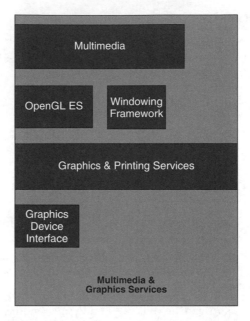

Figure 8.9   Multimedia and Graphics Services block

Audio data too was supported from the beginning, with a built-in recorder application forming part of the original application set for the system, something many phones still cannot match. (A Psion Series 5 running the early version of what became Symbian OS beat a cassette player hands-down for nailing that hard-to-master guitar lick.) Symbian OS moved onto phones when the state of the art was a polyphonic ring tone; in contrast, it allowed users to launch a complete sound clip as a phone ring tone (leading to offices full of baaing sheep and baby-gurgle effects).

Increasingly, as Symbian OS has driven into the phone market from its base as a more generic OS for the family of connected, PDA-like devices from Psion and as phones themselves have become more sophisticated, support for full-scale multimedia has become essential.

The Symbian-based Nokia 7650 was the first camera–phone outside Japan. The Symbian-based Sony Ericsson P800 played movie clips and the P900 shipped with a built-in MP3 player. The Symbian-based Nokia nGage integrated an FM radio along with game graphics and stereo sound.

The new device trends include full-stereo sound, multimegapixel cameras with true optics (true camera lenses and optical zooming) for both still and movie images, hardware-accelerated graphics, multiple high-pixel-density displays, and onboard high-definition broadcast TV.

Symbian OS v9 supports these hardware features with a new Multi-media Framework that is lightweight and flexible and aims to provide consistent and hardware-independent interfaces at the application level, while providing flexible support to device makers wanting to integrate the available multimedia hardware and support new multimedia applications and services.

Graphics Design Goals

Graphics is central to the goals of the OS to provide an easy-to-use, consumer-oriented operating system capable of driving a wide range of devices but offering a sophisticated and, above all, open (to third-party application developers) platform for general application development.

From the beginning, the system has been optimized to produce fast graphics on low-power devices. The importance of rich font support was recognized early on, including support for exotic scripts. Increasingly, the focus for graphics has moved towards multimedia applications and games, making device graphics generally more critical, as well as specifically making it more important to support open graphics standards.

On the one hand, Symbian offers a much more integrated graphics architecture optimized for its device class than, for example, Linux, which requires licensing an application-level graphical toolkit, for example, Trolltech or GNU, on top of which to either license or implement a bespoke user interface. On the other hand, the Symbian graphics solution aims to be more carefully architected, more modular, and better-scaled than a system such as Windows Mobile, which has a monolithic user interface implementation (with its origins in the PC-centric, legacy design of Windows itself).

Multimedia Design Goals

The first implementation of a multimedia server was introduced in Symbian OS v7. It was enhanced and substantially re-architected in Symbian OS v8 and has evolved significantly in Symbian OS v9. Partly, its evolution is the result of the rapid pace at which multimedia hardware and services have migrated to mobile phones and the push from both licensees and operators to integrate sophisticated new hardware and support new media services, and partly it is a natural evolution enabled by other enhancements in Symbian OS (including the adoption of the real-time kernel, which opened the way for a significant change in phone hardware complexity, and platform security, which makes Symbian OS an ideal platform for a movie and music player, including DRM-protected media cards and downloads).

The Multimedia Framework therefore provides a single extensible framework for integrating support for audio, video, MIDI, automated speech recognition, cameras, and integrated broadcast tuners. Its purpose is to consolidate and standardize the multimedia APIs, so that they are common across all devices based on Symbian OS, while also providing a flexible foundation for extension and customization.

The framework is designed around the concept of controllers that provide a full range of standard multimedia functions (such as audio and video recording and playback, as well as more advanced functions such as speech recognition) and define standard APIs, allowing uniform client access across all Symbian OS devices regardless of their different capabilities, and a standard plug-in interface.

The framework is implemented as a lightweight, multithreaded, ECOM-conforming plug-in framework, which may run as one or more threads in the client application process, and consists of a number of components that implement the application-level interfaces.

The actual implementation on any device is provided by controller plug-ins which are supplied by licensees and used by client applications to access multimedia functions. The media capabilities of a given device therefore depend on the available hardware and the supporting controller plug-ins, and are ultimately determined by the underlying device hardware. Multiple controllers may be available for any given format. Applications can choose whether they want to select a controller or to leave selection to the framework.

The Image Conversion Library provides an extensible plug-in framework, also conforming to ECOM, and a standard set of conversion plug-ins supporting conversions between standard image formats. The Camera component defines a standard API for onboard cameras and provides a reference implementation. The Broadcast Tuner component provides a standard API for onboard radio tuners.

The camera and tuner APIs are implemented as frameworks into which custom plug-ins are loaded to support specific hardware available on different Symbian OS devices but provide a standard API both for plug-in writers and for client applications, so that applications can work consistently across different phones including phones from different vendors.

The framework also defines the lower-level interface to the Media Device Framework (see Chapter 10), which defines device-level plug-in interfaces including support for hardware accelerators, and provides a standard set of device drivers.

The framework is heavily plug-in dependent. From a client application perspective, it provides a rich and consistent set of interfaces for all kinds of multimedia. From a system perspective, it provides a number of different plug-in interfaces to support writing of plug-ins that implement chosen levels of support (from logical to physical) for the onboard hardware.

By default, Symbian supplies only a simple audio plug-in, supporting WAV, AU and RAW formats. Codec implementations are provided for a number of encodings including various PCM encodings, A-Law and u-Law, and GSM6.10. Licensees are expected to supply the full set of plug-ins required on a particular phone, providing custom controllers, codecs, and format support.

OpenGL ES

OpenGL ES is an open standard for 2D and 3D graphics, specifically targeted at embedded systems including consoles and phones. It defines application APIs for rendering, texture mapping, and other graphical effects, as well as a portable binding to native windowing systems, as a subset of the workstation- and desktop-oriented OpenGL standard.

OpenGL ES support in Symbian OS consists of a framework that implements the API binding and a standard client API definition but not a concrete implementation. A stub implementation is supplied by OpenGL ES Headers and the OpenGL ES component is a reference implementation of a third-party OpenGL ES renderer. The API includes Display Properties (optional in the OpenGL ES standard) that encapsulate drawing properties (e.g. displaying rectangles and clipping regions), enabling drawing to be delegated to threads that don't have access to an RWindow object.

The framework is provided to implement the OpenGL ES binding and ensure compatibility between different devices.

Windowing Model

The Window Server is at the heart of the graphics architecture of Symbian OS and it is central to the event-handling model that drives applications. Unlike many other operating systems, in Symbian OS there is no notion of character-based applications or devices (there are no teletypes or green-screen terminals in mobile phones). All applications in Symbian OS are intrinsically graphical and the screen is where application events are realized, as well as being an important source of application events. The Window Server is at the heart of screen control and is, therefore, central to applications.

The Window Server uses the concept of application-owned windows onto the display device to serialize access to the display by multiple concurrent applications. A window on a Symbian OS device is a rectangular screen region that can be drawn to on behalf of its owning application in response either to system or application events, and which receives focus events as well as keyboard and pointer (pen) events. The Window Server owns the screen as a resource and owns the single event queue through which all device events, whether system- or application-originated, are handled, managing kernel and application events as well as events generated by the Window Server itself and distributing them to applications or system user interface components (status bars and so on). The Window Server implements a classic Symbian OS pattern – serializing access to shared resources, which in this case include the physical display and interaction and other events. (Note that devices may also have multiple physical displays.)

A window is an abstraction for making a screen region available to an application for interaction. A window abstracts a region of the physical screen. From the perspective of an application developer, a window is a screen region in which an application view can be constructed. To draw into the window area, and to receive user input, applications create controls inside windows and the controls become the units of interaction.

Applications are, by definition, window-owning processes. Applications may create and destroy windows, may have many windows, and may switch between them. Application windows form a window group. The first application window in a group is the top client window and an application must have at least one of these in order to display. Windows allow applications to display and have screen modes (e.g. color depth), a drawing area, and so on.

Logically, windows are maintained in a window stack, implying that they have a 'Z' order which is enforced by clipping; windows higher in the stack hide windows lower in the stack. Windows come in and out of focus (i.e. have the focus of the user) and, typically, the window at the top of the stack is the window which currently has focus. (A window group, however, may choose not to receive focus.)

By default, windows are the size of the full screen (less any area reserved for device status bars, control button arrays, or similar system-owned screen resources). Therefore, windows do not overlap but hide each other, a design optimized for small-screen devices (but a policy which is ultimately determined by the variant user interface running on a given device).

In implementation terms, a window is an 'R' class (RWindow) object returned by the Window Server to a client opening a window subsession in a Window Server session.

From an application perspective, most of the Window Server functionality is abstracted through the Control Environment and is made available to applications through the APIs provided to create and manipulate controls. Thus while applications need some knowledge of the windowing model, the window methods are provided through the Control Environment, not from the Window Server explicitly.

The Window Server is a system server that is started by the System Starter at boot time and runs until system shutdown. Only when the Window Server is running and providing access to the screen and events, can the Application and UI Frameworks be started, and only then can the phone application be run.

The Window Server is also responsible for starting some servers at startup and it provides the plug-in interface for animation (see Chapter 7). (It also provides other plug-in interfaces, for example the RSoundPlug-in interface. The Keyclickref plug-in is an example implementation of a

Window Server key-click plug-in library, which provides the audible clicks for keystroke events.)

The Window Server has a number of responsibilities:

- implementing client-side buffering of windowing commands to minimize calls across process boundaries between client and server, while enabling fine-grained control by the client (which can flush the buffer)

- managing bitmapped drawing via the Font and Bitmap Server

- managing the clipping and valid or invalid regions of screen, for example, when part of the screen becomes uncovered by some window

- managing system-initiated redraw events, window stacking (Z-order), etc.

- providing backed-up windows as a special case for applications that lack the ability to manage their own redrawing efficiently

- handling special effects including shadowing and animation.

Event handling is based around the RequestEvent active object, which the Window Server uses to get events from the kernel, once it has registered itself as the default event handler. Event types include digitizer and pointer events, keyboard events, and some other hardware events (including switch off, case open and case close, which are legacies of the early device architecture of the Series 5, a clamshell device in which closing the case caused the device to suspend and opening it caused the device to resume operation).

Window Server processes these events and passes processed events to clients. Thus, for example, pointer events may be translated into focus events or other logical events. Window Server may also perform rotation and other logical processing or scaling of screen coordinates for generated events (some devices support multiple screen orientations; others support full and 'flip' mode sizes). It performs key events and logical key events, including translations that are implemented by Front End Processor (FEP) plug-ins, for example to translate pointer events on an on-screen soft keyboard to logical keyboard events; to translate scan codes to character codes for physical key events; and to interpret hotkey events and combinations. Window Server also initiates events, for example redraw events, and manages the event queues for clients. Each client has its own queues.

The Window Server also supports a direct access (DSA) drawing mode, which bypasses the server itself but still enables an application to determine which screen region it owns (so that it does not overwrite other applications or system components which may have visible elements on the screen). The DSA framework notifies Window Server when it is invoked (but otherwise the Window Server is not directly involved).

The Window Server has been a central part of Symbian OS since the beginning but has seen many enhancements in subsequent releases, including semi-transparent windows, multiple screens, double buffering, and a configurable origin and scaling factor for windows (supporting rotated screens and flexible screen size).

Fonts and Bitmaps

From the perspective of the graphics system, all graphics devices are bitmap devices. All bitmapped graphics services and font services, including printing support, are managed by the Font and Bitmap Server. It owns the graphics devices and serializes client access to them (whether clients are applications or other system services). All access therefore to the screen or to printers and all bit-oriented screen operations, including font operations, are conducted through a client session with the Font and Bitmap Server, within a bitmapped device context. The Font and Bitmap Server also ensures that screen operations are efficient by sharing single instances of fonts and bitmaps between its multiple clients. It also provides the framework for loading bitmap and vector fonts.

While the Font and Bitmap Server owns the graphics devices, the Bit Graphics Device Interface (GDI) actually rasterizes drawing to bitmapped devices. Bit GDI implements the concrete instances of bitmapped graphics contexts, from the basic device abstractions providing hardware-independent access to display devices and screen attributes using a variety of graphics primitives. Bit GDI also provides transparency support (alpha blending).

Font management is delegated to Font Store, which manages all fonts in the system, both native Symbian OS format bitmapped fonts (glyph fonts) and open vector fonts, and performs closest-fit matching of font requests. Font Store provides APIs for storing, querying and retrieving bitmapped fonts and all properties of glyph fonts. Vector fonts are drawn by the FreeType Font Rasterizer and the available vector fonts are vendor-dependent. (Symbian OS includes a reference implementation of the FreeType rasterizer and reference bitmap fonts. Licensees may choose to replace the FreeType implementation with one of production quality or may omit it and replace the reference bitmap fonts with their own bespoke fonts.)

FreeType supports FreeType 2 TrueType fonts. On small-display devices, and on phones in particular, carefully optimized bitmap fonts offer a more optimal font solution than standard vector fonts.

WYSIWYG printing is provided by the Printer Driver Support printing framework, which stores and manages printer drivers and manages access to and mapping of printer ports, and for which reference implementations of concrete printer 'driver' plug-ins (type: PDR) are provided. Printer drivers are not device drivers in the standard sense of controlling physical hardware; rather they are printer-driver information files that provide translations from device-independent bitmap-based graphics descriptions to printer page descriptions. More precisely, a printer driver implements the GDI-defined bitmapped device abstraction and is a DLL plug-in to the framework. Printer ports are virtualized over the available device hardware, typically serial or short link. As well as loading driver plug-ins, Printer Driver Support creates printer driver lists. WYSIWYG printing support is considered legacy functionality for a modern phone. (See the earlier discussion of application-level support for printing, for example, based on Bluetooth profiles.)

The close coupling of drawing and printing and the inclusion of support for line breaking and margin calculation alongside polygon and ellipse rasterization among the Bit GDI primitives shows the legacy of the early implementation of Symbian OS. For example, on a modern phone, fast rendering of games and smooth rendering of streamed video on a device where many other things may be going on (calls being received, music being played, and so on) is more relevant than margin calculation for office-style documents.

Graphics Contexts and Color Palettes

The lowest layer of the graphics system provides the abstract interface to the device hardware (the physical interface is managed by the logical and physical device drivers in the Kernel Services and Hardware Interface layer).

The GDI abstracts the physical graphics device (a bitmap display or raster device) as a Device Context containing settable drawing and font properties (pen and brush settings for line styles, character and font information and metrics), all drawing methods (for lines, polygons, circles, rectangles, as well as text and bitmaps), and the clipping region defining the drawable rectangle.

Since GDI pixels and font metrics are device dependent, methods are provided to map from twips values (Symbian OS device-independent units)[3] to pixels and to zoom fonts by a specified zoom factor.

Text rendering supports bi-directional text, that is, both right-to-left and left-to-right as well as mixed text, and line-breaking algorithms. GDI also manages color value, handling mapping RGB values into display-mode color spaces.

The Color Palette supports color-array handling and conversion between RGB values and palette indices, and supports dynamic palettes, that is, color palettes may be supplied by external classes, allowing clients to control the palette capabilities depending on the available device hardware.

---

[3] A twip is a decimal variant of a typographical point. A point is 1/72 of an inch; a twip is 1/20 of an inch.

Font metrics and selection (matching a device-specific font to the font request) were significantly improved in Symbian OS v9 to support higher-resolution screens and to better support screens with non-square pixels. Calculation algorithms for font metrics (ascender and descender sizes, capital heights, etc.) were added and there are methods that offer choices based on maximum height to guarantee that the supplied font fits the given screen space.

Graphics Architecture

At the heart of the graphics architecture are the Window Server, the Bit GDI and the GDI components. Together they provide the services required to write to bitmapped physical displays from within a system or application graphics context and support the windowing abstractions that allow multiple clients to independently manipulate the display.

The Window Server abstracts the key ideas of event-driven pro-gramming for graphical applications and applies object-oriented design principles (and native idioms of the operating system, for example, active objects) to provide a straightforward programming model for native applications.

From an application perspective, a window is a secondary object that is created from the application view (the top-level application control; every application needs at least one control that owns or controls a window, that is a view). Once associated with a view by a Set() operation, a window is abstracted to the top-level graphics context in which all subsequent drawing, clipping and similar operations are performed.

While the graphical architecture of Symbian OS is central to the user interaction and application model, so that, in effect, nothing can happen on a device (from a user perspective at least) without the involvement of the Window Server, from a system perspective graphics is well isolated from the kernel and the basic system services. Thus to implement a base port, a text-only version of the Window Server and a Text Shell replace the complete graphics infrastructure with a simple event handler and a console shell. The resulting bare-bones system has no application support, communications or other 'higher' services but, from a kernel perspective, it is fully functioning.

The graphics system therefore is (from the kernel perspective) just another user-side process; it runs user-side (i.e., in non-privileged mode) and uses the standard machinery of client–server inter-process commu-nications to communicate both with the kernel (which is a server and to which it is a user-side client) and with its own clients.

The newer additions in the graphics area, for example, vector fonts and the OpenGL ES interface, as well as the Multimedia Framework itself, build on top of the basic window and graphics system.

Component Collections

Multimedia Collection

This framework defines application-level APIs for multimedia support of all kinds and provides a number of standard implementations as framework plug-ins.

Table 8.3 Multimedia Components

Component Name	Development Name
Multimedia Framework	MMF, COMMON
Multimedia Framework Plug-ins	MMFAUDIOCONTROLLER, MMFRAWFORMAT, MMFAUFORMAT
Image Conversion Library	ICL, ICL_IMAGEDISPLAY, IMAGETRANSFORM
Image Conversion Library Plug-ins	ICL_GIFSCALER
Camera	ECAM
Broadcast Tuner	TUNER

- The Multimedia Framework component provides a high-level extensible framework for multimedia support of all kinds, providing client utilities for common tasks, for example audio, tone, video, and MIDI playback and recording, as well as speech recognition. The framework is designed to accept controller plug-ins, which in turn provide the interface to lower level plug-ins (supplied by the Media Device Framework, see Chapter 10) that interface to hardware and provide acceleration APIs.

- The Multimedia Framework Plug-ins component provides controller plug-ins to the framework; reference controllers are supplied for standard audio formats.

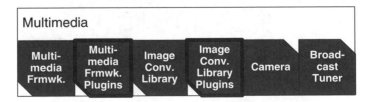

Figure 8.10 Multimedia components

- The Image Conversion Library component provides an extensible framework for integrating still-image conversion codecs into the Multimedia Framework. It recognizes picture file formats by providing a MIME-type recognizer plug-in to the MIME Recognizer Framework.

- The Image Conversion Library Plug-ins component provides default reference codecs for common still-image formats including GIF, JPEG, PNG, BMP and MBM.

- The Camera component provides an implementation for an onboard camera, allowing a camera object to be created and controlled and imagery data to be requested and received from it.

- The Broadcast Tuner component provides an implementation for an integrated broadcast tuner.

OpenGL ES Collection

These components comprise a framework supporting the OpenGL ES 2D- and 3D-graphics standard. OpenGL ES provides multi-client access to screen, keyboard, and pointer or digitizer for GUI applications and includes a keyclick reference plug-in that produces key or pointer clicks.

Table 8.4 OpenGL ES Components

Component Name	Development Name
OpenGL ES Headers	OPENGLSHEADERS
OpenGL ES Display Properties	OPENGLESDISPLAYPROPERTY
OpenGL ES	OPENGLES9.X

- The OpenGL ES Headers component provides standard OpenGL ES headers and binary definition files to encourage compatibility between OpenGL ES implementations for Symbian OS. The headers bind the OpenGL ES API to the underlying graphics model and support a plug-in renderer implementation.

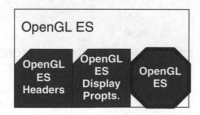

Figure 8.11 OpenGL ES components

- The OpenGL ES Display Properties component encapsulates display-drawing properties (e.g. display rectangles and clipping regions), enabling window surface access, that is, drawing, to clients from threads that do not own a window.

- The OpenGL ES component provides a reference implementation of an OpenGL ES renderer implemented as a plug-in, which is replaced by licensees.

Windowing Framework Collection

The Window Server owns and manages access to the screen as a drawable resource, which is made available to applications through the abstraction of windowed screen areas. It also provides access to the keyboard and pointer or digitizer for GUI applications, including the keyclick reference plug-in that produces key or pointer clicks.

Table 8.5 Windowing Framework Components

Component Name	Development Name
Window Server	WSERV8.1

Windows are at the top of the abstraction hierarchy for screen elements; all applications must own (or control) a window in order to display or to receive events. The Window Server receives and interprets events on behalf of applications, as well as generating events based on received application events (focus events, for example).

Graphics and Printing Services Collection

These components support all bitmapped graphics operations on display and printer devices, including all font and drawing operations. The principal components are the Font and Bitmap Server, through which all operations are made within a client-side server session to a bitmapped

Figure 8.12 Windowing Framework components

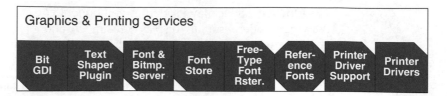

Figure 8.13 Graphics and Printing Services components

Table 8.6 Graphics and Printing Services Components

Component Name	Development Name
Font and Bitmap Server	FBSERV
Text Shaper Plug-in	ICULAYOUTENGINE
Bit GDI	BIT GDI
Font Store	FNTSTORE
FreeType Font Rasterizer	FREETYPE
Reference Fonts	FONTS
Printer Driver Support	PDRSTORE
Printer Drivers	PRINTDRV

graphics context, and the Bit GDI, which implements the bitmapped graphics context abstraction.

- The Font and Bitmap Server owns all bitmapped graphics devices and provides the framework for other graphics components. The server manages system-wide shared access to single-instance fonts and bitmaps, providing bitmap and font services for native bitmap fonts and vector fonts through its client-side APIs. It is responsible for loading the plug-in font rasterizer for vector fonts.

- The Text Shaper Plug-in component to the Font and Bitmap Server enables improved glyph placement for Hindi (i.e. Devanagari script).

- The Bit GDI component provides a polymorphic interface independent of device and display modes to bitmaps and the screen device via graphics primitives that implement the concrete device context for bitmaps.

- The Font Store component provides font storage and font file loading, using plug-in font rasterizer libraries if required. It also performs closest-fit matching of font requests.

- The FreeType Font Rasterizer component provides a reference implementation and library wrapper for the FreeType font rasterizer, supporting FreeType 2 TrueType font descriptions.

- The Printing Support component provides a framework that manages and loads printer drivers as bitmapped device context implementations and manages access to printer ports. It is considered a legacy component on most modern devices and is only relevant to PDAs.

- The Printer Drivers component provides reference implementations of concrete printer drivers that implement the polymorphic interface defined by GDI. It is considered a legacy component on most modern devices and is only relevant to PDAs.

Graphics Device Interface Collection

This is the lowest level of the graphics services, providing low-level graphics abstractions and color palette support.

Table 8.7 Graphics Device Interface Components

Component Name	Development Name
GDI	GDI
Color Palette	PALETTE

- The GDI component provides a device-independent graphics context abstraction, which supports drawing to various devices including screens and printers (which are treated as specialized graphics contexts). Normally all drawing, text display, and so on, is performed on a graphics context.

- The Color Palette component supports color-array handling, conversion between RGB values and palette indices, and dynamic palettes. Color palettes may be supplied by external classes, allowing clients to control the palette capabilities depending on the available device hardware.

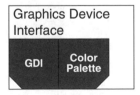

Figure 8.14   Graphics Device Interface components

8.8   Connectivity Services Block

Connectivity Services in Symbian OS (see Figure 8.15) consist of dedicated service and transport frameworks designed to support basic device or host connectivity functions, including backup and restore, remote file browsing, remote software installation, and so on.[4]

The first releases of Symbian OS based their connectivity on the proprietary PLP serial and infrared-based protocol. Symbian provided basic software for both PCs and devices, enabling backup and restore, synchronization of PIM-application engines, remote software install, and remote access to the file system. Licensees mostly provided basic customizations.

While Symbian OS v6.0 retained PLP, Symbian OS v6.1 moved to a TCP/IP-based framework (based on m-Router, licensed from Intuwave) and also introduced Bluetooth as a bearer, thus extending support to include cable, infrared and Bluetooth. m-Router also adds a service-loading framework and can load custom services.

From Symbian OS v8, there has been significant re-architecture of the Connectivity Services, principally on the host-side (in other words, on the host computer to which the device is connecting) but including the introduction of the Bearer Abstraction Layer to improve standardized access to connected phones.

Figure 8.15   Connectivity Services block

---

[4] The best introduction is [MacDowell 2005].

Design Goals

Good connectivity is a vital feature for any mobile device and especially for consumer-oriented devices. Symbian OS provides good device-side support for generic connectivity services based on configurable, standards-based technologies (such as SyncML), and the drive towards more consumer-oriented devices will hopefully see licensees (or opportunistic third-parties) providing good solutions for connecting to all host platforms, including Macintosh and Linux, for example. Easy connectivity based on standard technologies and compatible between devices from different licensees across multiple host operating systems is vital to support migration of data between devices (from an old device to a new device, for example).

Interestingly, while Symbian OS makes TCP/IP the standard protocol for its connectivity services, OBEX is more common on phones not based on Symbian OS. OBEX is optimized for simple transfer of small objects, for example, contact records and SMS messages. While OBEX is supported by Symbian OS and while some licensees may provide their own support for OBEX-based connectivity, it is not part of the standard connectivity solution.

Overview

The connectivity architecture provides a framework within which the device-side of TCP/IP-based device-to-host services can be created. Since the actual bearer is abstracted, such a service runs on any bearer. Implementations are provided for the basic device–host connectivity services of device backup, remote software installation and remote file browsing.

Windows PC desktop-side implementations are supplied as part of the Connectivity Services implementation but, in principle, the services on the Symbian OS device are agnostic about the host operating system. Since the services are based on TCP/IP, host-side implementations can be written for any operating system. Typically, all licensees provide a host connectivity suite of some kind; most support only Windows, some support Mac OS/OS X. Third-party freeware packages provide varying degrees of support for Linux or Unix connectivity for Symbian OS devices.[5]

The device-side framework is extensible, so that new (device- and host-side) services can be written, and open, so that host-side services

---

[5] *http://symbianos.org/~malm/SymbianLinuxHowTo.html* documents connectivity solutions for legacy releases up to Symbian OS v7.0; for current Symbian OS phones, data synchronization with other SyncML supporting systems should be possible but may require configuration. Alternatively, *www.scheduleworld.com* provides a web-based SyncML service, which should enable synchronization between Symbian OS and other SyncML-supporting systems.

can be written for platforms (e.g. Linux/Unix) that device vendors do not support out of the box. The framework is intended for use by developers of host-side software to access the device and its applications and is customizable by extension.

As supplied, the PC-side connectivity application uses Windows Winsock over RS232 serial, USB, Bluetooth, and infrared connections. On the device-side (i.e. Symbian OS), the chosen bearer propagates (through the Sockets Server) to a Connectivity Services Server Socket. Bearer-level components interoperate with the Sockets Server (see Chapter 9) to provide services to the framework.

Services

Connectivity Service Providers are device-side services that support basic interactions with a desktop host to perform device backup to the host, file browsing and transfer (in both directions; typically, browsing the device file system from the desktop and copying files between the device and desktop host) and software installation (from desktop to device).

The basic supported services are:

- backup and restore of a drive on the device to a desktop host

- file management (e.g. copying files to and from the device, renaming and deleting files and directories on the device, and formatting device drives)

- installation of software from the desktop host.

Additionally, the infrastructure supports starting named services on the device from the desktop host and managing the connection between the device and the host.

Data synchronization functions are not supported by the Connectivity Services but are provided elsewhere (for the device side, see Chapter 7; on the host side, there are various third-party offerings as well as licensee-provided software packages).

All the supplied services use the Service Broker framework. The Remote File Server provides an interface, via the Service Broker, to the device file system for a host-side client. Similarly, the Software Install Server enables a host-side client to interact with the device Application Installer to install SIS, JAR and JAD files over TCP/IP or OBEX. Similarly also, the Secure Backup Server enables a host-side client to interact with the Secure Backup Engine, which performs the interaction with the device-side file system and other processes to back up data from the device to the host.

Framework and Transport Abstractions

The Service Broker framework is the core of the Connectivity Services implementation, allowing device-side services to register a port number

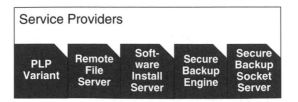

Figure 8.16 Service Providers components

for use by host-side clients, allowing host-side services to be started. The Service Broker protocol requires a TCP/IP connection to the host, for which it relies on the Bearer Abstraction Layer (BAL). Named services (supplied by the connectivity component) use the Service Broker. Port-number registration is based on XML-defined configuration files.

The Bearer Abstraction Layer (introduced in Symbian OS v9) provides a bearer-abstraction framework and a connection-management API to PC-link-type applications, allowing selection and configuration of connection bearers. Typically, the link application is provided by a licensee as part of a connectivity suite for a particular product. The framework supports plug-ins that encapsulate actual bearers (for example, m-Router).

Server Socket is a helper library that allows TCP/IP services based on port numbers to be created for use by the Service Broker, which is simpler and more ROM-efficient than creating bespoke named services from scratch.

Architecture

The Connectivity Services block provides device-side support for connectivity services. Services are organized around a central Service Framework component, the Service Broker, with named services, which are clients of the framework and use it to propagate service port numbers to remote clients, and bearer services, which are used by the framework to provide TCP/IP-based services over a variety of available bearer technologies.

The Bearer Abstraction Layer provides a common platform on top of the m-Router TCP/IP-based transport, independently of the actual bearer. Bearer support is provided to the Bearer Abstraction Layer as a plug-in, interfacing to a networking Sockets Server socket connection.

Component Collections

Service Providers Collection

These components provide named services which run on the device side to provide service interfaces to remote (host-side) clients. All use the Service Broker as an intermediary to propagate their port numbers to the remote client.

Table 8.8 Service Providers Components

Component Name	Development Name
Remote File Server	REMOTEFILESERVER
Software Install Server	SWINSTALLSERVER
Secure Backup Socket Server	SBSERVER
Secure Backup Engine	SECUREBACKUPENGINE
PLP Variant	PLPVARIANT, PLP, BRDCST

- The Remote File Server component provides on-device file-management functions to a remote client over TCP/IP, including access to backup and restore functions provided by other system components.

- The Software Install Server component interacts with the software installation components on the device to enable remote installation of SIS, JAR and JAD files over TCP/IP or OBEX. Installation events can propagate to a connected host, passing progress information and errors and allowing user interaction.

- The Secure Backup Socket Server component provides backup/restore functions to a remote client over TCP/IP.

- The Secure Backup Engine component manages backup and restore of device-side data, including private data and installed software, as controlled by the Secure Backup Socket Server. This component exposes an API and can be used by other components to carry out a remote backup and restore (for example, to a connected PC) or a local backup and restore (for example, to a removable memory card).

- The PLP Variant is a deprecated legacy component that returns fixed-device information, for example the device ID and required free memory, to applications running on other devices or connected hosts. It is retained only for compatibility with third-party components that use some of its APIs. It is implemented as a DLL to which applications link, not as a plug-in.

Service Framework Collection

This service based on configuration files and port registration enables device-side services to register a port number for use by PC-side clients, which can query for and start device-side services. The configuration files have an XML-based format.

Figure 8.17   Service Framework components

Table 8.9 Service Framework Components

Component Name	Development Name
Service Broker	SERVICEBROKER

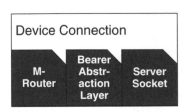

Figure 8.18   Device Connection components

Table 8.10 Device Connection Components

Component Name	Development Name
Bearer Abstraction Layer	MROUTER-PLUG-IN
Server Socket	SERVERSOCKET
m-Router	MROUTERSECURE

Device Connection Collection

This is the lowest (bearer-level) layer of the phone's Connectivity Services.

- The Bearer Abstraction Layer component is a framework for plug-ins, which encapsulates actual bearers (for example m-Router), providing a connection-management API to PC link-type applications.

- The Server Socket component is a helper library that supports creating (new, unnamed) port-number-based TCP/IP services for use by the Service Broker for device–host communications, for example with a PC. It communicates service port numbers and manages messages and commands.

- m-Router is a licensed, PPP-like data-communications protocol and framework, which provides a TCP/IP-based connection between two devices (typically, a Symbian OS device and a desktop computer; it runs on both sides of the connection). The connection may run over Bluetooth, infrared, USB, or serial cable connections. m-Router provides a proprietary framework for loading custom services.

9

The Comms Services Block

9.1 Introduction

The system model represents Comms Services as a major, self-contained block within the OS Services layer of Symbian OS.

'Comms' (or communications), in this context, really means 'data communications' – the art, science and technology of moving data between different devices over direct connections or networks. See Figure 9.1.

What connections are available depends both on the hardware architecture of a given device and on what services happen to be accessible through the hardware at any particular time. A typical modern mobile phone includes a data cable connector of some kind for connecting to a desktop computer (typically for data synchronization and backup), infrared or Bluetooth radio or both for more transient connections (to other phones or devices such as printers) and, of course, the telephone radio hardware itself. Typically the data connector is proprietary but, increasingly, mini-USB ports have begun to appear on phones. On devices such as PDAs they are standard, as they are on other digital devices such as cameras and music players. Most recently, Wi-Fi has begun to appear on high-end phones.

Figure 9.1   The Comms Services block within the OS Services layer

Whatever the physical connections and whatever their purpose, the issues from a communications perspective are essentially the same.

- Two-way communications requires *protocols*; to successfully exchange data requires a surprisingly complex set of shared assumptions between two parties: getting the other party's attention, agreeing who speaks when, agreeing what counts as 'data', keeping up with each other, and so on.

- As well as protocols to manage the conversation between the end parties, protocols are required to relay or *transport* data between them, if the two parties are not directly connected to each other.

- Finding a *route* that connects the parties can also be complex; even where the parties are directly physically connected, an appropriate interface to the connection must be selected and configured (there may be multiple interfaces available, even for the same physical connection) and where there is no direct connection, a network route must be found.

- Specialized hardware requires appropriate *drivers*, to push data through and manage the hardware state (powering hardware down when not in use is especially important in a low-powered or battery-powered device, for example).

- In a multitasking system, contention for hardware between multiple clients is likely so that hardware needs to be *shared* (for example, if two applications are trying to use a serial port at the same time).

- Finally, at the application level, *settings* may need to be saved, shared, updated and managed.

Communications is complex because the task is complex. Communications also continues to evolve at an explosive rate, not least because it is also where computing and telephony converge, where wired and radio technologies converge and where personal and enterprise usages converge. For all these reasons it is usually considered to be at the technological leading edge.

Symbian OS supports a wide range of communications technologies including conventional serial communications, short link technologies such as USB, Bluetooth and infrared, as well as networking technologies, from standard Internet protocols to newer protocols such as SIP (which are designed to support services from VoIP to the latest packet-based data services) and, of course, telephony voice, data and messaging services for 2G, 2.5G and 3G networks, whether GSM/UMTS or CDMA/CDMA2000.

Symbian's communications support has evolved not just to track new technologies but also in response to their rapid convergence and, in particular, in response to the increasing importance of packet-based technologies for 2.5G and 3G telephony services.

9.2 Purpose

Comms Services in Symbian OS provides the support for a wide variety of communications protocols and services:

- Serial protocols including RS232, IrDA and USB
- Bluetooth radio
- Networking protocols including TCP/IP (both IPv4 and IPv6), network security (TLS and IPSec) and dial-up protocols (PPP and SLIP)
- Wi-Fi
- 2G, 2.5G and 3G mobile telephony voice, data (including fax) and messaging services for GSM/UMTS and CDMA/CDMA2000 networks.

These protocols in turn enable the infrastructure for higher-level services including:

- networking including browsing and VPN support
- SIP session support
- email, SMS, MMS, WAP and OBEX messaging
- SyncML data synchronization
- WAP browsing
- Fax.

These services are supported over physical hardware including cable serial ports, infrared, USB connectors, Bluetooth radio and GSM/UMTS or CDMA/CDMA2000 phone–air interface.

The system model divides the Comms Services block into four distinct sub-blocks: the Comms Framework, which provides the overall supporting infrastructure for data communications, and Telephony, Short Link and Networking sub-blocks, each of which defines the dedicated services required for its respective technology.

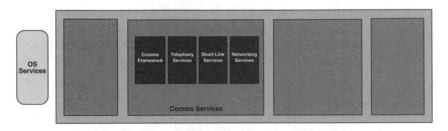

Figure 9.2 The Comms Services sub-blocks

Figure 9.3   Comms Framework sub-block

Comms Framework

The Comms Framework provides the generic infrastructure that supports all communications services.

Most importantly, it includes the Comms Root Server, which is the 'meta' process server for all communications services and the ESock Socket Server which provides the generic, sockets-style interface used to access all communications services. See Figures 9.2 and 9.3.

Telephony Services

The Telephony Services are based on the ETel Telephony Server (and its extensions) that provides support for 2G, 2.5G and 3G mobile phone networks, including GSM/GPRS/EDGE/UTMS (2G/2.5G/3G) and CDMA/CDMA2000 (2G/2.5G/3G North America).

GPRS and EDGE are the incremental packet data and 'go faster' increments to GSM; UMTS and CDMA2000 are the respective GSM and CDMA evolutions to 3G. See Figure 9.4.

Networking Services

Networking Services provides packet-based network services with Ethernet emulation and includes the TCP/IP stack implementation, secure

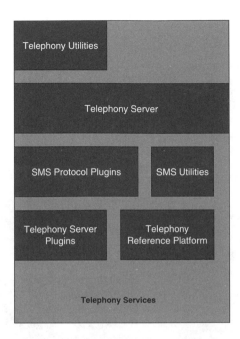

Figure 9.4 Telephony Services sub-block

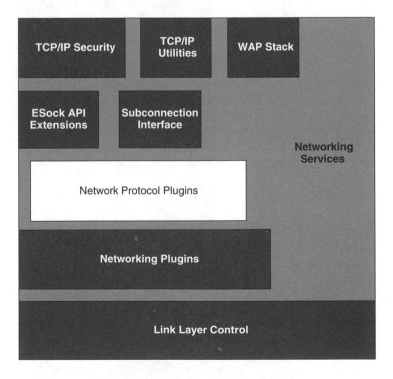

Figure 9.5 Networking Services sub-block

networking extensions including TLS/SSL and IPSec, which support secure browsing and VPN gateways, together with a variety of application-level Internet services including FTP and HTTP. (FTP does not expose public APIs.) All networking services are designed to be virtualized over telephony, serial or short-link bearers.

Support for Wi-Fi appears for the first time in Symbian OS v9 (although licensees have introduced Wi-Fi-enabled phones based on earlier releases). See Figure 9.5.

Short-link Services

Short-link services provides USB, Bluetooth and infrared services including support for the OBEX binary object protocol, USB class support that enables a Symbian OS phone both to use and serve as a USB host, and full implementations of the IrDA and Bluetooth protocol stacks. See Figure 9.6.

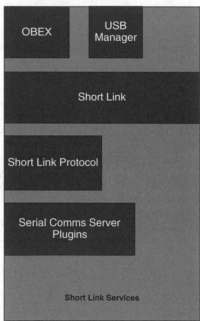

Figure 9.6   Short-link services sub-block

9.3   Design Goals

A phone is an extreme case of a mobile, connected device, which was the original design point for Symbian OS. While the ER5 release was explicitly targeted at PDA-style devices, even as the first Symbian OS devices reached market, convergence with mobile telephony was beginning to drive the company strategy. Symbian OS has been a leader in the trend which has seen PDA functions largely absorbed into mobile phones.

Compared with the original Symbian OS devices, current mobile phones (even low-end ones) make vastly greater demands on communications support. On the Psion Series 5, for example, the communications hardware consisted of a single UART, which could be switched between the serial port and the infrared port but could not be used by both simultaneously. Despite the simple hardware, a full set of integrated communications applications was envisaged, from email and web clients to network news readers and multiplayer games (network Doom, for example) and, of course, including infrared printing and beaming.

By the time the Series 5 came to market, communications support in the operating system had already been extended to include basic telephony. However, following the logic of the simple communications hardware design of the Series 5, the early networking and telephony use cases envisaged a Symbian OS device as one half of a two-box solution, using a conventional serial modem or a GSM mobile phone as a dial-up modem to connect to an ISP for network access (including Internet) or driving a GSM mobile phone (sending AT commands over a serial link), for example to dial directly from a phonebook on the Symbian OS device or writing and sending SMS messages from the Symbian OS device via a GSM phone. In each case the physical link was serial (either cable or infrared).

Even when Symbian OS migrated onto devices with onboard phone hardware (even before the release of the Series 5 in July 1997, phone projects were underway with licensees), the connection between the phone-side hardware (a dedicated second processor running a GSM stack) and Symbian OS was serial. Thus even true telephony functions, such as setting up lines and answering and making calls, ultimately went through the serial server and a serial port.

Each subsequent release of Symbian OS has taken it a further step away from this early legacy to support the evolving reality of data-enabled phones capable not just of full network access (browsing or email, for example, over a VPN tunnel into a company network) but also of running real-time communications applications, for example, video conferencing which requires two-way, real-time video streaming.

As Symbian OS has evolved, it has become capable of real-time processing and thus capable of directly hosting the telephony baseband stack, making single-core phone designs possible. (The real-time kernel first appears as an option in Symbian OS v8 and is standard in Symbian OS v9.)

In Symbian OS v8 and Symbian OS v9, Comms Services has evolved significantly. In particular, the Root Server was introduced as the primary communications server, responsible for starting and stopping the dedicated communications servers on demand and providing the common context within which all communications servers run. (In earlier releases, the C32 serial server provided this service.) The goal is to support more seamless interoperability between services and the faster data throughput required by new high-data-rate services.

In another significant change, from Symbian OS v9 the Comms Database has been integrated into the Central Repository, which provides a single point of storage for all system settings and a single common interface to all settings and service configuration. (The legacy CommDB interface is retained as a 'shim' layer providing backwards compatibility for existing applications.)

9.4   Overview

Symbian OS is designed for devices that typically do not have permanent or predictable connections (unlike a networked desktop computer, for example) and which have also typically not had even transient Ethernet connections (although this is beginning to change as Wi-Fi starts to appear on phones). The key requirement for communications services is therefore the ability to virtualize almost any service required at the application level over whatever transient connections are available at the time.

The hallmark of Symbian's communications implementation is the high degree of integration between the services at application level and the high degree of interoperability of technologies at a system level.

Logically, Comms Services is divided into sub-blocks, based on technologies. Each sub-block is organized around one or more primary servers and frameworks. Each server exposes client interfaces through its client-side APIs; implements system-level services by providing appropriate protocol implementations as Socket Server plug-ins; and defines a hardware adaptation interface through a framework for which it provides implementation plug-ins, while also enabling extension by licensees and partners (who can write their own plug-ins to support bespoke hardware).

9.5   Architecture

Servers and frameworks, characteristically for Symbian OS, provide the unifying architectural patterns for Comms Services.

The servers collaborate to provide the necessary level of interoperability essential for mobile devices which, by definition, rely on transient connections of varying kinds, depending on availability, rather than being a permanent part of a known, fixed infrastructure.

Each server exposes a client-side API. Each server is implemented as a framework. The frameworks supply the mechanisms for extensibility, which is designed in at a number of levels (see Figure 9.7):

- new protocols can be added to the system by server extensions

- new hardware types at the lower level can be supported by adding

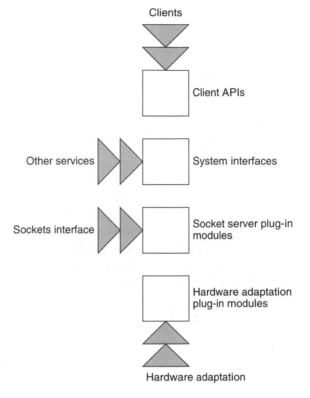

Figure 9.7 Logical layering of Comms Services

supplier module extensions (plug-ins which provide the hardware abstraction).

The architecture has proved its flexibility and adaptability over time, as it has evolved to support technologies such as Bluetooth and USB, as well as almost continuous evolution in telephony.

The Comms Server Model

In Symbian OS, each dedicated communications service is organized around a principal server and a protocol implementation. The servers include the ETel Telephony Server that provides telephony services, the C32 Serial Server that provides data communications services (typically virtualized over short-link connections), Internet extensions to the ESock Socket Server that provides networking services, and Bluetooth and USB managers that provide short-link services.

In the original communications architecture, all communications services were virtualized over a simple serial connection, supported by a hardware architecture which provided a single UART switchable between a cable serial port and an infrared port. The C32 Serial Server was therefore the primary service provider, accessed directly by clients using its client-side APIs. The Serial Server also provided the framework which defined the low-level abstract API for communications modules (CSY files) which were implemented as plug-ins supporting the available serial hardware (the serial port and IrDA).

Networking services were designed around a server that provided a Berkeley-style Sockets API and a TCP/IP stack implementation, which was loaded as a server plug-in. At a lower level, however, all networking services were virtualized over serial connections (for example, an IrDA link to a network-connected computer).

When telephony services were introduced to the operating system, the design was quite closely modeled on the serial services architecture, with a primary server, the ETel telephony server, providing the client-side APIs and the abstract framework for hardware-facing telephony modules (TSY files), which were analogous to C32 Serial Server CSY communications modules. Interestingly, the addition of telephony services did not substantially change the earlier assumption of the primacy of serial communications, since the initial expected use for telephony was a two-box solution, using a serial connection (either cable or infrared) to connect a Symbian OS device to a modem or a mobile phone.

This was more or less the communications architecture of the first releases of the operating system and largely survived through the Symbian OS v6 and Symbian OS v7 releases. Over those releases, there were significant extensions, most obviously to telephony and networking to add the required packet capabilities for 2.5G and 3G data services, as well the addition of new short-link technologies such as Bluetooth and USB. However, the general principle of the serial server as the primary communications server (the 'first among equals') remained even though, as each release increasingly specialized the operating system for mobile phones, the primary communications use case was not serial communications but on-board telephony, with or without networking.

This architecture was unsatisfactory for a number of reasons, not least of which was the resource cost of having the serial server running all the time to support non-serial communications services.

Beginning with Symbian OS v8, therefore, some significant changes to the communications architecture were introduced as the foundation for further evolution to support the increased demands for high data rates. The key change was to introduce a primary communications server, the Comms Root Server, which is designed to provide a purpose-built, lightweight server for which the dedicated communications servers (C32 serial, ETel telephony and ESock sockets servers) act as service providers. The Serial Server is relieved of its privileged role and becomes just another dedicated service provider.

In this architecture, the Root Server becomes a communications process server, initiating a single communications process within which it runs the servers for individual services as threads, starting and stopping them in response to client requests and providing process, shared resource and common settings management including fast, low-overhead communication between the dedicated server threads. Each server is run in its own thread and only a single instance of any server is ever running. A number of supporting components implement the messaging abstractions and communications channels which allow passing of messages between running server threads, while the Comms Database provides the shared settings service. (From Symbian OS v9, the CommDB API is provided for compatibility only; the Central Repository should be used for all shared settings.)

The Root Server is responsible for running the following dedicated servers, which implement a common Comms Provider Module (CPM) interface defined by the framework:

- C32 Serial Comms Server
- ETel Telephony Server
- ESock Socket Server
- Resolver Server
- Fax server.

The individual services are described in more detail in the sections that follow. Each service provides a client-side session API, encapsulated in a single static DLL to which clients link. The general usage pattern is thus:

1. Create a client session with the appropriate server, for example the Serial Server or Socket Server; this exposes the server's client-side APIs to the client.

2. Create a client sub-session with an appropriate object, for example a communications port or a socket; this exposes the object APIs to the client.

3. Use the object.

4. Close the sub-session with the object when finished.

5. Close the session with the server when finished.

It is also worth noting that in Symbian OS communications services are provided user-side; in other words, communications services are not built into the kernel. This protects the kernel from resource failures or badly behaved processes originating from communications services or clients.

Frameworks

As well as implementing server functions, the principal communications servers also provide extensible frameworks, which are at the heart of the communications architecture.

Frameworks provide extensibility at a number of levels, including:

- at the client-interface level (for example, extending core telephony services to enable fax over mobile networks)
- within the protocol stacks at the protocol level (for example, adding the WAP stack or extending core TCP/IP services to enable packet-level security)
- at the network-interface level (for example, adding support for new technologies such as the Bluetooth Personal Area Networking (PAN) profile)
- at the hardware-abstraction-interface level (for example, extending the telephony baseband interface to support CDMA).

All implementations of communications framework plug-ins conform to the Plug-in Framework (i.e. ECom), in other words they are polymorphic DLLs that implement the standard interfaces which enable the Plug-in Framework server to find and load the appropriate modules at run time on behalf of the requesting framework, as well as the communications-specific interfaces required by the specific communications frameworks.

The Comms Services frameworks include:

- C32 Serial Server, which defines CSY virtual serial port modules
- ETel Telephony Server, which defines TSY baseband interface modules
- Socket Server, which defines PRT protocol modules
- Network Interface Manager, which defines AGT interface agent and NIF network interface modules.

In addition, the Comms Framework component defines the CPM interface which is implemented by all of the dedicated communications servers (but not by the Root Server itself).

9.6   Comms Framework

The Comms Framework components implement the infrastructure used by all communications services:

- The Comms Root Server is the primary communications server, responsible for starting and stopping the communications servers that provide dedicated services and for providing the process context in which all dedicated servers are run.

- The C32 Serial Server and the ESock Socket Server are, respectively, the data communications and socket servers that provide the two direct client interfaces for communications services (all communications services are accessed through sockets and serial communications services are also available directly through the Serial Server).

- The Network Interface Manager and Network Controller are, respectively, the network interface and connection managers that find and set up appropriate network connections requested by Socket Server clients and that are used (indirectly) by all communications services.

The Comms Framework also includes common utility and framework support, including the framework classes that define the Comms Provider Module (CPM) interfaces to which all communications servers conform and specialized messaging and memory management (Comms Channels and MBufs), designed to enable fast inter-thread communications within the communications process including thread-shared memory. (Communications servers run in their own threads inside the single communications process managed by the root server.)

Also included is the Comms Database, which supports the legacy interface used for storing shared communications settings. (New applications should use the Central Repository.) See Figure 9.8.

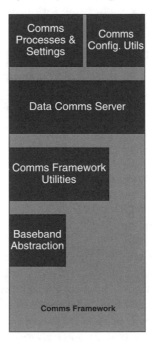

Figure 9.8 Comms Framework components

Design Goals

It is important to remember that on a typical device based on Symbian OS (a mobile phone, for example), all communications must be virtualized over an available, and usually transient, connection. Thus Internet browsing, for example, typically does not take place over a direct Internet connection (as it would on a PC) but is virtualized over telephony or short-link services. As Wi-Fi begins to appear on phones, direct network connections also become possible but very much as complementary options.

The Comms Framework has evolved to provide a generic infrastructure that enables the seamless interoperation of services while providing improved performance, ready for the next generation of high-data-rate services.

Architecture

The Comms Framework sub-block is less a self-contained architectural unit than the architectural glue that binds the different dedicated communications services together. It provides the frameworks that define essential, common communications abstractions, the Root Server that provides the runtime context within which all communications services operate, and the shared settings database and utilities, as well as utilities and libraries, such as the MBuf Manager and Elements components.

The Root Server and Framework Utilities

From Symbian OS v8, all communications servers are implemented as Comms Provider Modules and are run and managed by the Root Server, which loads, configures, runs and monitors CPMs as dedicated threads within the Root Server's own process. Starting the Root Server creates the single communications process and starts the server as the main running thread within in. The Root Server runs from device startup to shutdown.

In Symbian OS, a process is the fundamental unit of protection, with its own address space, while a thread is the fundamental unit of execution, running inside a process and sharing the process address space and any other resources (file handles, for example) with other threads running in the process.

The Comms Framework is the component that provides the abstractions needed to implement Comms Provider Modules including the CPM interface, common thread management and Comms Channels, the asynchronous message queue abstraction that provides an efficient communication mechanism between active CPMs. The CPM framework also defines a file-based configuration method that is used by the Comms Root Server to configure CPMs on loading.

To support implementation of new CPMs, the Comms Elements provides a reusable catalog of common design pattern implementations, for server startup, message passing and generally useful abstractions such as state machines. The MBuf Manager provides a memory management framework that allows direct sharing of data (for example, network packets) between CPMs without copying.

Serial Communications

The C32 Serial Server provides serial services for application and system clients. A key component from the first Symbian OS release, it has been re-architected and re-engineered to support platform security and the new communications infrastructure based on the Root Server. From Symbian OS v8, the C32 Serial Server is a CPM, run and managed by the Root Server. The CPM and supporting mechanisms provide data sharing and efficient inter-server communications without the overhead of running the Serial Server to support other communications services.

The Serial Server follows the standard Symbian OS server pattern, providing serialized access to shared resources. In the simplest case, and unlike other communications servers, clients can gain direct access to the serial hardware on a device by initiating a client session with the server (by making a serial service request to the communications configurator) and then from within the session loading, opening and configuring a (virtualized) serial port. This creates what is, in effect, a raw serial link over the chosen port (either an actual serial port, or virtualized over Bluetooth, infrared, or USB) to another, connected device. Clients can also access serial services through the Socket Server.

As well as providing a client API, the Serial Server defines the framework interface that communications plug-in modules (CSY files) implement. A CSY module is implemented as a polymorphic DLL (with a CSY extension, by convention) that exports a factory function for a CSerial-derived CPort class object. CSY implementations are supplied for true RS232 serial ports and serial port emulation over IrDA, Bluetooth and USB. At the level below the plug-in modules, logical and physical device drivers implement the hardware-level interfaces.

Sockets

Sockets were first introduced as a networking abstraction in Berkeley Unix (BSD), providing a generic mechanism to associate a communications protocol with a data pipe (dedicated communications channel) connecting two processes, transparently of where the processes were actually running and using a simple, file-type semantics. The ESock Socket Server provides sockets-based communications on Symbian OS through a client session API and an underlying framework for creating and loading protocol implementation plug-ins (PRT files) that determine the type of the socket and provide the underlying protocol implementations. Sockets-based protocol implementations are supplied allowing services to be run over a wide range of possible bearer protocols including Bluetooth, IrDA, TCP/IP and SMS.

The sockets abstraction provides a common client interface to networking, serial and short-link communications protocols, providing a sockets API plus name and address resolution and connection management.

ESock was originally provided as part of the networking implementation of the first Symbian OS release, but over subsequent releases it has evolved into a more generic mechanism for requesting any communications services. Since Symbian OS v8, the Socket Server presents itself to the Comms Root Server as a collection of CPMs whose purpose is to provide protocol sessions to requesting clients by finding and loading an appropriate protocol module, serving it through a client session to the client, transparently managing the shared data structures and channels used for socket communications, monitoring and cleaning up after thread panics and, generally, performing all necessary housekeeping functions and resource management.

Clients connect to the ESock server with a `Connect()` call and then open a sub-session by calling `Open()` on a socket of the chosen type. The socket type is based on the transport protocol. In response to a socket request from a client, the Socket Server loads an appropriate protocol module (PRT file) that implements the requested protocol.

In Symbian OS v9, the Socket Server is multi-threaded, improving performance.

Network Interfaces

The underlying interface to the network transport layers is provided by the Network Interface Manager, or NIFMan, and its supporting components, which load interface agents (AGT files) to establish network connections and then create an appropriate network interface (NIF file). Connections supported at Symbian OS v9 are either circuit-switched or packet-switched data connections running through telephony services, or an Ethernet implementation running over serial communications or short-link services. The chosen network interface is bound to the TCP/IP stack. NIFMan defines the plug-in framework (i.e. the base classes from which plug-ins must derive and hence the interfaces they must implement) for the network controller modules. A network controller owns both networks and bearers.

The Network Controller component is used by the Network Interface manager to select a suitable outgoing interface, for example from those pre-configured in the Comms Database. It loads first the appropriate agent to establish the physical connection and then the appropriate network interface. Thereafter, data can flow between the requesting client and the network interface through the loaded PRT stack module and the Socket Server that loads it.

At the lowest level of the networking services are the modules that implement the interfaces to the physical link layer, the plug-ins to the Network Interface Manager (NIF files) and other related low-level plug-ins. Supported interface types include Ethernet, PPP, SLIP and a tunneling NIF, each of which can serve as an interface to different physical link-layer carriers, for example physical cable or infrared implementations of serial communications, Bluetooth, GPRS, and so on.

NIFMan can be thought of as the server that manages the overall control of network and bearer selection, delegating the actual work to the Network Controller and the agents that plug-in to NIFMan. Agents are the workhorses that manage the pairings of networks to bearers. Typical bearers might include:

- a GSM radio network supporting circuit-switched data calls

- a CDMA95 radio network supporting circuit-switched data calls

- a GSM radio network supporting packet-switched data contexts

- a UMTS radio network supporting packet-switched data contexts

- a CDMA2000 radio network supporting packet-switched data contexts

- an Ethernet wired network connection to a LAN

- an 802.11 (Wi-Fi) radio network connection to a LAN.

With multi-homing, there may be multiple access technologies available to reach the same network destination. For example, a given network (an Internet ISP, say) may be reachable by all of the following: circuit-switched data (for example, a GSM data call), packet-switched data (for example, a GPRS connection) or WLAN (directly via Wi-Fi or perhaps via Bluetooth connection to a PC). In contrast, another network (the user's 3G mobile network, for example) may be reachable only by packet-switched data. Multi-homing enables each network and bearer combination to be separately defined, so that the relationship of networks to bearers is no longer 1:1 but one to many (i.e. multiple combinations for a given network, based on all the possible bearers).

Shared Settings

Historically in Symbian OS, the Comms Database, or CommDB, is the repository in which all communications-related settings and configuration information is stored. Settings are used, for example, by system-level components for default host-name resolution and to determine connection preferences, availability of physical modems, services, configured ISPs,

GPRS access points, LAN services, and so on, as well as by applications that, for example, may need to allow users to set or change settings.

As well as containing preferences and settings, CommDB provides the utilities needed to set, store and manipulate settings and to read and write settings into XML formats.

CommDB has been a part of the system since the first Symbian OS releases. In Symbian OS v9, however, its functions are replaced by the Central Repository, to which the CommsDat component provides a communications-specific interface for stored settings. Compatibility is maintained for old-style CommDB requests.

Component Collections

Comms Process and Settings Collection

The Comms Root Server provides the main thread in the communications process and is responsible for starting and managing all other communications process threads. These are started at device boot, rather than on demand, as in previous operating system releases. See Figure 9.9.

Table 9.1 Comms Process and Settings Components

Component Name	Development Name
Comms Root Server	ROOTSERVER

It provides client-side APIs for loading, configuring and binding provider modules; polices any relevant security policies; and publishes a Publish & Subscribe property to notify thread death of provider modules.

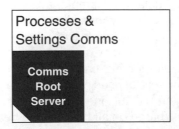

Figure 9.9   Comms Process and Settings components

Comms Configuration Utilities Collection

Communications-related settings and configuration information are used to set and determine the host name, connection and service provider defaults. The Comms Database (CommDB) is the legacy repository that,

Figure 9.10 Comms Configuration Utilities components

Table 9.2 Comms Process and Settings Components

Component Name	Development Name
Comms Database	COMMSDAT, COMMDB_SHIM, COMMDB_COMPAT

from Symbian OS v9, is replaced by the CommsDat interface to the Central Repository, although the CommDB API is preserved for compatibility. See Figure 9.10.

Data Comms Server Collection

This collection contains servers and supporting components that provide the key client interfaces for data communications. See Figure 9.11.

Table 9.3 Data Comms Server Components

Component Name	Development Name
C32 Serial Server	C32
ESock Server	ESOCK
Network Interface Manager	NIFMAN, DIALOG
Network Controller	NETCON

- The C32 Serial Server provides the client session APIs and server implementation for serial type communications and the framework

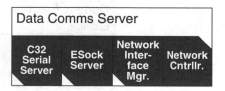

Figure 9.11 Data Comms Server components

for creating and loading the communications plug-in modules (CSY files) that implement the serial-port abstractions, enabling clients to access virtual serial ports independently of the underlying hardware.

- The ESock Socket Server provides the client-session APIs and server implementation for sockets-based communications and the framework for creating and loading protocol implementation plug-ins (PRT files).

- The Network Interface Manager provides the bearer-level support for the Socket Server, providing the framework for creating, loading and managing interface agent (AGT file) and interface plug-ins (NIF files). Interface agents find and load network-interface implementations and bind them to the TCP/IP stack to create the bearer-level connections over which the socket protocols served to clients by the Socket Server actually run.

- The Network Controller is the component that selects a network interface agent to create an appropriate network interface. It reads connection preferences for the client from stored communications settings, based on which it chooses both a network and a bearer (i.e. an access technology). Having made its choice, it loads the appropriate agent. It is implemented as a plug-in library loaded by the Network Interface Manager.

Comms Framework Utilities

These utilities provide framework support for the Root Server and for Comms Provider Module mechanisms. See Figure 9.12.

Table 9.4 Comms Framework Utilities Components

Component Name	Development Name
Comms Framework	COMMSFW
Comms Elements	ELEMENTS
MBuf Manager	MBUFMAN

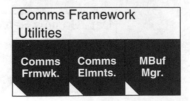

Figure 9.12   Comms Framework Utilities components

- The Comms Framework provides the framework base classes and utilities that support the communications architecture based on the Comms Root Server, including base classes that define Comms Provider Modules (the addressing and binding mechanism used by the Comms Root Server to identify and load modules), the message definitions and communications channel queue abstraction used to communicate between modules and the Comms Root Server, and thread-creation support.

- The Comms Elements are an internal library of ready-made programming patterns, for example state machines and message parsers, that are used within communications services and are made available as reusable objects.

- The MBuf Manager defines and manages MBufs, a communications-specific shared-memory mechanism allowing provider modules (i.e. multiple threads within the primary communications process) to share memory buffers and therefore avoid unnecessary copying of messages and data. For example, MBufs can contain data packets as well as arbitrary C++ objects.

Baseband Abstraction Collection

The Baseband Channel Adaptor (BCA) provides an abstraction of the actual channel used to communicate with the baseband processor, for use by communications components (which, therefore, don't need to understand the actual channel implementation) and a plug-in framework for a hardware-specific interface implementation module. The actual channel is dependent on the hardware design and may comprise a physical fast serial link, USB or other fast bus, a shared memory or even a shared register protocol. See Figure 9.13.

Table 9.5 Baseband Abstraction Components

Component Name	Development Name
Baseband Channel Adaptor	BCA

Figure 9.13 Baseband Abstraction components

9.7   Telephony Services

The telephony architecture was designed to provide flexible support for a wide variety of possible phone types, including conventional analog modems, GSM phones and even desktop phones containing an integrated Symbian OS device.

Like other communications services, the Telephony Services block is organized around a primary server and framework, the ETel Telephony Server, supported by protocol implementations for specific services, low-level plug-in modules implementing hardware adaptation interfaces defined by the framework and some assorted high-level utilities.

The design principle for ETel was to abstract a small core set of universal telephony functionality as the Core API, while providing a flexible extension mechanism to enable support to be added for specific service and network types at both the client interface, enabling support for custom services and at the hardware interface, enabling support for different telephone-baseband implementations. The straightforward goal of the Core API is to enable telephony clients to pass information over a generic phone link.

From this starting point, support has evolved from basic Hayes modem control (AT commands) through GSM 2G standards, to 2.5G (GPRS, EDGE) and CDMA (for the North American market and other markets, such as Korea, that initially adopted CDMA rather than GSM), and to

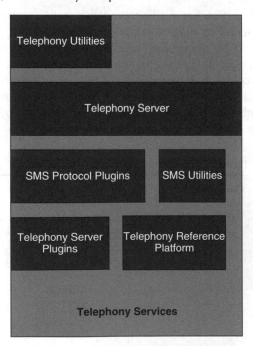

Figure 9.14   Telephony Services components

3G UMTS and CDMA2000 (respectively the 3G evolutions of GSM and CDMA).

From an initial emphasis on dial-up and modem connections, providing a fully integrated telephony service became important as Symbian OS moved onto mobile phones. More recently as mobile telephony has evolved towards packet-based networks, support for high-bandwidth data services has become important. See Figure 9.14.

While the basic phone services in Symbian OS are quite mature and were well established by Symbian OS v6, incremental enhancements have been introduced with almost every release since.

Architecture

Telephony Services are structured around the ETel Telephony Server. ETel provides a core set of common, network-independent telephony services that abstract control of telephony devices either connected to or integrated into a Symbian OS phone and enable client access to phone services. ETel is implemented as an extensible framework into which modules can be added to extend the core functionality at the client level.

The ETel framework also defines the low-level, hardware adaptation interfaces and provides the mechanisms that support hardware adaptation plug-in implementation modules (TSY files).

Conceptually, the ETel core API is extensible in two directions: in the direction of hardware, supporting new networks, baseband implementations and other hardware evolutions, and, in the client API direction, enabling new services to be supported.

While extensibility implies flexibility, it also implies a significant division of labor between Symbian and licensees to extend the telephony support appropriately for a given phone or family of phones:

- On the Symbian side, the ETel server core and Multimode framework support extensions for new standards (which is how the initial GSM-only support has been extended first through CDMA and then to 3G UMTS and CDMA2000) and expose the TSY provider module plug-in API.

- On the licensee side, the Telephony Application and low-level TSY provider modules support platform- and device-specific customization.

Licensees implement a custom TSY and any additional custom APIs they choose to add to support unique features of their own telephony hardware.

In addition, licensees provide engine support and custom UIs, for example, for phone security (such as PIN-based locking of the phone application), and the phone application itself, which must include a platform-specific user interface and must also support comprehensive user-interface-independent functions including handling networks, audio, contacts, logging and call handling, number parsing, and so on. Telephony Services includes a number of libraries and utilities that provide basic support for such functions.

Note that the 'licensee' may be either a platform vendor such as S60 or UIQ, providing a pre-integrated user interface and application suite solution to its customers, a phone vendor (or consortium such as FOMA) creating a bespoke UI and applications, a third-party developer of a phone application, or a hardware partner providing a packaged phone hardware solution.

Evolution of Mobile Services

Mobile phone services and technologies have evolved rapidly, as has the global market for mobile phones, including significant cycles of boom and bust. Basic mobile network technologies have evolved from 'plain old' GSM through GSM Phase 2+, otherwise known as 2.5G (GSM, GPRS, EDGE), to UMTS 3G, with similar evolutions from CDMA to 3G CDMA2000. Symbian OS has tracked these evolutions. It enables control of landline and mobile phone modems and supports wireless telephony standards for all markets.

GSM uses a packetized but synchronous Time-Division Multiple Access (TDMA) approach to sharing available bandwidth between multiple users. Voice is digitally encoded and transmitted as digital packets in timeslots (frames) at a data rate approximating 19 200 baud (equivalent to modem speeds from around the late 1980s).

Support for basic GSM services requires support for receiving and making voice calls, receiving and sending SMS messages, showing that SMS messages have been received, and receiving and making circuit-switched data calls, for example fax calls. GPRS adds the requirement to support making and receiving packet-switched data calls.

EDGE and 3G networks extend these requirements to include, for example, both one-way and two-way audio and video calls including support for two-way tele-conferencing; streaming of audio and video to a phone; interactive, session-like two-way request–response (for web browsing or remote database query); and background data delivery for example of SMS messages.

GPRS and EDGE add packet data services by stitching together multiple GSM voice channels to create a higher bandwidth channel. GPRS provides data rates up to 170 kbps, which EDGE improves by a factor of three (either in speed or in the number of simultaneous subscribers supported at GPRS data rates).

Both GSM and CDMA remain circuit-oriented, voice-centric technologies. UMTS evolves GSM to use Wideband CDMA to gain higher data rates. Unlike GSM or CDMA, UMTS is fully packet-switched, not circuit-switched.

Historically, CDMA has dominated the North American market, while GSM originated as a European standard that has had widespread global uptake. GSM has also recently increased its market share in many CDMA-dominated markets to become a second-line network technology.

Telephony Server

The ETel Telephony Server manages access to telephony functions on a Symbian OS device, regardless of the details of the available phone hardware. Indeed, there may be no onboard phone hardware, as was the case in the first Symbian OS devices. As well as supporting fully featured mobile phones, ETel supports the use of data ports thus enabling two-box solutions, for example using a mobile phone as a modem via infrared or Bluetooth, which was an early use case.

The server implements the standard Symbian OS client–server framework, providing a client-side API (as a separate DLL to which clients link). The server also implements the CPM interface and is thus a communications provider that is managed and run by the Comms Root Server and which provides thread management and communications channels for fast communication with other communications server threads.

The basic abstractions made available by the Telephony Server are phones, lines and calls. The server also provides an extension framework, which is used to add extended client services and a low-level hardware adaptation interface that is implemented by hardware adaptation plug-in modules. Clients open a server session with the Telephony Server and then open sub-sessions with phone, line and call objects.

The Core API includes generic functions for requesting the capabilities or status of the phone hardware and making and managing voice, data and fax calls.

Basic telephony extensions supporting GSM/GPRS are implemented by the ETel Multimode extension and other extension modules supply further CDMA, messaging, 3GPP packet data and fax-specific extensions. Collaborating components are all realized inside sub-sessions or the root ETel server session to a client, that is created when the ETel server is started by the Comms Root Server in response to a client request.

The ETel Server and Core API, together with the Fax Client–Server, formed the basis of the original telephony implementation in ER5. The ETel Core API was rearchitected in Symbian OS v7 when the other extensions were introduced and most were further enhanced in Symbian OS v8.

ETel Third-Party API

The ETel third-party API was introduced in Symbian OS v7 to provide a restricted but common 'safe subset' of telephony functionality to third-party (i.e. non-licensee and partner) application developers. It was significantly extended in Symbian OS v8, adding support for multiple voice calls, better access to onboard and network status information and system notifications and events, and access to IMEI and IMSI numbers.

Telephony Messaging

The ETel Multimode extension includes generic support for telephony messaging, with specific implementations (for example for GSM, CDMA and WAP) implemented as Socket Server protocol-module plug-ins, providing a common sockets-based interface to messaging clients. The protocol modules perform the actual encoding and decoding of messages, support SIM card message store management functions and interact with the Telephony Server (via ETel Multimode) for transmission and reception of message.

SMS messaging clients include the messaging application support components at the application-services level, for example the SMS MTM, CDMA MTM and Java messaging components.

Similarly, the WAP Stack is a client for WAP messaging, typically to expose a Wireless Datagram Protocol (WDP) service to a WAP client. The WAP protocol module in turn directly cooperates with the SMS protocol module, which undertakes the transaction with the Telephony Server.

Note that the Telephony Server may not be the ultimate provider of the message service, for example if an SMS is requested to be sent over a Bluetooth link. In this case, the Telephony Server creates a further Socket Server session requesting the appropriate bearer and the messaging interface is a serial port plug-in for the appropriate bearer rather than the TSY interface to onboard phone hardware.

Part of the messaging support consists of utility classes that implement encoding and decoding functions and streaming, logging and backup-server interface classes. Utilities are provided as standalone DLLS (linked to by clients at compile time).

Interfacing to the Baseband

TSY modules are the telephony equivalent of the Serial Server's CSY virtual serial-port implementation modules and are defined and loaded by the ETel Framework. A TSY is an ECom (plug-in framework) compliant plug-in that provides the glue between Symbian OS Telephony Services and the phone baseband (the telephony stack).

The Telephony Server passes client requests made to it on sub-session objects (which may be based on the Core API, Symbian-supplied extension APIs, or custom APIs created by licensees by extending the core framework) to the TSY, which translates them into proprietary requests the baseband understands. The TSY plug-in model is a direct borrowing of the CSY model used by the Serial Server.

The Telephony Server framework provides the abstract base classes for each of the objects implemented by a TSY, representing phones, lines, calls, faxes and extensions.

Symbian OS supplies four TSYs as reference implementations. The Multimode TSY shipped for the first time in Symbian OS v7, as an upgraded and renamed version of the original GSM.TSY that shipped with ER5, incorporating GPRS support. The CDMA and SIM TSYs also shipped for the first time in Symbian OS v7, as did a first version of a Telephony Reference Platform (TRP) TSY. The Symbian OS v9 TRP TSY runs on Texas Instruments H2 development board hardware but is designed to be easily ported to other platforms.

Component Collections

Telephony Utilities Collection

This collection contains helper components that use the telephony server but which are not used by it. See Figure 9.15.

Table 9.6 Telephony Utilities Components

Component Name	Development Name
Telephony Watchers	TELEPHONY_WATCHERS
Phonebook Sync	PHBKSYNC
Dial	DIAL

- The Telephony Watchers are watcher Framework plug-ins that monitor telephony conditions and report them as Publish and Subscribe properties, including current signal strength, battery level and whether a call is in progress. They were introduced in Symbian OS v8.

- The Phonebook Sync server enables synchronization of contacts between a phonebook application and entries stored in the Integrated Circuit Card (ICC) or 'SIM' card of a device. It was originally introduced as part of the ETel Multimode extension 3G support for UMTS and CDMA2000.

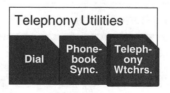

Figure 9.15 Telephony Utilities components

- The Dial component consists of dialing utilities the use of which is deprecated from Symbian OS v9.

Telephony Server Collection

The ETel telephony server and core framework implements the basic telephony functions and is extended by the Multimode framework into a uniform generic API for all mobile telephony independent of the underlying network, with additional packet-data extensions for 2.5G and 3G packet services, CDMA-specific extensions, plus SIM Toolkit utilities, fax support and a third-party API that opens a common subset of telephony functions to third-party application developers. See Figure 9.16.

Table 9.7 Telephony Server Components

Component Name	Development Name
ETel Server and Core	ETEL
ETel 3rd Party API	ETEL3RDPARTY
Fax Client and Server	FAX
ETel Multimode	ETELMM
ETel Packet Data	ETELPCKT
ETel SIM Toolkit	ETELSAT
ETel CDMA	ETELCDMA

- The ETel Server and Core API provides clients with access to telephony functions on Symbian OS. It implements the standard Symbian OS client–server framework, providing a client-side API (as a separate DLL to which clients link). The server in turn translates these into TSY requests, which are passed on to a TSY module. The server dynamically loads and unloads TSY modules at client request. The TSY implements a customized interface to the onboard hardware (although in a two-box case, it would route back through an appropriate socket to

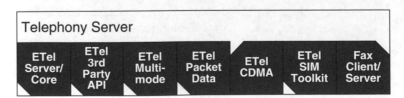

Figure 9.16 Telephony Server components

use the requested communications port). Like other communications servers, ETel is a CPM that runs as a thread within the communications process.

- The ETel Multimode component extends the Core ETel API to provide, as far as possible, a uniform API, for making voice, fax, data or multimedia calls, that is independent of the underlying mobile network and phone architecture (e.g. 2G, 2.5G, or 3G).

- The ETel 3rd Party API is implemented as a sub-session providing a subset only of the Core ETel, ETel Multimode and ETel Packet APIs. Unlike the other main telephony APIs, which are restricted to licensees, the ETel 3rd Party API is open to third-party developers, enabling them to make applications that can use the telephony features or create dedicated, phone-aware applications.

- The ETel CDMA component extends the ETel multimode sub-session to implement a high-level API for CDMA-specific telephony applications.

- The ETel Packet Data component is an ETel Telephony Server extension framework enabling access to GPRS Release 97/98, CDMA/CDMA2000, Release 99 (GPRS and UMTS) and Release 4 (UMTS) packet services. It enables clients to configure, modify and activate a PDP context for a network packet-switched service and to control a packet-switched connection.

- The ETel SIM Toolkit provides the functionality of a GSM/WCDMA (U)SIM Application Toolkit, implemented as a sub-session.

- The Fax Client and Server is not an ETel extension. The server is a DLL that provides a framework for adding fax functionality to applications and is driven by the Fax Client via ETel and a suitable TSY. The Fax Client is accessed by applications through the Messaging Send-As API.

SMS Protocol Plug-ins Collection

These protocol modules and other plug-ins implement telephony-based messaging for GSM and CDMA SMS and WAP messaging. See Figure 9.17.

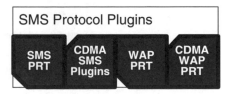

Figure 9.17 SMS Protocol Plug-ins components

Table 9.8 SMS Protocol Plug-ins Components

Component Name	Development Name
SMS PRT	SMSSTACK
WAP PRT	No unit
CDMA SMS Plug-ins	CDMASMSSTACK
CDMA WAP PRT	No unit

- The GSM SMS PRT protocol module enables its clients to send and receive GSM SMS messages, enumerate and delete messages from phone stores and read and write SMS parameters on the SIM. It is implemented as an ESock plug-in protocol module, therefore clients interact with it though an instance of RSocket; all operations are initiated by IOCTL calls on RSocket.

- The CDMA SMS protocol implementation conforms to IS-637 and supports Wireless Paging, Wireless Messaging, Voice Mail Notification, Broadcast SMS, Service Category Programming, Wireless Enhanced Messaging and Card Application Toolkit Protocol.

- The WAP PRT protocol module is used by the WAP Stack for sending and receiving SMS messages.

- The CDMA WAP PRT provides functional equivalents of the GSM WAP protocol module.

SMS Utilities Collection

The GSM Utilities and SMS Utilities components are used by the SMS protocol modules (the SMS stack) to assist in creating and processing SMS messages. For example, GSM Utilities includes encoding and decoding routines and SMS Utilities includes streaming classes (to stream message objects across the Socket Server), logging classes and interfaces to the backup server. They are implemented as utility DLLs that are linked to by clients. See Figure 9.18.

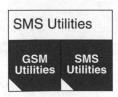

Figure 9.18 SMS Protocol Utilities

Table 9.9 SMS Protocol Utilities

Component Name	Development Name
GSM Utilities	GSMU
SMS Utilities	SMSU

Telephony Server Plug-ins Collection

This collection contains reference telephony server plug-in modules (TSY files), loaded by the Telephony Server. See Figure 9.19.

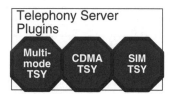

Figure 9.19   Telephony Server Plug-ins

Table 9.10 Telephony Server Plug-ins

Component Name	Development Name
MultiMode TSY	MMTSY
CDMA TSY	CDMATSY
SIM TSY	SIMTSY

- The MultiMode TSY provides the GSM and GPRS functionality. It uses the AT command interface to communicate with the phone or modem via standard AT commands over a serial or infrared link.

- The CDMA TSY is the CDMA equivalent of the MultiMode TSY for GSM. The ETel TSY reference plug-in for CDMA is replaced on an actual device by a hardware-specific licensee TSY.

- The SIM TSY is a simulator module designed to enable automated testing of a range of operating-system components in a simulated GSM, CDMA and WCDMA mode. It does not communicate with any real hardware (neither a modem nor a phone) but instead uses static configuration data and dynamic system-agent notifications to simulate the presence of phone hardware. It supports the Core ETel API, Multimode ETel API and ETel Packet API requests.

Telephony Reference Platform Collection

These components support a standard reference platform telephony implementation. See Figure 9.20.

Figure 9.20   Telephony Reference Platform components

Table 9.11   Telephony Reference Platform Components

Component Name	Development Name
TRP TSY	TRP
TRP CSY	TRP
Baseband Channel Adaptor for C32	C32BCA

- The TRP TSY is a reference TSY designed to run on development-board hardware, as part of a wider effort to make easier licensee development on phones using Symbian OS.

- The TRP CSY is used to manage the internal channel between the telephony hardware (a dedicated phone-side core running the TI phone stack) and the application hardware (an ARM core running Symbian OS). The logical driver on the TI H2 board presents the internal serial bus as a standard serial port.

- The Baseband Channel Adaptor for C32 is a reference plug-in providing a serial communications implementation of the Baseband Channel Adapter interface (see Section 9.6). For example, the Telephony Reference Platform provides a serial communications BCA plug-in implementing the BCA interface.

9.8   Networking Services

Web browsing and email were the functions that motivated the inclusion of networking services in the first releases of Symbian OS, although the potential for more exotic applications such as network news readers and multiplayer games was.

It is worth remembering in this context that the devices at which the first ER5 release was targeted were not phones, although they were connected and they were telephony-enabled in the sense that they were designed to interoperate with phones. That interoperation, however, was understood in terms of phone-as-modem, dialing up an ISP to access an email account, a corporate intranet or the Internet, or even the web. Neither email nor the Internet were ubiquitous in the way that they both now are and the web was still very much a novelty.

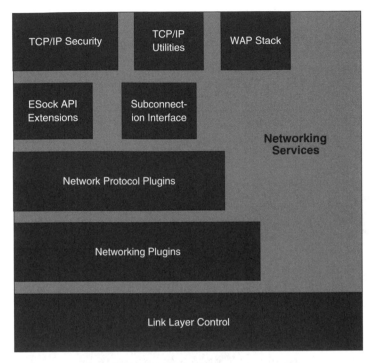

Figure 9.21   Networking Services components

The core of the networking implementation remains the TCP/IP v4/v6 networking stack, implemented as a PRT Socket Server plug-in module and the network interface plug-ins that support it and which are, in turn, supported by link-layer plug-ins.

While the ESock Socket Server and Network Interface Manager have migrated out into the Comms Framework to provide generic socket support for all communications services (and not just for networking), networking services have expanded to encompass TCP/IP enhancements such as IPSec, telephony-driven networking enhancements including packet-data services (for GPRS and UMTS) and Quality of Service (QoS, required for 3G services), as well as completely new technologies such as WAP and most recently (in Symbian OS v9) Wi-Fi. See Figure 9.21.

Networking Stack

Symbian OS Networking Services are based on a TCP/IP protocol implementation, TCP/IPv4/v6 PRT (in effect the transport and network layers of the OSI seven-layer model), together with IP extensions that implement various packet-level services including QoS and IPSec. See Figure 9.22.

However, TCP/IP packets and the stack itself are not directly available. TCP/IP packets are encapsulated within the stack and there are no visible TCP/IP packet classes, for example. The stack is implemented as a Socket

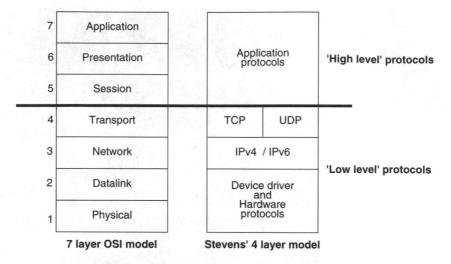

Figure 9.22   The OSI Seven-Layer model and simplified layer model

Server PRT protocol plug-in, and network services are made available through the sockets interface, by requesting a TCP/IP socket.

The stack does however support a hook mechanism provided by IP Hook to enable packets to be accessed within the stack on the inbound and outbound paths, for example to allow pre- and post-processing and other packet transformations; that, for example, is the mechanism used to implement IPSec packet-level encryption and decryption.

A socket is a session-based abstraction that sits logically above the networking protocol implementation, which provides the transport and network layers implementation.

The bottom interface of the stack relies on the Network Interface Manager to select a suitable outgoing interface, which in turn relies on the Network Controller to find a network agent to negotiate the chosen connection (for Network Interface Manager and the Network Controller see Section 9.6).

A number of network agents (AGT files) are available: CSD.AGT, to establish circuit-switched data connections; PSD.AGT, to establish packet-switched data connections; and NULL.AGT, which implements a minimal agent that is used with Ethernet.

A number of network interface implementations (NIF files) are available, including for PPP and Ethernet, as well as a QoS Test NIF that is used in conjunction with the QoS Framework.

At these levels, the architecture has evolved quite significantly since Symbian OS v7, which implemented a rather simplistic view of networks and connections. Particularly with UMTS packet-switched 3G networks, the networking world becomes more complex. For example, multi-homing means that devices can have multiple IP addresses (multiple network interfaces may be active, each with its own IP address and potentially each on a different network) and packet-switched phone data services mean that multiple interfaces and networks may provide access to a single network destination. However, from an application perspective these changes are mostly invisible and impact only systems developers.

Note that Bluetooth and wireless LAN are not supported by default but require comparable drivers to abstract the hardware for the (overlying) Ethernet NIF implementation. The NIF can support one lower-layer packet driver, loaded during initialization.

PPP NIF has been part of the networking delivery since the first releases of Symbian OS but was significantly enhanced in Symbian OS v8 to improve interoperability with MS Windows, for example by supporting Microsoft extensions to CHAPS dialup authentication.

The Tunneling NIF was introduced with VLAN support and IPSec reimplementation in Symbian OS v8.

Network Security

Network Security protocols operate at different levels in the overall networking stack. TLS and SSL (the two can be considered synonyms) operate at the transport level, providing per-packet encryption and decryption. IPSec on the other hand operates at the network level and is principally designed to support secure networks, for example Virtual Private Networks (VPN) based on policy.

The TLS component implements TLS v1.0 (Transport Level Security) and SSL v3.0 (Secure Sockets Layer), providing more or less transparent, per-packet encryption-based security to client applications, for example HTTPS or SyncML. TLS is implemented as a number of separate DLLs exposing client APIs to applications, which enable sockets to be secured and internal APIs used by networking and security components. TLS was first introduced in Symbian OS v7 and was redesigned and enhanced in Symbian OS v8. A typical use of SSL is to enable secure browser-based transactions.

The IPSec implementation provides security policy management, including support for multi-homed clients (so that different security policies can be associated with the different IP addresses in use by the device) and multiple active policies. IPSec is implemented as a policy server and supporting libraries, as well as a protocol-level PRT Core IPSec PRT plug-in. In effect, it sits between the Socket Server and clients requesting secure sockets. The IPSec PRT does not implement the full interface required by the Symbian OS v9 Socket Server architecture based on the Comms Framework and therefore is considered to be a 'pseudo-PRT'.

IPSec uses the networking stack Hook interface to inspect all incoming and outgoing packets and apply the required cryptographic transformations. The actual security algorithms and libraries are implemented by cryptography and security services components.

The VPN component uses IPSec to manage VPN policies and connections, including VPN password management and is implemented as a VPN manager server and supporting libraries.

IPSec was first introduced into Symbian OS in v7 and was redesigned and enhanced in Symbian OS v8.

Quality of Service

Quality of Service enables performance characteristics to be specified for a communications channel in a packet-based network, to ensure that the required data rates for a given application are met.

The QoS Framework PRT implements QoS policy setting for open sockets (Socket Server sub-sessions) that are treated as QoS channels, providing generic and UMTS-specific APIs for use by applications. While GPRS supports general QoS principles, UMTS defines four traffic classes (Conversational, Streaming, Interactive and Background). Like IPSec, QoS is considered to be a 'pseudo-PRT'.

Networking Daemons

A number of standard networking daemons are implemented as part of networking support.

- DND (i.e. DNS) and DHCP are implementations of the Internet standard protocols for domain-name resolution and dynamic host-address assignment.

- DND makes DNS queries to the (remote) network and listens for and responds to local queries. Like its Unix counterpart, it is implemented as a server 'daemon'. It is accessed through the `RHostResolver` class by Socket Server clients (i.e. as a Socket Server sub-session) and supports `GetByName` and `GetByAddress` queries.

- DHCP enables a device to obtain an IP address and network parameters dynamically from the network, so that having a fixed IP address becomes unnecessary. The DHCP implementation consists of a server daemon and a client-side interface and provides a limited API sufficient for the Network Interface Manager to configure an appropriate network interface. (It is not intended for other users.)

WAP Support

Wireless Application Protocol (WAP) evolved out of work started by the Unwired Planet consortium, which evolved into the WAP Forum in 1998 at around the time that Symbian joined it and into the Open Mobile Alliance (OMA) in 2002. WAP was a deliberate attempt to create a Web standard targeted at mobile devices in general and phones in particular, to make web content browsable on those devices.

WAP defines a protocol stack, much like TCP/IP, with transport and datagram layers defined over a variety of possible mobile phone network bearers. At the top of the stack, the wireless session protocol (WSP) behaves like a binary-encoded HTTP. Unlike HTTP, WAP enables both pull and push models. In the pull model, clients make requests to a WAP gateway that responds by sending data. In the push model, the gateway pushes data to the client, without a client request.

While the full WAP stack consists of multiple protocols, WAP Datagram Protocol (WDP) is the critical underlying mechanism, defining a binary encoding for datagrams over a bearer network, which can be any of GSM, CDMA, SMS, GPRS or 3G network protocols.

Symbian supplied a full WAP-stack implementation in Symbian OS v7. However, where licensees supplied a WAP browser it was generally tightly coupled to a particular WAP-stack implementation and where they didn't the Symbian stack was redundant in any case. Therefore from Symbian OS v8, Symbian OS implements a 'short' stack that only supports WAP messaging features (which, for example, are used by the Multimedia Messaging Service), providing connectionless WAP Push, connectionless WSP, and WDP.

The implementation consists of the Messaging API and the WAP Short Stack, which is supplied as a reference implementation only. These provide the client APIs and implementation for WAP messaging over GSM SMS bearers. (Note that the mapping of WDP to CDMA SMS not implemented.)

An important use case for WAP is as the carrier for MMS delivery. Unlike SMS, which is transmitted over network-signaling channels, MMS uses data-traffic channels and, hence, requires a transport technology (SMS relies on network-specific signaling mechanisms as its bearer). An advantage of WAP is that it provides a uniform transport protocol regardless of the underlying network type (GSM, GPRS, CDMA or 3G).

WAP push is based on notification sent to the terminal over SMS, followed by a WSP 'get' call to fetch the message. MMS is therefore independent of the network type (because WAP implementations run on all network types) and interoperable (because TCP/IP is used on the network side to link WAP gateways and can link gateways on different networks, for example, GSM and CDMA or UMTS).

Architecture

The general structure of Networking Services in Symbian OS will be recognizable to those familiar with the standard OSI 7 layer networking model and corresponds roughly to a utility layer plus the lower four layers. See Figure 9.23.

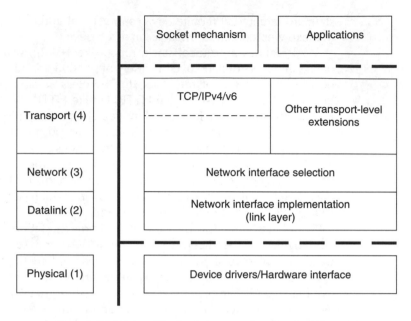

Figure 9.23   Networking Services mapped against the OSI model

The higher layers of the OSI model are mapped by components in the higher layers of Symbian OS, particularly in the Application Services layer.

The OSI model is a generic abstraction and not a rigid specification but understanding the mapping helps to understand the Symbian networking implementation. Roughly, the Symbian OS layers are as follows, working top down:

- networking services and utilities including network security, network daemons, plus the WAP stack implementation

- network-specific extensions to the Socket Server and Network Connection Manager

- core network protocols, the TCP/IP stack and its extensions

- network interface management agents

- network interface implementations.

The higher-level components provide application-level interfaces, the middle-layer protocol implementations are tightly bound to the sockets abstraction, through which all networking services are accessed, and the lower levels provide the interfaces to the available communications bearer technologies.

From a client perspective, the complex interactions between the networking components, the communications framework and the short-link and telephony bearer services are hidden behind the Socket Server. However, it is useful to have at least a general picture of the pattern of interaction.

When a Socket Server sub-session is opened by a client requesting a TCP/IP protocol socket, the request is passed to the TCP/IP stack, which tries to start an outgoing connection. If the stack fails to find an interface that will allow it to reach the selected destination, it reports its failure back to the Socket Server, which then requests the Network Interface Manager to load and start a connection agent of suitable type. Depending on its type, the agent requests a connection from either the Telephony Server or the C32 Serial Server. When the connection is established, the Network Interface Manager loads and starts a NIF module, which implements the required Network Interface and negotiates authentication and other link characteristics (for example, encapsulation and compression) and finally acquires an IP address. The Network Interface manager then binds the NIF to the TCP/IP stack.

Design Goals

The original design goals of Symbian OS Networking Services were based on dial-up access to a network via either a fixed-line modem or a mobile phone. The expected networking applications were standard Internet and web applications, for example email and browsing. Adequate data throughput and the ability to virtualize networking services over an available serial bearer were the key considerations.

Network protocols were also considered important as a way of standardizing support for connectivity with desktop computers for data synchronization and backup.

The increasing specialization of Symbian OS for mobile phones, and the evolution of mobile phones into true network devices as packet services have begun to dominate, has required almost continuous evolution in the architecture and implementation of Networking Services, to keep in step with rapid technological advance, rapid adoption of advanced technologies into mobile phones and the push to provide infrastructure for new services and applications.

Networking has become mainstream for telephony as the basis for high data throughput services such as two-way video conferencing and audio and video streaming. Direct network connection over Wi-Fi is also rapidly becoming a support requirement for mobile phones.

VoIP pushes these trends to their logical step, in effect subsuming telephony into networking.

Component Collections

TCP/IP Security Collection

These components implement secure networking, supporting transport-level security (security at the level of individual IP packets) and connection-oriented security (which is used, for example, to provide VPN support services to application clients running on Symbian phones). See Figure 9.24.

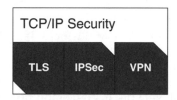

Figure 9.24   TCP/IP Security components

Table 9.12   TCP/IP Security Components

Component Name	Development Name
TLS	TLS, TLSPROVIDER
IPSec	IPSEC
VPN	VPNAPI, VPNCONNAGT, VPNMANAGER

- The TLS component is an implementation of Transport Level Security (TLS) including SSL Secure Sockets, which provide encryption per packet, supporting application-level encryption and authentication-based security, for example for secure web services. Client authentication is based on key management and certificate handling, including support for external cryptography modules ('secure tokens'), for example based on a phone smart card.

- The IPSec component operates at a lower level (i.e. network level) and is principally designed to enable secure networks, for example VPN, based on policy.

- The VPN component provides policy-based connection management and gateway interoperability for VPN connections, i.e. it enables users to connect to VPNs.

TCP/IP Utilities Collection

This collection contains implementations of standard networking 'daemon' server utilities. See Figure 9.25.

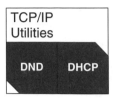

Figure 9.25   TCP/IP Utilities

Table 9.13   TCP/IP Utilities

Component Name	Development Name
DND	DND
DHCP	DHCP

- DND is a DNS implementation that makes DNS queries to the network and listens for and responds to local queries.

- The DHCP component is a Dynamic Host Configuration Protocol (DHCP) implementation used by the PAN Profile and other networking components.

WAP Stack Collection

Symbian provided a full WAP stack implementation in Symbian OS v7. Versions later than Symbian OS v8 implement only a 'short' stack, providing client APIs for connectionless WSP, connectionless Push and WDP. See Figure 9.26.

Table 9.14   WAP Stack Components

Component Name	Development Name
WAP Message API	WAPMESSAGE
WAP Short Stack	WAPSTACK

- The WAP Short Stack component is a cut-down WAP stack supporting WAP messaging, that is, WAP datagrams, WAP Push messaging and

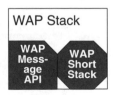

Figure 9.26   WAP Stack components

WSP but not full WAP browsing. It is supplied only as a reference implementation. Vendors replace it with their own short or full WAP stack implementations.

- The WAP Message API implementation provides APIs for WAP Push, connectionless WSP and WDP datagrams.

Sockets API Extensions Collection

The Internet Sockets component is a DLL that provides a library of utility classes and generally useful constants, which specifically support using Internet sockets, to store and manipulate IP addresses, routes, and so on.

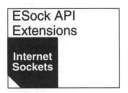

Figure 9.27   Sockets API Extensions

Table 9.15   Sockets API Extensions Components

Component Name	Development Name
Internet Sockets	INSOCK

Clients access the Internet Sockets through the generic sockets client API and use the TCP/IP-specific utility classes to perform the IP-specific manipulations. Clients link against the Internet Sockets library. See Figure 9.27.

Subconnection Interface Collection

This is a utility component used by QoS clients to create and package the QoS parameter list. Parameters are set using the RQoSChannel class. See Figure 9.28.

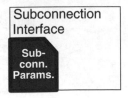

Figure 9.28   Subconnection Interface

Table 9.16 Subconnection Interface Components

Component Name	Development Name
Subconnection Parameters	

Network Protocol Plug-ins Collection

This collection contains the core TCP/IP functionality including the TCP/IP stack, which supports both v4 and v6 standards, the hook mechanism that allows access to packets for inline processing (for example allowing packets to be encrypted 'in place' as they flow through the stack), and IPSec and QoS implementations. See Figure 9.29.

Table 9.17 Network Protocol Plug-ins

Component Name	Development Name
IP Event Notifier	`IPEVENTNOTIFIER`
TCP/IPv4/v6 PRT	`TCPIP6`
IP Hook	`INHOOK6`
QoS Framework PRT	`QOS, QOSLIB, PFQOSLIB, SBLPAPI`
Core IPSec PRT	`No unit`

- The IP Event Notifier PRT is implemented as an IP Hook and raises events to clients based on state changes in the TCP/IP stack. It is principally used by DHCP to determine when and how to perform address negotiation.

- The TCP/IPv4/v6 PRT supplies the core protocol implementations for TCP/IP networking including the IPv4 and IPv6 stacks, TCP, UDP, ICMP and ARP protocols, a Hook interface allowing access to packets, IPSec and QOS protocol modules, and an event notifier service.

- The IP Hook PRT defines an interface to which modules bind to perform transformations on inbound and outbound packets, respectively

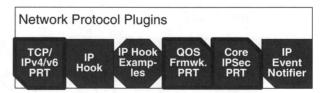

Figure 9.29 Network Protocol Plug-ins

upon receipt from or before delivery to the Network Interface. IPSec is such a Hook, inspecting all incoming and outgoing packets and applying cryptographic transformations as specified in the Security association database.

- The QoS Framework PRT is a Hook module, implementing QoS channels through which it schedules packets. Additional plug-ins map the desired QoS characteristics to relevant link technology.

- The Core IPSec PRT implements core functionality for IPSec in a multi-homed context, that is multiple active network interfaces, for simultaneous use by multiple applications, providing tunnel modes and various high-level APIs. It includes a cryptographic library module, policy managers and parsers.

Networking Plug-ins Collection

This collection contains the network interface agents (AGT files). Two additional components are also included, the Bluetooth PAN profile and the GPRS/UMTS QOS PRT (which is considered a 'pseudo PRT'). See Figure 9.30.

Table 9.18   Networking Plug-ins

Component Name	Development Name
Connection Provider Plug-in	IPCPR
CSD AGT	CSDAGT
PSD AGT	PSDAGT
NULL AGT	NULLAGT
GPRS/UMTS QOS PRT	GUQOS
Bluetooth PAN Profile	BLUETOOTHPAN
Secondary PDP UMTS Driver	SPUD

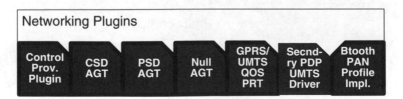

Figure 9.30   Networking Plug-ins

- The Connection Provider Plug-in provides IP connections to clients, supporting bearer mobility.

- The CSD AGT plug-in to the Connection Agent framework negotiates a circuit-switched data connection, for example to GSM or CDMA networks, supporting dial-up networking services.

- The PSD AGT plug-in is deprecated and its functionality is replaced by other components. It is an agent plug-in to the Connection Agent framework that negotiates packet-switched connection for example to GPRS networks, supporting 'always on' networking services.

- The NULL AGT plug-in implements a minimal agent used to pass straight through to an Ethernet connection that is provided by the Ethernet packet driver.

- The GPRS/UMTS QOS PRT is a plug-in helper module to the QoS Framework that gets and validates QoS parameters from the QoS framework at the request of a loaded NIF and is used to implement 3GPP parameters.

- The Bluetooth PAN Profile plug-in is an agent-like module that implements the Bluetooth Network Encapsulation Protocol (BNEP), as an Ethernet Packet Driver module. It serves as the network interface agent used to create PAN connections, enabling PAN to behave like a regular Internet access provider.

- The Secondary PDP context UMTS Driver (also called the PDP NIF) supports multiple primary PDP contexts (multi-homing over GPRS) on the telephony reference platform. It is not a production component.

Link Layer Control

Link-layer components of the networking stack, Network Interface modules (NIF files) are selected by the Network Controller and loaded, started and stopped by the Network Interface Manager to implement the interface to the physical link layer (which is, in turn, provided by networking device drivers, serial communications CSYs, or telephony TSYs). See Figure 9.31.

NIFs implement the polymorphic plug-in interface defined by the Network interface manager (NIFMan).

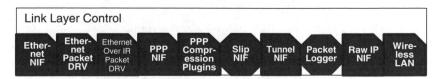

Figure 9.31 Link Layer Control components

Table 9.19 Link Layer Control Components

Component Name	Development Name
Ethernet NIF	ETHER802
Ethernet Packet Driver	ETHERDRV
Ethernet Over IR Packet Driver	IRLANPACKETDRIVERS
PPP NIF	PPP
PPP Compression Plug-ins	PREDCOMP, MSCOMP, STACCOMP
SLIP NIF	SLIP
Tunnel NIF	TUNNELNIF
Packet Logger	PACKETLOGGER
Raw IP NIF	RAWIPNIF
Wireless LAN	802.11

- The Ethernet NIF component provides a generic Ethernet layer network interface, that manages Ethernet framed packets. It is designed to sit below any number of supported Protocol modules and on top of more specialized Ethernet framing interfaces, called packet drivers.

- The Ethernet Packet Driver is an Ethernet framing interface, the driver-level component (DRV files, that is, lower-layer packet drivers) that supports the Ethernet NIF.

- The Ethernet Over IR Packet Driver is an Ethernet framing interface, the underlying networking interface driver for infrared.

- The Serial Line IP (SLIP) NIF component is supplied as a reference component that licensees can choose to remove or replace with a production implementation. SLIP was the earliest (and simplest) protocol for relaying IP packets over dial-up lines and has largely been replaced by PPP.

- The Point to Point protocol (PPP) NIF provides TCP/IP over serial communications (i.e. over a point-to-point link). It allows a device to connect to a phone and use it as a gateway to the Internet. Once the link has been established, optional facilities such as data compression may be negotiated.

- The PPP Compression Plug-ins supplies the implementation of common PPP compression algorithms as dynamically loaded DLLs. It includes Microsoft Compression (MSCOMP), Stac Electronics Compression (STACCOMP) and Predictor Compression (PREDCOMP) implementations.

- The Tunnel NIF component implements the IPSec tunnel to enable IPSec to operate in tunnel mode, for example, as used by VPN clients.

- Wireless LAN supports IEEE 802.11 wireless networking.

9.9 Short-link Services

Short-link services enable individual devices to communicate directly with each other ('peer-to-peer'), either over a physical cable connection such as serial or USB, or using short-range radio, either line-of-sight such as infrared, or unseen paired, such as Bluetooth. (Note that, by this definition, Wi-Fi, which is fast becoming important on phones, is considered a network access technology not a short-link connection technology, although Wi-Fi hardware supports a peer-to-peer mode.)

Symbian OS supports the principal short-link technologies: RS232 serial, USB, infrared/IrDA and Bluetooth, as well as the higher-level OBEX object transfer protocol, which is supported over both IrDA and Bluetooth.

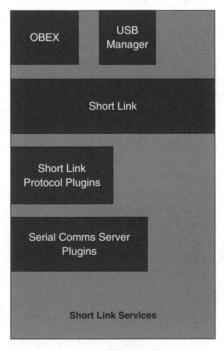

Figure 9.32   Short-link services

The short-link-services block includes managers, utilities, protocol implementations and serial-hardware-adaptation plug-ins. Associated device drivers are located lower down in the system model, at kernel level. See Figure 9.32.

For network-capable mobile devices (mobile phones and PDAs, for example), short-link connections are also important for network access. Typically, they provide the connection alternative to using the onboard phone. In Symbian OS, short-link services act as bearers for higher-level communications services, including both networking and telephony. This enables some interesting scenarios, for example, remote use of a phone in one Symbian OS device from another over a short-link connection.

Although continuing to evolve to enable increased data rates, short-link technologies are relatively mature and Symbian's support for them is relatively mature. RS232 serial has a long history and IrDA, Bluetooth and USB have all been standardized since the mid-1990s.

However, there are interesting and significant evolutions in all the technologies. In terms of connection speeds, while serial cable is limited to data transfer rates of 115 kbps, Bluetooth offers data rates closer to 1 mbps with a range of 10 meters, while 'newer', 'faster' IrDA standards increase rates beyond 16 mbps and even up to 100 mbps. USB began as a 12 mbps standard, before increasing 40-fold (with USB 2.0) to 480 mbps.

The application possibilities are also interesting and extend beyond basic data management and data synchronization. After a slow start, Bluetooth has become ubiquitous on phones, in particular for hands-free and headset peripherals, including stereo headsets. USB offers much more than just a physical link protocol. USB is both a link technology and a transport protocol definition with extras such as support for powering unpowered devices and hot-plugging ('plug and play' notification to the host). In a Symbian context, it allows a Symbian OS device to plug into a USB host (for example, a desktop computer) and offer multiple services.

Both IrDA and Bluetooth specify a complete protocol stack defining link, transport and application layers, which offers significantly more than just serial-like setup for a simple physical link.

Because Bluetooth allows ad hoc, 'promiscuous' connection between any devices within range, security is potentially an issue. The Bluetooth standard therefore includes security protocols (which Symbian OS implements).

As well as conventional serial communications, over a physical serial link or virtualized over IrDA or Bluetooth, Symbian OS supports a number of higher-level short-link services:

- Higher-level IrDA protocols are supported, for example including IrTranP for beaming camera images.

- IrDA Object Exchange (OBEX), a binary protocol for data exchange, is supported over IrDA, Bluetooth and USB connections.

- A number of Bluetooth profiles including security profiles are supported, with support for licensee extension.
- USB device management is supported.

Architecture

While short-link services forms a natural logical and functional block, it does not form a cohesive architectural unit. While the supported short-link technologies are designed to interoperate extensively and implement the overall architectural patterns of communications services (server- and framework-based, protocol module plug-ins to the Socket Server, serial port plug-in implementations to the Serial Server framework), the detailed architecture of each is distinct and should be understood independently of the serial architecture.

IrDA is implemented as a Socket Server plug-in module, loaded by the Socket Server when an IrDA socket is requested (either directly by an application, or by other components in the Comms Services). Within the Socket Server session, the protocol module communicates with the infrared port through the Serial Server and its IrDA serial plug-in CSY module, which ultimately drives the logical and physical device drivers for the onboard infrared hardware.

The OBEX implementation is designed as a wrapper for either a socket style API (RSocket for IrDA and Bluetooth) or a USB client API (RDevUsbcClient for USB). OBEX is implemented as a static DLL to which clients link at compile time, with the OBEX code running in the client thread.

Bluetooth is implemented as a Socket Server protocol plug-in module. Clients request a Bluetooth socket from a Socket Server session. The Bluetooth socket communicates with the firmware controller via the Bluetooth HCI implementation. Symbian OS implements the mandated v1.2 Bluetooth stack.

IrDA and OBEX

Symbian OS has supported IrDA since the first ER5 release, providing line-of-sight infrared data exchange between devices. IrDA is more than a simple connection protocol and, in fact, comprises a complete set of protocols from application level to link level, including IrTranP (Infra Red Transfer Picture, for devices with cameras), IrCOMM (IrDA serial port emulation) and TinyTP (TinyTransfer Protocol, providing flow control), as well as lower-level protocols including FIR (Fast Infrared). All are supported by Symbian OS.

IrDA also provides the underlying support for OBEX over infrared (Infrared Object Exchange, IrOBEX). OBEX is a protocol and not a service but application-level services can be created that use the protocol to send and receive data. At the application level, Symbian OS provides OBEX-based services including SendAs messaging, SyncML data synchronization, installer services, and so on. Symbian OS has supported OBEX since the first ER5 release. Since the introduction of Bluetooth support in Symbian OS v6, it has supported OBEX over Bluetooth and, since Symbian OS v7, OBEX over USB (but with server functionality only).

Bluetooth

Bluetooth also defines a complete protocol stack and not just a radio link technology. The Bluetooth services that run on top of the stack are defined as Bluetooth profiles. Symbian OS provides Serial Port, PAN (Personal Area Networking) and Generic Access profiles, as well as Remote Control (since Symbian OS v9), that enables a Symbian device to control Bluetooth peripherals, for example headsets. Licensees may add additional profile support.

Bluetooth components include:

- The Bluetooth Manager is the information store (implemented over Symbian OS DBMS) used to manage details of local and remote Bluetooth devices.

- Bluetooth SDP (Service Discovery Protocol) enables Bluetooth devices to find each other and store information about discovered devices. (The SDP database is not persistent.)

- The Bluetooth HCI (Host Controller Interface) interfaces the Bluetooth stack to the onboard controller hardware and is provided as a reference plug-in.

Symbian OS has supported Bluetooth since Symbian OS v6, with incremental support added over subsequent releases.

USB Manager and Classes

USB classes are analogous to Bluetooth profiles and represent the use cases that a device supports when it connects to a USB host. The USB Manager on a device enumerates, starts and stops the USB classes implemented on the device and provides a query interface for their status, providing a central control point and an on–off switch.

Symbian OS provides a USB Manager and implements USB CSY (serial over USB), Mass Storage and OBEX (OBEX over USB) classes. The USB Manager implements a server interface for USB class implementations and for clients requesting information or services from USB classes (typically the user is the USB host) and provides the underlying mechanism for application-level class configuration and querying of the USB host (the other connected device) across a USB connection.

Component Collections

OBEX Collection

This collection defines the OBEX (Object Exchange) session protocol. OBEX is a binary protocol and is therefore compact and can support application-level services from simple beaming of vCard and vCal entries to full-scale synchronization, for example, as a SyncML bearer.

Table 9.20 OBEX Components

Component Name	Development Name
OBEX Protocol	OBEX, IROBEX
OBEX Extension API	OBEX_EXTENSIONAPIS

In Symbian OS, OBEX is supported over IrDA infrared, Bluetooth and USB, providing session-style APIs, that is, Connect and Disconnect and basic Get and Put commands. See Figure 9.33.

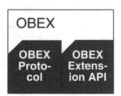

Figure 9.33 OBEX components

USB Manager

This collection comprises the manager for the USB classes present on a device, for example providing the mechanism beneath a configuration application like a control panel to switch on and off the available USB classes on a Symbian OS device and to query a USB host (not a Symbian OS device) application across a USB connection. See Figure 9.34.

Figure 9.34 USB Manager components

Table 9.21 USB Manager Components

Component Name	Development Name
USB Manager	USB

Short Link Collection

These higher-level components support the Bluetooth protocol implementation and Bluetooth profiles. See Figure 9.35.

Figure 9.35 Short Link components

Table 9.22 Short Link Components

Component Name	Development Name
Bluetooth Protocol Client APIs	No unit
Bluetooth Manager	BLUETOOTHMANAGER, BLUETOOTHBTEXTNOTIFIERS, BLUETOOTHCONFIG, BLUETOOTHGAVDP, BLUETOOTHROM, BLUETOOTHUSER
Bluetooth SDP	BLUETOOTHSDP
Bluetooth Profiles	BLUETOOTHAVRCP
Remote Control Framework	BLUETOOTHREMOTECONTROL
HCI Framework	BLUETOOTHHCI

- The Bluetooth Protocol Client APIs are used by Bluetooth socket clients
  and provide support for low-level control of protocol parameters
  (packet sizes, for example) and hardware (power modes, for example).

- The Bluetooth Manager provides an information store for managing
  details of the local and remote Bluetooth devices, implemented over
  Symbian OS DBMS, allowing entries to be stored, retrieved, modified
  and deleted.

- The Bluetooth Service Discovery Protocol (SDP) is the mechanism used by connected Bluetooth devices to query each other and exchange information about the Bluetooth services they support.

- The Bluetooth Profiles include Generic Access Profile (GAP), Personal Area Networking (PAN), since Symbian OS v8, and (from Symbian OS v9) Audio and Video Remote Control (AVRCP).

- The Remote Control Framework enables sending and receiving of remote-control commands to and from remote Bluetooth devices. (It is supported from Symbian OS v9.)

- The HCI Framework is a reference implementation of the Bluetooth Host Controller Interface as used by the Bluetooth Stack to interface to the onboard controller hardware. It provides a full range of HCI commands, accessed indirectly via L2CAP and RFComm layers. Licensees can replace the supplied implementation.

Short Link Protocol Plug-ins

This collection implements the Bluetooth core stack, including the Bluetooth protocols and the HCI firmware implementation and the IrDA protocol suite as PRT Socket Server plug-in-in protocol modules. See Figure 9.36.

Table 9.23 Short Link Protocol Plug-ins

Component Name	Development Name
Bluetooth Stack PRT	BLUETOOTHSTACK
Bluetooth HCI	BLUETOOTHHCIPROXY
IrDA PRT	IRDA, INFRA-REDCONFIG

- The Bluetooth Stack PRT component implements the Bluetooth stack as a Socket Server protocol plug-in, providing a complete implementation including L2CAP, RFCOMM and SDP.

- The Bluetooth HCI is a reference implementation of firmware-specific support for the standard Bluetooth Host Controller Interface (the

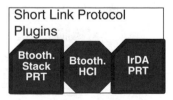

Figure 9.36 Short Link Protocol Plug-ins

stack-side implementation of the interface forms part of the standard Bluetooth support provided by Symbian OS).

- The IrDA PRT is an implementation of the IrDA protocol stack as a Socket Server protocol plug-in, provides a complete IrDA implementation including IrTranP (for sending pictures) and FIR (Fast Infrared).

Serial Comms Server Plug-ins Collection

CSY modules are implementations of serial ports virtualized over different bearers (RS232, USB, Bluetooth, IrDA) and are loaded by the C32 Serial Server in response to clients to provide ports of the types requested. See Figure 9.37.

Table 9.24   Serial Comms Server Plug-ins Components

Component Name	Development Name
Serial Port CSY	ECUART
USB CSY	ECACM
Bluetooth CSY	BTCOMM
IrDA CSY	IRCOMM

The Serial and IrDA CSY components were both present in ER5.

- The Serial Port CSY component implements an RS232 virtual serial-port abstraction for conventional serial communications and directly drives the ECOMM.LDD and ECOMM.PDD logical and physical device drivers.

- The USB CSY component was introduced in Symbian OS v7.0 supporting a single-port configuration and extended to support multiple virtual ports in Symbian OS v7.0s. It provides a multiple serial-port-like interface over a USB connection and directly drives the EUSBC.LDD and EUSBC.PDD logical and physical device drivers. Note that this is an implementation of USB intended for legacy applications that require conventional serial support, rather than for USB-aware applications.

Figure 9.37   Serial Comms Server Plug-ins

- The Bluetooth CSY component was introduced in Symbian OS v6.1 with the first Bluetooth implementation for Symbian OS. It is a plug-in to C32 Serial Server and implements an RS232-like virtual serial port over a Bluetooth link using an RFComm socket. Port configuration is performed using the Bluetooth Manager APIs.

- The IrDA CSY component implements the IrDA standard for serial communications, IrComm, emulating a serial port over an IrDA link. Internally, it uses an IrDA socket (IrDA.PRT), through a Socket Server session, which in turn drives the ECUART.LDD and UCUART.PDD logical and physical drivers to drive the infrared hardware.

10

The Base Services Layer

10.1 Introduction

To get Symbian OS up and running on new hardware, whether on a reference board (from a supplier such as Intel or Texas Instruments) or on the hardware for a new phone, you need to port the base layers of the system.

The lowest level of the system contains the operating system kernel, device drivers, and the device-driver framework support, which provide operating system primitives and hardware abstraction frameworks. Sitting just above them are the low-level libraries, servers, and frameworks that build on the kernel layer to create a programmable and usable operating system. Because Symbian OS is a microkernel system,[1] the 'kernel side', which runs in protected or privileged mode on the host processor ('supervisor' mode on ARM processors), is kept as small as possible. The kernel-side/user-side distinction roughly divides the base of the system into two layers.

The Base Services layer is the higher of the two layers and it contains the user-side servers, frameworks, libraries and utilities that build on the kernel layer to provide the basic operating system services. Together, the two layers constitute the minimal system which can be booted, run and programmed on real hardware. In a monolithic operating-system design, most (and possibly all) of the Base Services would form part of the kernel implementation. See Figure 10.1.

10.2 Purpose

The Base Services layer extends the bare kernel into a basic software platform that provides the foundation for the remaining operating system

---

[1] In fact the design is not 'pure' microkernel, but borrows from both microkernel and monolithic design principles (see Chapter 11).

Figure 10.1   Base Services layer in the system model components

services, and effectively encapsulates the user side of the 'base' operating system. It also provides the minimum services required to enable a complete and self-contained basic build of the lower-level system, which supports only text-mode program execution and is used to create the first stage 'base-port' to new hardware.

As well as providing foundational frameworks and utilities which are used both by system components and by applications, it also provides the operating system libraries that support the programming model, in other words, which support the creation, loading, and running of programs on the operating system and which implement many of the signature Symbian OS idioms, for example the cleanup stack, active objects and descriptors.

10.3   Design Goals

In many (but not all) respects, Symbian OS offers a textbook example of a microkernel operating system architecture.[2] The most significant exception is the inclusion of the two-level device-driver framework, and

---

[2] See the rationale for and description of the microkernel pattern in [Buschmann *et al.* 1998].

device drivers themselves, on the 'kernel side' of the system. A true microkernel design would move these into user space.

The microkernel principle is to keep the kernel small;[3] core functionality which is, however, above the level of the basic operating system primitives, is kept out of the kernel itself and instead is located in system servers. System servers extend the microkernel to provide necessary services, and also encapsulate any lower-level software and hardware dependencies. In Symbian OS, the core system servers that are required to create a complete but minimal running system on real hardware are located in the Base Services layer; the remaining system servers, which are not essential for a basic hardware port but which are required to engineer a complete product based on Symbian OS, are located one layer up, in the OS Services layer.

The goals of the Base Services layer therefore are to provide efficient and effective extensions to the basic kernel functionality, which are in a concrete sense complete (i.e. they enable a complete but minimal system to be built), while being both portable and extensible.

10.4   Overview

The Base Services layer includes a number of essential frameworks and libraries on which almost all higher-level services, as well as applications, have some direct or indirect dependencies.

- The User Library provides the basic programming model for Symbian OS, including system-specific types (such as the CBase class and manifest constant[4] definitions), as well as the APIs that define the unique native idioms, for example active objects, descriptors and UIDs, libraries which provide DLL and executable entry point stub classes, and so on.

- The File Server includes file-system utilities and the concrete file-system implementation plug-ins in use on a particular device.

- The Store is a persistent storage framework. The Base Services layer also includes the DBMS implementation, as well as more recent additions such as the Central Repository, which provides a single location and set of APIs for managing all system settings.

---

[3] The most significant immediate benefits of 'small' are portability, because all essential hardware dependencies are encapsulated within the small core of the system, and small memory footprint, a small system consuming less ROM (where the system is ROM-based) and RAM (put simply, there is less system to load at runtime). The additional goal of simplicity is also more likely to be realized in a small system than in a large one.

[4] Those are named constants whose underlying definition can be varied at compile time for different platforms; in Symbian OS, they include TInt, TReal, TBool and TAny.

- Other essential frameworks and libraries include the Plug-in Framework (ECOM), cryptographic libraries, Application Utilities (such as the Basic Application Framework Library, BAFL), character encoding and conversion libraries, XML parsers,[5] the power management and shutdown framework, as well as the low-level framework support used by multimedia services to communicate with hardware-accelerator adaptor plug-ins. (The actual adaptors and the device drivers with which they interact are located lower down, at the kernel level.)

- Components such as the Text Window Server and Text Shell are required to make the base system complete and to avoid dependencies on higher-level services, for example, graphics.

Put simply, from a programming perspective, many of the most basic characteristics of the operating system are realized in the Base Services layer.

10.5   Architecture

The Base Services layer of Symbian OS is in many ways the foundational layer of the system, extending the microkernel and the lowest level hardware-abstraction services provided by the kernel layer into a basic but complete system. A number of critical services which in monolithic architectures would be included in the kernel itself, for example the file system and the user libraries which provide the programming model for the operating system, are found here. The key boundary which defines the separation of these services from the kernel is the division between kernel (supervisor or privileged) and user (non-privileged) processes. In a monolithic system, most of these services would run as privileged kernel processes.

The design decision to separate these services from the kernel and to implement them as user-side services is a distinguishing feature of the operating system, separating it from monolithic systems (Unix/Linux, Windows-derived systems) and putting it squarely in the tradition of microkernel operating system design.

From the perspective of applications and higher-level operating system services, the Base Services layer libraries and frameworks provide the logical interface to the basic low-level operating system. The Base Services layer extends the raw hardware support and the basic kernel abstractions of the low-level system and adds file-system support and the File Server,

---

[5] XML is considered an essential service since XML is increasingly used as the basis for internal configuration files and other essential data formats, for example, Central Repository entries.

the User Libraries that support the programming model, a simple text-window server and a text-based shell, and an assortment of other low-level frameworks and utilities. Together, this is enough to support, test and validate a first-stage port to new hardware and it provides the foundation for creating complete support for all device hardware. The boundary between the Base Services layer and the higher-level services in the layers above it, therefore, is a concrete one: nothing above the Base Services layer is required get a port running on specific hardware.

The system model organizes the Base Services layer components into a number of collections, divided broadly between the low-level components that interact closely with the kernel to provide basic services (the User Library, file-system support) and higher-level components that build on these services (for example, Store, which provides the persistence model, the Cryptography Library and the Text Shell).

The User Library

It is through the User Library that the fundamental abstractions implemented by the kernel, which together define the native programming model for Symbian OS, are made available to clients. These include processes, threads and memory chunks and mutexes, semaphores and message queues. The User Library also implements many other programming idioms specific to Symbian OS, including active objects and descriptors, the cleanup stack, the client–server framework, and the Publish-and-Subscribe mechanism. It supplies an assortment of utility classes, including timers, date and time services and locale definition and collection classes, including arrays, lists and binary trees. It defines the native data types, both class-based and manifest constants, and supplies the libraries that implement the low-level system and language bindings, including DLL and executable entry point stub classes.

In the original kernel architecture of Symbian OS (EKA1, before Symbian OS v9), the User Library was called from both user-side and kernel-side code. In order to guarantee time bounds, the EKA2 kernel-side code does not link to the User Library but instead uses a small utility library (incompatible with the user-side library) accessible only by the kernel side.

The User Library includes the following APIs:

- the native types used in the system which include C++ base classes (including `CBase`) and manifest constants (`TInt` and others)

- collection classes (buffers, arrays and lists), descriptors, Unicode-character support, raw-memory management (copying and filling) and geometric concepts (points, sizes, rectangles and regions)

- math libraries including 64-bit integers and floating-point math

- idioms specific to Symbian OS including the cleanup stack, descriptors, active objects, UID manipulation, and implementations of memory allocators, named and reference counted objects and bitmap allocators

- other useful classes supporting lexical analysis, bitstreams, Huffman compression, timers and timing services.

In addition, it supplies libraries that provide DLL global data and static data and thread local storage; and executable and DLL entry point support (for example calling static constructors).

The User Library also provides the Publish-and-Subscribe mechanism (since Symbian OS v8), as a means of storing system-wide global variables and a platform-security safe IPC mechanism (again, since Symbian OS v8) for peer to peer communication between threads in the operating system. Publish and Subscribe is based on the notions of properties (data values), publishers (threads with rights to update given properties), and Subscribers (threads interested in changes to given properties). Because it is available on both user-side and kernel-side, it also provides a possible asynchronous communication mechanism between user-side and kernel-side code.

The File Server

The File Server provides the framework architecture supporting the implementation of file systems as custom plug-ins and the default plug-in implementations for FAT file systems, the native format for externally visible drives, for example, those implemented on removable media, as well as internal-only formats such as Read Only File System (ROFS), the internal file system to which ROM code is copied for execution in hardware architectures that do not support execute-in-place memory.

File-system plug-in implementations may in turn be further extended via extension DLLs to support specific hardware differences, for example FAT on NAND flash, which implements a NAND flash translation layer transforming requests coming from the FAT file system into a format suitable for a NAND flash-media driver. Note that the file server is multithreaded (since Symbian OS v9), using one thread per storage medium used.

The File Server also provides some file-related utility functions, for example FAT filename conversion which supports translation from full Unicode file names to ASCII. (While EKA1 supported Unicode strings internally, the real-time EKA2 kernel uses only ASCII strings internally; note that there is no impact on the full, system-wide support for Unicode.)

The file server has traditionally had an additional role in Symbian OS, as the first of the system services to be started by the final stage of the kernel boot process. In Symbian OS v6 and v7, the file server was responsible for starting the Window Server, in effect completing the boot process. From Symbian OS v8, the File Server instead launches the System Starter, which performs final initialization of the File Server including adding and mounting file systems on appropriate local drives, and then initiates start-up of the rest of the system, including implementing the customizable server start-up policy (which defines which servers should be started and in which order).

Essential System Frameworks

The Base Services layer includes some essential system frameworks, including the Plug-in Framework, which underpins the Symbian OS framework–plug-in architecture, and the persistent storage model.

Plug-in framework

The Plug-in Framework, known as ECOM, has two principal purposes: to make it easier to design and implement new services or features as framework plug-ins by providing a standard (and best-practice) pattern together with ready-made run-time support. Framework plug-in architectures improve the overall modularity, extensibility, and customizability of the system, thus improving usability (from a system perspective) as well as improving design consistency. As importantly, it provides an evolution path for already conforming framework plug-in components to migrate relatively painlessly to the platform security model introduced in Symbian OS v9, making it easier for components to adopt the required security policies (i.e. to ensure trust between frameworks and the plug-ins they load and to avoid plug-in loading being exploited to subvert platform security).

ECOM defines an interface to which all plug-ins conform (plug-ins derive from the ECOM base classes) and provides the dynamic discovery and instantiation mechanisms which find, create, and load them on demand.

ECOM's original design was evolved from the design of the WAP browser framework plug-ins. Broadly, it provides:

- methods for defining and implementing interfaces as DLL plug-ins

- plug-in registration and methods for managing multiple interface implementations, including plug-in 'upgrades' (later versions)

- fast dynamic discovery and instantiation methods for plug-ins, as well as static registration for known system plug-ins

- capability policing, that is, enforcement of the security restrictions of its clients

- other features including support for easy localization of plug-ins and start-up state awareness (to improve system boot-up performance).

ECOM was first introduced in Symbian OS v7 and was then significantly enhanced in Symbian OS v8, to support and conform with the new platform security model. Initially it offered an optional, standard mechanism for frameworks to define plug-in interfaces and a standard plug-in registration and loading mechanism. Subsequently it was elevated from an optional to an obligatory mechanism; from Symbian OS v8, it is the standard interface used by all frameworks to define how plug-ins interact with and extend the framework and the global runtime binding mechanism that finds and loads plug-ins into requesting frameworks on demand, while conforming to the Platform Security requirements and limitations on processes.

Security issues

ECOM ensures that frameworks are only able to find plug-ins they have the capability to load and which pass the platform security check, that is, matching of the DLL UID field from the RSC resource file to the SID (secure identifier) of the corresponding DLL.[6] Plug-ins are loaded into the requesting client framework's process, allowing the kernel to police the capabilities of the plug-in DLL. (Although if the plug-in's capabilities do not match those of the client process, then it could be loaded into a separate process.)

ECOM is implemented with a standard client–server architecture, based around a central registry (of interface implementations) and a server client-side API that handles inter-process communication (IPC) between servers and their clients (wrapping the invocation parameters, passing the wrapped request over the IPC boundary and unwrapping any return parameters when a call completes). Client frameworks use a session object as the interface to the ECOM server for finding, creating and destroying plug-in providers of the framework interface.

Calls to the ECOM server are translated into registry or load calls to perform:

- addition and removal of interface implementations (registrar functions)

- access and persistence mechanisms (registry data functions)

---

[6] Strictly speaking the UID3 of a DLL is not really the SID, since SIDs are only assigned to executables or processes (based on the executable's UID3) and not to DLLs. Also any single DLL can potentially contain multiple different implementations of a given interface, which would share interface UIDs but differ in implementation UIDs. [Heath 2006] is the best reference for following up the details.

- resolution and searching mechanisms returning 'best fit' results (resolver functions)

- loading and unloading (load manager functions).

A single instance of the registry exists. Registry data is held in two forms, an internal format for fast access, consisting of a subset of the full registry data, and persisted data, consisting of the registration set stored in file form, divided into branches with one branch per available drive (branches may be transient, supporting removable media).

Client frameworks (i.e. interface definers) may supply custom resolver implementations to ECOM to implement custom criteria.

Full discovery of plug-ins occurs at ECOM server start-up, that is, at device boot time. Additional discovery of non-read-only internal drives occurs when a drive is added or removed and when a secure plug-in is added to or removed from a writable drive.

Persistence model

The Symbian OS persistence model is based on the Store architecture, which defines abstractions of streams and stores.

A stream is an abstract interface that translates between internal and external object representations, that is, between bit layouts in RAM and bit layouts saved onto storage media or sent over a network. As well as encryption and decryption streams, four alternative stream implementations are provided, suited for different underlying storage media:

- fixed-size memory streams

- variable-size memory streams

- file streams

- store streams.

A store is an abstract interface that allows a network of streams to be manipulated, including Externalize and Internalize operations, which allow complex data structures (e.g. whole documents or databases) to be stored or restored from external media or from a network.

As well as secure stores (which provide encryption and decryption) and supporting store dictionaries (used to locate the various streams inside a store), the Store architecture provides alternative implementations suited for different underlying storage media or uses:

- stores using RAM as the underlying storage media (for example, used as undo buffers by some applications)

- stores using files as underlying storage media, either *direct* file stores used by 'file-based applications', which keep all their data in RAM when running (in other words, which create and manipulate conventional documents), or *permanent* file stores used by applications that only part-load their data (for example, database applications)

- stores that can be embedded into other stores thus allowing document embedding to create compound documents (e.g. pictures in a text document)

Streams and stores provide the native, object-oriented persistence model for Symbian OS. Both the DBMS relational database interface and the Central Repository are implemented on top of store mechanisms.

DBMS

The DBMS component defines a general relational-database-access API and provides implementations either for small client-side databases or for client–server-based multiple-client implementations. Client–server databases are stored in files. Client-side databases can either be a whole file or a single file stream (enabling multiple single stream databases to reside in a single file).

Databases can be manipulated either through a native API or a subset of SQL. Basic database functions are supported, including table creation, manipulation and deletion, database queries and transactions.

From Symbian OS v8, where required, DBMS supports security-access-control policies for databases, including shared-access policies. For system-supplied databases, it allows additional finer-grained policies to be specified for named tables within a database (for databases created within the DBMS private data-cage).

Central repository

The Central Repository provides a single persistent store for global settings as well as a notification mechanism allowing clients to register to be notified when specific settings change.

The Central Repository is designed as a collection of repositories, where a repository is a collection of settings. A setting is represented by a data value (a 32-bit integer, a real number, a byte-array or a text string). Repositories are created from a definition file based on a standard template and may be compacted into a binary format. Each repository has an owner and is required to declare an access-control policy, which is set in the initialization file and cannot thereafter be changed. Access control may be specified at the level of the whole repository, for individual settings or for ranges of settings, and may include settings which have not yet been created.

Depending on access control, individual settings may be created, searched for, have their values set, or deleted. Range operations are supported and a notification registration mechanism is provided allowing clients to register interest in settings changes (including creation and deletion). 'After-market' repositories (e.g., for user- or network-installed applications) are supported by the Application Installer. Backup, restore and caching of repositories is also supported. Access to repository settings is restricted based on the capabilities of the client making an access request together with the repository security policy.

In general, settings replace the use of INI files to store application and system defaults and other information, for example default file names, locale settings and user preferences. Similarly, settings replace the use of the Comms Database for storing communications-specific defaults and settings, although the Comms Database interface is preserved for compatibility.

The earliest releases of Symbian OS included a Registry, but it was removed (as it was not portable) in Symbian OS v6 and replaced by solutions based on INI files and the Comms database. In Symbian OS v8, the Central Repository was introduced to provide more efficient and consistent settings management.

Other Services and Utilities

The Base Services layer contains a number of additional frameworks, libraries, utilities and servers.

Application Utilities

The Application Utilities, known to developers as the Basic Application Framework Library (BAFL), provide an assortment of utility classes organized as a single library DLL:

- resource-file handling including loading and reading of legacy formats (before Symbian OS v7) and Unicode-compressed and Unicode-and-dictionary-compressed formats (since Symbian OS v7), including robust reading classes able to handle corrupt resource files

- file utilities, including file finding based on file type as defined by UID and file matching to select between files based on the current locale

- string pools, a storage mechanism allowing for fast string comparisons

- dynamic arrays for descriptors, supporting mixed 8-bit and 16-bit descriptors

- incremental text-matching comparing two text buffers (reading left-to-right)

- support for showing localized names of 'user-showable' plug-ins

- clipboard copy–paste support implemented as a direct file store with stream dictionary, allowing applications to retrieve clipboard data by UID

- system sounds for messages, events, errors and so on, specified by UID

- minimal support for spreadsheet-style 'cell' and 'range' data types

- legacy change notifier (derived from active objects) wrapping the `RChangeNotifier` for system environment changes relating to time, locale, power and thread death.

Character Encoding and Conversion Framework and Plug-ins

The Character Encoding and Conversion Framework provides an API for converting text between Unicode and other character sets based on an extensible converter plug-in architecture.

In Symbian OS v9, conversion is supported for a variety of ASCII formats (including common ISO codepages), UTF-7 encodings (including Shift-JIS and JIS) and UTF-8 encodings. Conversion is performed by specifying the Unicode character set of interest (for conversion to or from) and then requesting the conversion.

As well as text conversion, text utilities are provided to manage character sets (create character-set arrays, find the character-set UID from the character set name and vice versa) and to detect character sets automatically based on sample texts.

XML Framework and Parser Plug-ins

The XML Framework provides an extensible framework for XML parsing based on a parsing model similar to SAX 2.0, into which custom parser-implementation plug-ins (as well as validator, DTD and auto-correction plug-ins) can be loaded. Default plug-ins are provided for non-validated parsing of XML 1.0 and for WAP Binary XML (WBXML).

Parsers are selected based on a document's MIME type and other criteria supplied by clients when using the framework. The parser class defines methods that parse XML data from descriptors (all in one go or incrementally) and from files. Internally within the parser, text is stored in UTF-8 format to ensure preservation of extended characters.

The WBXML parser plug-in can be extended to support additional document types by providing WBXML token-to-string translation tables ('String Dictionaries'). Default tables are supplied for SyncML, WML and Service Indication.

The design goal for the framework is to provide a single, standard, platform implementation of a flexible and capable XML parser to replace the various task-specific and ad hoc parsers provided locally in the system. The framework also provides sufficient extensibility for likely future uses (including generating capability).

So-called 'processor' plug-ins (for example, validators and auto-correctors) may be chained with parsers to provide multiple processing stages.

String Dictionaries are implemented using string pools that make string comparison almost instantaneous (at the expense of string creation; however, this supports parsing cases where string constants are known at compile time particularly well, as is the case where documents follow standard DTDs such as SyncML, SMIL or WML).

Media Device Framework and Plug-ins

The Media Device Framework provides hardware-abstraction interfaces for audio and video accelerators to the Multimedia Framework (see Chapter 8) and its clients. Typically, accelerators are hardware devices (codecs) but they may also be software emulations. The framework defines APIs for sound, video, MIDI, and ASR (Automatic Speech Recognition) accelerators, and the architecture for loading the lower-level adaptor plug-ins (DevSound, DevVideo, DevMIDI, and DevASR; see Chapter 11). A client utility API for speaker-independent speech recognition is also supplied as a plug-in and is available to any client wanting to interface to ASR hardware (or software emulations).

The framework also includes a policy server that manages access to the underlying audio and video hardware, deciding which clients can access the hardware and when. Licensees can customize access policies.

The Media Device Framework evolved from the earlier Symbian OS v6 and Symbian OS v7 MediaServer. Previously, Multimedia Framework controller plug-ins were able to directly interface to audio and video codecs via adaptor plug-ins. By defining a standard interface between controllers and adaptors, the Media Device Framework enables portable adaptors to be developed to support specific accelerator hardware. The framework has evolved significantly over subsequent releases compared with its first implementations, which supported only audio.

Cryptography Library

The Cryptography Library provides system-level support for a wide-range of non-RSA cryptographic algorithms including symmetric and asymmetric ciphers, hash functions and a cryptographic strong random-number generator. The cryptographic algorithms are supplied in two variants: export-restricted (strong) and non-export-restricted (weak). Note the change since Symbian OS v7, which provided an export-restricted and an RSA-based library, with no non-export-restricted variant.

Subcomponents of the library include:

- random-number server, an implementation of a random-number generator

- random-number library DLL, providing an API for generation of cryptographically strong random numbers

- hash library DLL, providing an API for generating cryptographic hashes, supporting MD2, MD5, SHA1 and HMAC

- password encryption API DLL supporting key generation from password (PKCS#5 key-derivation function) and key-based encryption and decryption

- cryptographic library, providing non-RSA cryptographic algorithms, supplied in weak and strong versions (depending on possible export restrictions) and implementing symmetric and asymmetric ciphers, padding schemes, and big integers.

Clients link against the Cryptography Library for all functions. Calls are transparently forwarded to whichever version of the library implementation is present at run time (strong if present, weak if not; this is determined at ROM build time). The weak version is limited to symmetric cryptographic operations with a maximum key size of 56 bits and asymmetric cryptographic operations with a maximum key size of 512 bits.

Zip Compression Library

Port of the zlib compression library (see relevant RFCs, for example, RFC1950) used to support compression and decompression of SIS files (Symbian native installable-application format) and Java Archive (.JAR) files, and for PNG decompression.

Shutdown Server

The Shutdown Server provides a notification service to clients to provide 'save data' and 'release resources' notifications in case of switch-off or low memory and similar events, enabling a client to save data (for example, if it is an application) and possibly also close itself (to free up resources).

It consists of a client-side library that clients use to request notifications, the Shutdown Server that provides 'save data' notifications and which may be derived from to create bespoke shutdown servers (for example, Uikon implements a customized shutdown server), and a server launcher (executable) that launches the service.

Feature Registry

The Feature Registry (introduced in Symbian OS v9.2) provides an API enabling run-time queries to discover whether known but optional features are supported on the particular running platform (device or emulator).

A 'feature' is a Symbian OS or user interface variant API (or set of APIs) identified by a Feature UID.

A configuration file listing features present is generated at ROM build time (on real devices) or provided as part of the emulator support in the licensee SDK and is held in a Publish and Subscribe property queried by the Query API. A Notify API is also provided but not currently enabled, with the intention that in future releases the feature set will be updatable at run time (the Symbian OS v9.2 implementation fixes the feature set at ROM build time).

Text Shell and Text-Window Server

Together, the Text Shell and the Text-Window Server that supports it make the base layers of the operating system independent of higher-level services (for example graphics and windowing support as well as the GUI-based application support), allowing functional text-mode-only builds of the base to support porting and other low-level development.

This enables a minimal but functional system to be built for and run on new hardware as a first step to providing full hardware support. In principle, all hardware dependencies are encapsulated within the base layers of the system; once the base port is complete, the rest of the system can be moved over to run on top of it without any further adaptation being required. In practice, the situation is a little more complex; Comms Services, in particular, are hardware-dependent at the lowest level of hardware abstraction and interface. In practice therefore porting is a two-stage activity: once the base port is complete, a communications port is needed to interface the communications stacks to the device hardware. When the communications port is complete, the remaining system services can be moved over.

The Text Shell provides a console-like (command-line) interface to basic operating system services, for example navigating the file system and launching executables, when standard graphics, application, and GUI support are not available. The Text Shell is also available on the emulator, where it is used, for example, when developing servers that run without a user interface.

The Text-Window Server supports the Text Shell, using a text-mode display driver to provide standard VGA/LCD screen displays on local hardware as well as VT100 terminal emulation over a serial line.

Figure 10.2   Component collections in the Base Services layer

10.6   Component Collections

The Base Services layer (see Figure 10.2) contains several collections of components.

- The User Library and File Server and User-Side Hardware Abstraction collections contain essential system services providing file-system support and essential user libraries.

- The Text-Mode Shell provides character-based text services that enable the lowest two layers of the system to be built independently of graphics frameworks.

- The Low-Level Libraries and Frameworks, Character Conversion, Persistent Storage and XML collections contain frameworks and libraries useful to applications, as well as to other system components.

User Library and File Server Collection

The User Library and the File Server implement essential basic functionality that should be considered central to the operating system. They interface to the kernel in a uniform way using the standard client–server model. Because they run user-side, the kernel is protected both from programming errors by users of the basic libraries (including resource exhaustion) and from timing latencies introduced on the user side enabling real-time guarantees to be met. See Figure 10.3.

- The User Library component provides much of the signature functionality of Symbian OS to (system) programs and to applications,

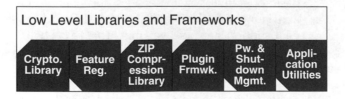

Figure 10.3   User Library and File Server components

Table 10.1 User Library and File Server Components

Component Name	Development Name
User Library	EUSER
File Server	F32_EKA2
Filesystem Plug-ins	FILSYS
FAT Filename Conversion Plug-ins	FATCHARSETCONV

including native data types, clean up and clean-up-aware base classes, active objects, descriptors, as well as the system–language binding including DLL stub mechanisms, IPC and similar mechanisms, and generally useful low-level services including (since Symbian OS v8) Publish and Subscribe.

- The File Server component manages all file access through client-side file-server sessions. It is a framework of file-system plug-ins and extensions which supports the implementation of custom file systems. The server is responsible for brokering client requests and passing them through to the file system, where the real work is performed. The file server includes an embedded ROM file system.

- The Filesystems component provides file-system plug-in implementations of LFFS and FAT file systems. FAT is the native format for externally visible drives, for example those implemented on removable media.

- The FAT Filename Conversion Plug-ins support filename conversion from and to Unicode.

User-Side Hardware Abstraction Collection

This API provides Get and Set functions to query and set information about specific hardware features from the user-side, providing a way to access and control many device-specific features independently of the hardware platform. See Figure 10.4.

Figure 10.4 User-Side Hardware Abstraction components

Table 10.2 User-Side Hardware Abstraction Components

Component Name	Development Name
User HAL	HAL_EKA2

This component is deprecated for application use. The intended users are system components running on the user-side and needing to access hardware properties, for example fault and exception, memory-page size, timer-tick period, screen properties (whether a screen backlight is present or not, setting the display contrast), and so on.

Text-Mode Shell Collection

Together, the Text Shell and the Text Window Server that supports it make the base layers of the operating system independent of higher level services (for example graphics and windowing support as well as GUI-based application support), allowing functional builds of the base to support porting and other low-level development. See Figure 10.5.

Table 10.3 Text Mode Shell Components

Component Name	Development Name
Text Window Server	EWSRV
Text Shell	ESHELL

- The Text-Window Server supports the Text Shell, using a text-mode display driver to provide standard VGA/LCD screen displays on local hardware as well as VT100 terminal emulation over a serial line.

- The Text Shell provides a console-like (command-line) interface to basic operating system services, for example navigating the file system and launching executables, for use in porting, testing, and low-level development in which only the base layers of the system are built.

Low-Level Libraries and Frameworks Collection

This collection contains a number of basic system frameworks and libraries which are used throughout the system as well as by applications.

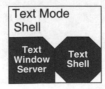

Figure 10.5 Text-Mode Shell components

It includes the Plug-in Framework, which provides a uniform and secure plug-in definition and loading mechanism, Store, which implements the Symbian OS persistence model, and a varied collection of system utilities. These include a cryptography library, which implements both weak and strong versions of standard cryptography algorithms, a Zip compression library, and a basic application utilities library (BAFL).

Table 10.4 Low-Level Libraries and Frameworks

Component Name	Development Name
Cryptography Library	CRYPTOGRAPHY
Zip Compression Library	EZLIB
Plug-in Framework	ECOM_ONGOING
Power and Shutdown Management	DOMAIN
Application Utilities	BAFL
Feature Registry	FEATREG

Among the more recent components (new in Symbian OS v8) is the Central Repository which is provided to store state and settings information that need to be persistent for clients, for example default filenames, locale settings, user preferences, etc. See Figure 10.6.

- The Cryptography Library implements (since Symbian OS v7) non-RSA-based cryptographic support for symmetric and asymmetric ciphers, hash functions, random number generation, and password encryption.

- The Zip Compression Library is a port of the zlib compression library (see relevant RFCs e.g. RFC1950) used to support compression and decompression of SIS files (the native Symbian OS installable application format) and Java Archive (JAR) files, and for PNG decompression.

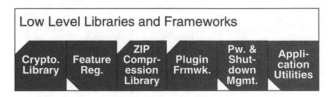

Figure 10.6 Low-level libraries and frameworks

- The Plug-in Framework is a framework and server for plug-in interface implementations. It defines the standard base classes used by conforming plug-ins and a client-side API used by framework clients to locate and load plug-ins on demand. It manages a registry of available plug-ins and implements security policy mechanisms (e.g. capability policing).

- The Power, Memory and Disk Management component is a customizable user-side power manager supporting policy-driven power management via power domain 'profiles' at device switch-on and switch-off. It includes a notification service (the so-called 'Shutdown Server') to clients to provide 'save data' and 'release resources' notifications in case of switch-off, low memory and similar events.

- The Application Utilities component, known to developers as BAFL, provides an assortment of utilities organized as a single library DLL including utility classes for resource-file handling and file finding, and implementations of string pools and descriptor arrays.

- The Feature Registry (introduced in Symbian OS v9.2) provides an API enabling run-time queries to discover whether known but optional features are supported on the run-time platform.

Character Conversion Collection

This collection provides a character-code conversion framework and plug-ins. See Figure 10.7.

Table 10.5 Character Conversion Components

Component Name	Development Name
Character Encoding and Conversion Framework	CHARCONV_ONGOING
Character Encoding and Conversion Plug-ins	CHARCONV

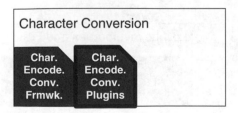

Figure 10.7 Character Conversion components

- The Character Encoding and Conversion Framework supports conversion of text between Unicode and non-Unicode character sets. Symbian OS native text formats are Unicode.

- The Character Encoding and Conversion Plug-ins provide conversion between a variety of ASCII and UTF-7 and UTF-8 text formats. The Unicode text format is UTF-8.

Persistent Data Storage Collection

The persistence model, plus the DBMS abstraction implemented as a layer around it, provides an SQL-interface for database applications. It also includes the Central Repository that provides a uniform approach to persistent settings management. See Figure 10.8.

Table 10.6 Persistent Data Storage Components

Component Name	Development Name
Store	STORE
DBMS	DBMS
Central Repository	CENTRALREPOSITORY

- The Store component defines the Symbian OS persistence model based on the two abstractions of streams and stores, providing an application data-storage model which shields applications from the underlying File Server implementation.

- The DBMS component defines a general relational database access API and implementations for fast client-side-only exclusive access and slower client–server-based shared-access databases. Databases can be manipulated either through a native API or a subset of SQL.

- The Central Repository component provides a single persistent store for global settings as well as a notification mechanism allowing clients to register interest when settings change. The Central Repository was introduced in Symbian OS v8.

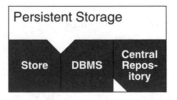

Figure 10.8   Persistent Data Storage components

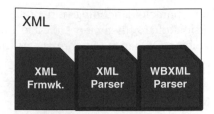

Figure 10.9   XML components

Table 10.7 XML Components

Component Name	Development Name
XML Framework	XML
XML Parser	XMLPARSERPLUGIN
WBXML Parser	WBXMLPARSER

XML Collection

XML support includes an extensible framework and parser plug-ins for parsing and validating XML documents (see Figure 10.9).

- The XML Framework provides an extensible framework for XML parsing based on a parser model similar to SAX-2.0 and supporting DTD and processing plug-ins (for example, validators and auto correctors) as well as parser plug-ins.

- The XML Parser component is a non-validating parser plug-in for XML 1.0.

- The WBXML Parser component is a parser plug-in for WAP Binary XML (WBXML).

Media Device Framework Collection

The Media Device Framework (see Figure 10.10) defines standard hardware acceleration APIs which are used by the Multimedia Framework and its clients, enabling multimedia controller plug-ins to communicate with hardware accelerator adaptors through standard interfaces.

- The Media Device Framework contains standard acceleration APIs for audio, video, MIDI, and Automatic Speech Recognition (ASR).

- Media Device Framework Plug-ins is an ASR Client Utility API that provides speaker-independent speech recognition to the Multimedia

Framework and directly to other clients wanting to interface to speech-recognition hardware (or software emulations).

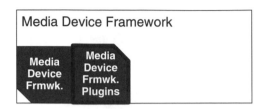

Figure 10.10 Media Device Framework

Table 10.8 Media Device Framework Components

Component Name	Development Name
Media Device Framework	MDF
Media Device Framework Plug-ins	AUDIODEVICE, MDFAUDIOHWDEVICEADAPTER, VORBISDECODERPROCESSINGUNIT, VORBISENCODERPROCESSINGUNIT, MDFVIDEODECODEHWDEVICEADAPTER, MDFVIDEOENCODEHWDEVICEADAPTER

11

The Kernel Services and Hardware Interface Layer

11.1 Introduction

The Kernel Services and Hardware Interface layer (see Figure 11.1) is the lowest layer of Symbian OS. It contains the Symbian OS kernel and supporting components.

These include the kernel-level components which must be customized in order to bring up a minimal build of the operating system on new hardware (although a typical port entails customizing other components too).

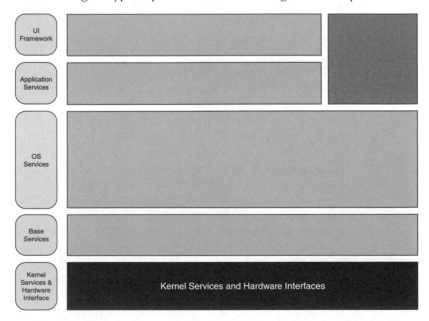

Figure 11.1   Kernel Services and Hardware Interface layer in the system model

The layer boundary also marks the 'kernel side' boundary; all components which run in privileged mode in the runtime system are included within the layer.

11.2 Purpose

The Kernel Services and Hardware Interface layer is the foundational layer of Symbian OS. It includes the kernel and all the supporting infrastructure needed to boot and run the kernel on the underlying hardware platform. It is responsible for fundamental operating system services:

- bootstrapping the physical or emulated device to provide the basic initialization of the hardware

- creating and managing the fundamental operating system kernel abstractions, for example, threads, processes, memory address spaces, and other resources including timers, mutexes, and so on

- scheduling, pre-emption and interrupt handling

- access to devices, providing the device-driver framework and device drivers that abstract device hardware and implement the two-tier logical and physical device driver model

- encapsulating the kernel–user boundary; all processes which run in privileged mode originate from this layer

- encapsulating the lowest level of an operating system port ('base port') to new hardware

- insulating all higher layers from actual hardware.

The system model collects the kernel and kernel extensions, device drivers, and the other hardware abstraction components which are required for hardware porting, into a single Kernel Architecture block. (Versions of the system model for Symbian OS v8 have two Kernel Architecture blocks, for each of the EKA1 and EKA2 kernel versions.)

Two small collections sit within the layer but outside the block. These collections each have a single component which is independent of the kernel version but which requires customization in a new port. These components implement locale support, which is used by the kernel, and the screen driver.

From Symbian OS v9, the Kernel Architecture block is organized to reflect the basic architecture of the kernel side of the system, as well as the recommended structure of a base port. The kernel includes extensions that implement the device-driver framework, providing a two-layer logical–physical device-driver model in which logical device drivers abstract a generic device interface and physical device drivers drive the actual hardware. Below the kernel, abstraction of device hardware is divided between the Application-Specific Standard Part (ASSP), an off-the-shelf integrated CPU, and the Variant components. The ASSP component contains ASSP-specific code that is otherwise hardware-agnostic (it supports the specific silicon package used in a product, typically a standard part containing the CPU core and custom chips). The Variant components contain hardware-dependent code which is specific to a product, for example hardware-specific flash-memory translation.

11.3 Design Goals

Releases up to and including Symbian OS v8 shipped with the original kernel, EPOC Kernel Architecture 1 (EKA1). Symbian OS v9 and later releases ship with the new kernel architecture of the EKA2 'real-time' kernel.

At the highest level, the design goals of the kernel layer of Symbian OS are common to both kernel versions:

- provide an operating system kernel optimized for its device class – palmtop and smaller

- optimize for ROM-based execution – XIP- or RAM-shadowed execution

- optimize for mobile – no fixed wires

- optimize for battery operation – anything from the two 'AA' batteries of the original Psion Series 5 to the latest mobile phone rechargeable battery

- target consumer-oriented devices – for 'ordinary' non-technical users.

Immediate performance goals follow:

- meet the requirements of the device class – in terms of the operating system image size, start-up time, task-loading and task-switching times, its ability to run forever, and overall robustness

- meet consumer-device goals – robustness in the face of typical failure scenarios, for example out-of-memory, no signal, low battery or sudden battery removal, media card removed in mid-write, disk full but camcorder still running, and so on

- provide a highly portable operating system kernel – to enable porting to multiple hardware architectures in as pain-free a way as possible

- support typical licensee product models, that is, the product line or product family principle – multiple minor hardware revisions follow from an initial 'lead product' and porting effort should scale down significantly between a first port and subsequent incremental ports.

Compared with the original kernel, EKA2 is explicitly designed to make porting easier by improving the modularity of the kernel and the structuring (and packaging) of its supporting components. Thus the core kernel is independent of both ASSP and variant. In contrast, in EKA1 the separation between hardware-dependent and hardware-independent code was less clear-cut, and hardware support was less cleanly partitioned between the ASSP and variant.

It is also important to remember Symbian's origins as an application-centric operating system, which determines additional design goals:

- provide a fully programmable platform – enabling user-installable applications as well as a complete native application set

- provide a fully graphical system which is intuitive to use – with full interactive GUI, multitasking and instant task switching.

From the beginning, Symbian OS has also been strongly focused on international markets, with early support for non-Western scripts (for languages such as Chinese, Arabic, Thai and Hindi):

- Unicode multi-byte characters supported throughout the system

- easy localization

- non-Roman and multidirectional script display.

Increasingly, the application emphasis has evolved from PIM applications (calendars, contacts books and so on) toward high-data bandwidth applications, including camcorder applications and mobile digital TV, following the trend of increasingly multimedia capable devices.

As well as requiring the architecture to support ever higher data rates, this overall shift in the market away from PDA-style products towards mobile phones has led to an important evolutionary goal and, in particular, to specific requirements on the EKA2 kernel. The kernel was to be capable of supporting typical licensee phone hardware architectures, including one-core and dual-core variants, and Symbian-only as well as 'partner operating system' configurations (requiring cooperation with a real-time 'partner' operating system driving the baseband hardware and software).

Arguably the most critical design goal follows from the above: provide a highly adaptable and evolvable kernel architecture capable of change in a rapidly evolving technology, product and market context.

The strength of the kernel architecture is demonstrated by its stability and continued fitness for purpose in the face of rapid change – for example, the almost complete transformation of the mobile phone in less than a decade, from the pre-Symbian OS basic phone of the mid-1990s to the PDA–phone 'smartphone' hybrids with which Symbian OS entered the phone market to today's full multimedia devices.

11.4   Overview

Symbian OS has a microkernel architecture,[1] which means that the responsibilities of the kernel are kept to an essential minimum. The design approach is to implement a minimal set of operating-system primitives in the kernel, on which higher-level, generic operating system services can be built, the goal being to keep the kernel small, and therefore fast, and to keep its complexity low, to achieve high reliability and predictability.

Simplistically, kernel responsibilities are divided between implementing suitable primitives for use by the higher layers of the operating system and interfacing to the underlying hardware platform. Surrounding the kernel itself are the additional components required to provide complete hardware support.

Because the kernel layer is the interface to the hardware platform, it is dependent on the hardware. To port the operating system to new hardware entails porting the kernel layer. An important design consideration, therefore, is to optimize ease of porting by isolating hardware dependencies. The design of the kernel and its supporting components is highly modular, to make porting simpler.

An important distinguishing feature of Symbian OS is its optimization for ROM-based systems. Symbian OS was designed to be built into device ROM and executed in place without requiring loading into RAM, in contrast to more conventional systems (including Linux/Unix and Microsoft Windows), which are designed to be loaded from the file system into RAM before executing.

Supporting ROM-based systems has become more complex as memory technologies and hardware architectures have evolved to keep pace with the burgeoning requirements for storage capacity. The latest releases of Symbian OS are optimized for multiple hardware architectures and memory types, including the latest NAND-flash-based systems as well as more conventional NOR flash. On NOR-flash systems, Symbian OS is executed in place (XIP). On NAND flash, which is not byte-addressable, Symbian OS shadows itself to RAM from where it executes. In both cases, it provides a translation layer to interface the filing system to the flash drive.

11.5   EKA1 and EKA2

The origins of EKA1 go right back to the first releases of the operating system. The original architecture of the Symbian OS kernel was driven by the need to provide a robust platform for a PDA-centric (and, therefore, application-centric) operating system. Almost from its first release, however, Symbian OS has been evolving to meet the high data-throughput

---

[1] There are some aspects in which it is more hybrid than pure (see the detailed discussion below).

and real-time requirements of more communications-centric devices (in particular, advanced mobile phones).

As early as 1998 (i.e. two years before Symbian OS v6 was released) a 'real-time' project began in the kernel team to investigate the issues involved in providing real-time support and to prototype a solution.

The eventual result of that work was a new, real-time-capable kernel, EKA2 (also known as EpocRT), benchmarked in terms of its ability to directly support a full mobile phone signaling stack. It was intended for release in Symbian OS v7 and reached the market in Symbian OS v8, becoming the standard kernel in Symbian OS v9. (In Symbian OS v8.1, customers were offered a choice between the EKA1 and EKA2 kernels.) Even so, the new kernel's initial selling point for customers was probably less its real-time capabilities than its support for the new Platform Security architecture, which had become commercially necessary. Platform Security requires kernel support to police security policies as part of its inter-process communication (IPC) mechanism. While Platform Security was introduced in stepped phases to be compatible with the original kernel, the full features of Platform Security are only available in a system running EKA2.

EKA2 was designed to be closely compatible with EKA1. In important respects, the two are functionally equivalent, as evidenced by the choice of using either EKA1 or EKA2 in Symbian OS v8.1. The critical difference is that EKA2 is designed to offer true real-time behavior.

11.6   Singleton Component Collections

The Kernel Services and Hardware Interface layer consists of the Kernel Architecture block (or blocks, in the case of releases that include both kernel versions) and two singleton component collections (see Figure 11.2) containing components that, while they are not part of the kernel architecture proper, nonetheless can be counted as belonging on the kernel side of the kernel–user boundary.

Localization Collection

This component is a customizable plug-in that implements locale-specific settings including standard strings (for example, day and month names),

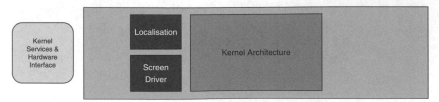

Figure 11.2   Localization and Screen Driver collections

Figure 11.3 Localization components

distance units, currency symbols, date and time formats, collation orders, and so on. Standard locales, including Japanese and several Chinese variants, are provided with the system.

Locale Support is included in the Kernel Services layer because it implements various strings used directly by the kernel (e.g. default system messages). It is loaded by the User Library.

Table 11.1 Localization Components

Component Name	Development Name
Locale Support	LOCE32_ONGOING, ELOCL

Screen Driver Collection

This component implements the generic operations defined by the Bit GDI to manipulate the physical memory map of the device display or bitmap memory map. (Typically, in-memory bitmaps and the display memory map are addressed in the same way in hardware, hence a common interface is provided to both.) It supports dual screens, which feature in flip-phone designs. The Screen Driver forms part of a base port to new hardware.

Table 11.2 Screen Driver Components

Component Name	Development Name
Screen Driver	SCREENDRIVER

11.7   Kernel Architecture Block

In one sense, the Symbian OS kernel has always been larger than a microkernel, since in both EKA1 and EKA2 it includes extensions and

Figure 11.4 Screen Driver components

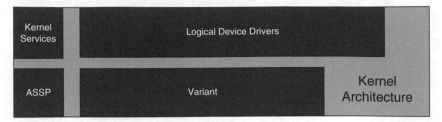

Figure 11.5   Kernel Architecture block in Symbian OS v9

device drivers. In another sense, in EKA2 (see Figure 11.5) it is even smaller, with a true nanokernel at its core.

However, both kernel architectures have true microkernel properties. For example, major services such as the File Server and the User Library, as well as all graphics and communications services, including networking and telephony, remain outside the kernel and are run as user-side processes. This is in contrast for example to the monolithic kernel architectures of both Linux and Microsoft Windows.

A microkernel limits the kernel responsibilities to a small set of core functions, and builds higher-level operating-system functions on top of a small set of kernel primitives. The microkernel can thus be kept small and fast. Another important feature of microkernel architecture is that kernel functionality is deliberately simplified; more complex higher-level behavior is moved out of the kernel onto the user side. The principles parallel those of Reduced Instruction Set Computer (RISC) versus Complex Instruction Set Computer (CISC) processor design, with broadly comparable arguments in favor.

From microkernel architectures, the Symbian OS kernel borrows the following features:

- a message-passing framework for the benefit of user-side servers

- networking and telephony stacks as user-side servers

- file systems implemented as user-side servers.

At the same time, for performance reasons Symbian OS compromises on microkernel purity by allowing kernel extensions and including the device-driver framework in the kernel. However, device drivers are not embedded in the kernel binary but follow the typical Symbian OS design pattern of being implemented as run-time loadable and unloadable plug-ins.

From monolithic kernel architectures, the Symbian OS kernel borrows the following features:

- kernel-side device drivers

- scheduling policy implemented in the kernel.

The test case for the success of the EKA2 kernel architecture is its ability to support the real-time requirements of a GSM/wCDMA or CDMA phone-signaling stack ([Sales 2005, p. 778]). To do so requires that real-time guarantees can be given for key services, most importantly interrupt latencies, thread latencies and context switches. ('Real-time' in this context means deterministic and bounded by a predictable and known time; which is not quite the full definition.[2])

The rationale for providing real-time support is two-fold. First, as phones become more complex and add more custom hardware, particularly to support multimedia functions, interrupt latencies become increasingly critical to data throughput. Secondly, there is a specific goal to enable the Symbian OS nanokernel to operate as a true real-time operating system capable of hosting the baseband software. The baseband (phone software stack or 'modem') in a mobile phone requires real-time support in order to respond to the timing requirements of the signaling stack.

Typical phone designs host the baseband stack on a real-time operating system. In a phone that also provides sophisticated application-side software including, for example, a full GUI, a second, application-centric operating system is dedicated to providing the application support. The 'dual operating system' design most commonly also implies a 'dual core' two-processor design, in which dedicated baseband and application-side processors host the respective operating systems and, in many cases, also own dedicated peripheral hardware including memory. Adding real-time capability to Symbian OS is intended to enable 'single operating system', and hence 'single core', designs.

The EKA2 architecture is based on a nanokernel which is designed to have sufficient functionality and the real-time properties required to directly host a GSM, wCDMA or CDMA phone stack. Phone stacks are not yet commodity items and most have been written to interface to an existing real-time operating system (RTOS), whether bespoke or off the shelf. The EKA2 nanokernel therefore supports a 'personality' layer mechanism that enables an interface layer to be written between a given RTOS and the software above it, while mapping calls into the interface to the underlying functionality of the nanokernel. This provides a way of running a baseband stack directly on Symbian OS without having to first port the stack. Writing a personality layer is a small task compared to that of porting an existing phone stack or writing one from scratch to the Symbian OS nanokernel interface.[3]

The kernel re-architecture had a secondary goal of improving the modularity of the kernel. The EKA1 architecture includes an undesirable

---

[2] See [Sales 2005, Chapter 17] for a detailed discussion of what real-time means and how Symbian OS meets real-time requirements.

[3] The task of creating a new phone stack, from design to full type approval, can take 10s to 100s of man-years of coding effort.

degree of hardware dependency. Although at the lowest level of the kernel an EKA1 'variant' DLL encapsulates many of the device-specific hardware dependencies, many ASSP-specific assumptions are contained in generic EKA1 kernel code, meaning that customization of kernel code is still required to move to a new ASSP architecture (or, at the very least, the kernel needs to be recompiled).

In the EKA2 architecture, all peripheral-related code (i.e. code which is ASSP-specific, but not specific to a particular licensee product) moved out of the kernel into a separate ASSP DLL. This provided better isolation of the kernel from the porting effort and enabled a more flexible approach to porting (since a licensee can choose how to structure the port between the Variant and ASSP DLLs, to better support families of similar but not identical devices; indeed, a licensee can even choose to dispense with the ASSP DLL for a one-off port).

Architecture

The project that led to the creation of EKA2, the 'real-time' kernel, had a number of goals:

- to enable the creation of single-core, single-operating-system products in which baseband software, for example a GSM protocol stack, executed on the same processor as the application software, supported by the same operating system

- to improve average overall performance, as well as portability and robustness

- better timer resolution, easier debugging, a better emulator (i.e. a more faithful 'virtual' port to Microsoft Windows), and general architectural housekeeping.

EKA2 meets all of those goals. In particular, it is highly portable, running on X86 as well as many flavors of ARM processor architectures (ARM720/920/SA1/Xscale); on systems with different Memory Management Unit (MMU) styles, including no MMU;[4] and on multiple ASSPs. Its architecture (see Figure 11.6) is highly modular and carefully layered to isolate hardware dependencies.

At the heart of the design, the nanokernel implements essential operating primitives and supports real-time guarantees for interrupt latencies, thread latencies and context-switching time bounds. The nanokernel is responsible for the most basic thread scheduling, synchronization and

---

[4] Realistically, the 'no MMU' option is intended as an aid to porting rather than a supported target architecture. The security model, for example, depends on an MMU being present to enforce memory protection between processes.

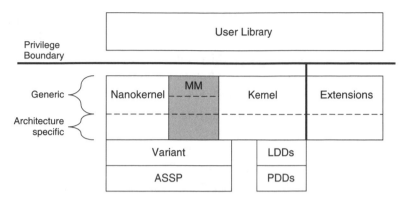

Figure 11.6 Kernel architecture for EKA2

timing functions. Extending the nanokernel, the kernel proper provides higher-level operating-system services compatible with the EKA1 kernel.

Both the nanokernel and the kernel are isolated from hardware dependencies by the Memory Model, ASSP, and Variant modules. The Memory Model provides per-process address spaces and inter-process data transfer. The Variant represents the specific, 'off-chip' system hardware, while the ASSP represents the core silicon package.

The device-driver model and extension mechanism control peripheral devices and provide client interfaces. (Extensions are statically linked device drivers.)

Kernel Responsibilities

The Symbian OS Kernel implements the operating-system primitives on top of which generic services are built by higher-level components (such as the File Server and the User Library). In particular, the kernel, including its extension mechanisms, implements:

- the thread and process models that provide the underlying basis for all code execution, including process creation and termination, code loading, thread scheduling and the scheduling policy

- memory management, including Direct Memory Access (DMA), which is an essential service underlying the process model

- process protection and IPC mechanisms that guarantee process independence while allowing processes to cooperate; IPC is at the heart of the client–server model and policing of the platform security model

- the device-driver model, which provides the device-level interface for system clients and applications

- interrupt management, which is exclusively the responsibility of the kernel

- the power model, which provides an interface for higher-level clients to manage and respond to the device power state

- logical device drivers (LDD), which are implemented as plug-ins to the device driver framework and provide a high-level device interface (i.e. LDDs support classes of device, e.g. Ethernet ports)

- physical device drivers (PDD), which are implemented as plug-ins to LDDs and provide the low-level interface to actual hardware present on a device (for example, a specific Ethernet card)

- various other high-level drivers (for example, accelerator plug-ins to the Media Device Framework), which are not implemented as LDDs, operate at an equivalent level of abstraction.

Kernel Executive Calls

Executive calls are the mechanism used to call into the kernel from user-side programs. While the interface is defined by the User Library, the underlying mechanism is a kernel-side software interrupt dispatch table that decodes software interrupts generated by invocation of the CPU software interrupt instruction into a specific executive call. From the user side, the mechanism is entirely wrapped by methods of the User Library static classes.

Inter-Process Communication

IPC is supported by an asynchronous message-passing mechanism, based on Executive calls, which is the basis for client–server communications, as well as inter-thread communications used in system-level programming (for example, device-driver programming). EKA2 also introduces IPC based on message queues, which is distinct from client–server IPC, and enables shared chunks.

Publish and Subscribe

EKA2 introduces Publish and Subscribe, a mechanism for defining global properties whose values may then be 'published', that is, updated by the property owner, to 'subscribers' who have a dependency on the value. Publish and Subscribe is, in effect, a system-wide asynchronous notification mechanism to which interested clients can subscribe to track the changing values of arbitrary properties. Broadly speaking, it is an asynchronous IPC mechanism although, more precisely, it is really an inter-thread mechanism, since both publishers and subscribers are threads.

Properties are single data values, that is 32-bit integers or 'string' values (strictly speaking, descriptors that contain byte or text data, including

Unicode text) of up to 512 bytes in size. Larger property arrays of up to 64 KB can also be defined but do not have the same deterministic operation.

Real-Time Processing

In EKA1, the kernel is single-threaded. In EKA2, the kernel is multithreaded and all threads are pre-emptible. Interrupt latencies and process switching are time-bounded (whereas they are potentially unbounded in EKA1).

Memory Model

In EKA2, all MMU-related code is moved into a separate module (the 'memory model'), which is linked against the kernel at build time. (EKA2 also uses a different memory-map base address.) The kernel itself is thus MMU-agnostic. The following memory models are supported:

- the moving model (ARMv4 and ARMv5 architecture) is functionally equivalent to EKA1; it uses a single memory-page directory within which entries are moved when address spaces are switched

- the multiple model (ARMv6) uses per-process memory-page directories

- the single model has no address-space paging but uses a single address space (it is used with CPUs without an MMU or to simplify the early stages of porting)

- the emulator model for memory management on a PC.

Device-Driver Model

In EKA2, the device-driver model is made more flexible to allow multiple user-side client device-driver requests to be handled by a single kernel thread, which in effect serializes access and therefore simplifies device-driver programming.

Device drivers are DLLs that allow code running in Symbian OS to communicate with hardware in the variant or kernel extensions. Device-driver DLLs are loaded into the kernel process by explicit load commands from the user side. (In contrast, kernel extensions are automatically loaded during the kernel boot process.)

User-side code accesses a device driver through a specific API provided by the kernel, which provides functions to open a channel to a device driver and to make requests. These functions are protected and the device-driver author provides a derived class to implement functions that are specific to the device driver.

Device-driver DLLs come in two types – logical (LDD) and physical (PDD) device drivers. Logical device drivers provide an abstracted representation of hardware and typically support functionality common to a class of hardware devices. Physical device drivers support specific devices. Thus, for example, there is a single serial communications LDD (ECOMM) that supports all UARTs, providing buffering and flow-control functions. On a particular hardware platform, different physical UARTS (for example, RS232 and infrared) are supported by device-specific PDDs.

Porting

Modularity is improved in EKA2 and more flexible porting strategies are supported. Changes to the ROM building tools also enable greater ROM build-time (rather than compile-time) customization.

Emulator

The EKA2 Emulator is written as a true port of Symbian OS to a (virtual) hardware platform and, in that respect, is like any other variant port. This is unlike the EKA1 emulator implementation, which mapped native Symbian OS system services to their best equivalents on the Microsoft Windows host (i.e. trying to make the Symbian OS kernel API work on Microsoft Windows, so that the underlying services, including the scheduler, are all Microsoft Windows services).

As a result, the EKA2 emulator is a much more faithful representation of Symbian OS although, since Microsoft Windows forces the emulator executable to be a single process, the emulator must use Microsoft Windows threads to emulate Symbian OS processes. A deliberate goal of these changes is to make it easier to implement an emulator on platforms other than Microsoft Windows.

Power Management

EKA2 introduces a new power management framework, which is intended to improve flexibility by supporting a wider range of hardware and by separating policy from mechanism. The new framework is based on the concept of power domains. (A domain is a set of processes that share the same power management characteristics.)

A user-side domain server provides a single point of interaction with the kernel. Policy (the definition of power states) is implemented in a customizable DLL.

On the kernel side, a power manager is embedded in the core kernel, which implements the power-management executive calls. An ASSP-specific power controller is implemented in a kernel extension and manages the different power states and sleep modes supported by the ASSP.

SDIO Support

The MMC/SD bus controller is extended in EKA2 to support SDIO cards (SD cards that provide an interface to a hardware device as well as memory). For example, an SDIO card could provide access to a camera and to some memory.

Unicode Support

Unicode strings are not directly supported within the EKA2 kernel, and therefore all kernel objects (processes, threads, and so on) have ASCII names, implying that user-side code should use only the ASCII subset of Unicode when creating such objects.

User Library

In EKA2, the User Library is available only on the user side and a kernel-specific utility library is used on the kernel side. In EKA1, both user-side and kernel-side code were linked against the User Library DLL.

Summary of Major Kernel Differences

- EKA2 offers real-time support.

- EKA2 supports alternative memory models.

- Device-driver implementation in EKA2 is simplified by supporting an alternative, serialization approach to handling multiple user-side requests in a single kernel thread (in the case where multiple device drivers share a kernel thread, which is optional).

- EKA2 includes improved modularity and greater flexibility for porting, a new power-management model, emulator improvements, and support for SDIO cards.

- EKA2 does not support Unicode strings inside the kernel and the User Library is available only on the user side. (The kernel implements its own User Library subset.)

Overall, there are some small system-wide impacts from these changes:

- EKA2 threads use more RAM than EKA1 threads (4 KB to support a per-thread kernel-mode stack)

- In EKA2, memory chunks are limited to 16 per process (for the Moving Memory model, in order to support deterministic operation although context-switching times remain nondeterministic). For the Multiple

Memory model, there is no chunk limit and, in addition, it provides fast, deterministic context switching.

Hardware Interface, Base Porting and Reference Hardware

Kernel extensions provide the interface to the platform hardware (the device-driver model provides the interface to specific devices). The modular, extension-based architecture is designed to allow for flexible customization when moving the operating system to new hardware. In particular, it is designed to enable generic platform dependencies encapsulated by the Application-Specific Integrated Circuit (ASIC) or ASSP, for example, processor type, MMU architecture and standard peripherals such as DSPs and LCD drivers, to be isolated from the device-specific hardware such as the flash-memory interface.

Platform dependencies are typically encapsulated as ASSP dependencies and device-specific dependencies as 'Variant' dependencies.

- An ASSP extension module implements hardware-dependent support for a given ASSP.

- A device-specific 'variant' extension module implements other device-specific hardware support, for example, for peripherals that are not standard to the ASSP.

- Other standard extensions include the power framework and the peripheral bus and USB controllers.

To provide a reference point for porting work, each release of Symbian OS is built and warranted against the hardware reference platform. The hardware reference platform for Symbian OS v8 releases was the Intel Lubbock development board. For Symbian OS v9, the hardware reference platform is based on the Texas Instruments OMAP development boards.

For Symbian OS v9.0 and v9.1, the hardware reference platform is the H2 development board with OMAP 1623 (ARMV5-based core):

- other versions of the H2 boards are not officially supported

- includes a DSP, SD/MMC/SDIO card, USB, camera, display (240×320, rotatable, 8 or 16 bits per pixel), NAND flash

- operating system installation from MMC or serial loads either into RAM or into NOR Flash on the board

- JTAG (IEEE 1149.1) is also supported.

From Symbian OS v9.2, the hardware reference platform is the H4 development board with OMAP 2420 (ARMv6-based core):

- adds USB for bootable image source to H2

- introduces some NAND-flash differences compared to the H2 board.

11.8   Kernel Architecture Component Collections

The Kernel Architecture block in the system model contains four separate collections (see Figure 11.7): Kernel Services, Logical Device Drivers, ASSP, and Variant.

Kernel Services Collection

The EKA2 component consists of the nanokernel, which is the real-time kernel core, and the operating system kernel that builds the basic threading, process and memory models on top of it.

Table 11.3 Kernel Services Components

Component Name	Development Name
Kernel Architecture 2	E32_EKA2

Logical Device Drivers Collection

Logical (see Figure 11.9) device drivers (LDDs) are plug-ins to the kernel device-driver framework that provide the logical abstraction of hardware

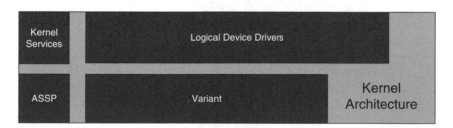

Figure 11.7   Kernel Architecture collections

Figure 11.8   Kernel Services components

Figure 11.9

Table 11.4 Logical Device Drivers

Component Name	Development Name
Ethernet Driver	ETHERDRV
USB Driver	USBC
Audio Driver	SOUNDDEV
MIDI Driver	DEVMIDI
Speech Driver	DEVASR
Video Driver	DEVVIDEO
Other LDDs	
Media Drivers	MEDUSII, MEDUSII_CRASHLOG, MEDUSIIS
SD Card Driver	SDCARD4C
Peripheral Bus Controllers	EPBUS

devices, and accept the physical device driver (PDD) plug-ins, which communicate with real hardware.

Symbian OS supplies specific Ethernet and USB drivers, as well as hardware accelerator plug-ins used by the Media Device Framework, which form part of the hardware abstraction for multimedia devices.

- The Ethernet Driver is a logical device-driver implementation for Ethernet cards, including the emulator.

- The USB Driver is a logical device driver for USB. The standard USB software architecture on Symbian OS supports dynamically configurable USB 2.0 device functionality.

- The Audio, MIDI, Speech and Video accelerator API plug-ins to the Multimedia Device Framework (MDF), the lowest-level framework supporting multimedia services, are used by MDF controllers. They all include hardware- or kernel-dependent components.

- o DevVideo is the hardware-abstraction layer for video decoding and encoding acceleration enabling playing and recording of video; it includes a client API that enables policy management (e.g. request contention and file-type matching).

- o DevMIDI is the API that supports hardware-accelerated MIDI engines.

- o DevSound is the hardware-abstraction layer for digital audio acceleration enabling Playing, Recording, Conversion and Tone generation of sounds; it includes a client API that enables policy management (e.g. request contention and file-type matching).

- o DevASR is the hardware-acceleration API for Automatic Speech Recognition, allowing the computationally intensive speech-recognition algorithms to be performed in hardware, where present.

- The peripheral bus controllers for supported variants are implemented as kernel-side DLLs that interface media and I/O device drivers to PC-card or MMC-card socket hardware.

ASSP Collection

Hardware dependencies divide between ASSP dependencies, based on properties of the CPU core and the peripherals packaged with it in the same silicon chip, and additional 'off-chip' hardware peripherals.

Table 11.5 ASSP Components

Component Name	Development Name
ASSP	OMAP 1623

Isolating the ASSP-dependent code allows it to be reused on multiple systems that use the same ASSP. OMAP 1623 is the ASSP supported in the Symbian OS v9 hardware reference platform, as used in the Texas Instruments H2 development board. (Other H2 ASSPs are not supported.) The ASSP module contains source code tailored to a range of different microprocessors (e.g. ARM720/920/SA1100/Xscale). See Figures 11.9 and 11.10.

Figure 11.10   ASSP components

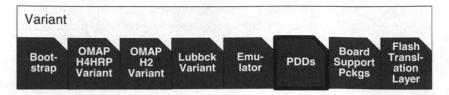

Figure 11.11   Variant components

Table 11.6 Variant Components

Component Name	Development Name
Bootstrap	`BOOTSTRAP`
Emulator	`WINS_VARIANT_EKA2`
Lubbock Variant	`LUBBOCK_EKA2`
OMAP H2	`OMAP_H2`
OMAP H4	`OMAPH4HRP`
PDDs	
Board Support Packages	
Flash Translation Layer	`UNISTORE2_DRIVERS`

Variant Collection

The Variant collection contains components which are associated with off-chip hardware, that is, which are independent of the ASSP.

- The Bootstrap component prepares hardware including memory and peripherals, maps virtual address space if an MMU is present, and starts the kernel. It includes processor-, MMU- and other device-dependent code, as well as a generic layer.

- The emulator component is treated as a hardware target variant. The Emulator runs on Microsoft Windows platforms and maps Symbian OS services and logical devices to Microsoft Windows APIs and local hardware. Single-process implementation uses Microsoft Windows threads to emulate Symbian OS processes. It is valuable for high-level programming, but the implementation creates practical issues for low-level and device-dependent programming compared to hardware targets.

- The Lubbock Variant component is Variant code for the Symbian OS v8 hardware reference platform.

- The OMAP H2 component is Variant code for the Symbian OS v9.1 hardware reference platform. (From Symbian OS v9.2 the reference hardware platform moves up to the OMAP H4 board.)

- The PDDs are physical device drivers, low-level plug-ins used by LDDs providing the device-dependent level of the two-tier device-driver model.

- The kernel requires additional loading and media translation support when running on NAND-flash devices. A special boot loader is required to move the kernel into RAM so that it can execute (since NAND flash is not byte-addressable), together with a translation layer that manages the flash device and presents a standard byte-readable and -writable interface to the file system. The Flash Translation Layer component is the original file system plug-in implementation of flash-driver support. Media drivers provide a newer implementation (implemented as conventional Symbian OS device drivers), supporting more NAND formats: small, large and OneNAND.

12

The Java ME Subsystem

12.1 Introduction

Symbian OS offers licensees an optional Java implementation. Since Symbian OS v7.0, this has been based on the Java 2 Platform, Micro Edition (Java ME) MIDP and CLDC specifications, which have become the standard for Java on mobile phones and other communicator-style devices.

Java ME is subdivided into configurations, profiles and optional packages. A Java ME configuration provides a basic[1] Java platform for a large class of constrained devices by defining a Java Runtime Execution (JRE) environment consisting of a Java language subset, a Java Virtual Machine (JVM) and a base set of necessary class libraries. It is commonly based on a subset of the J2SE APIs.

Currently, there are two Java ME configurations:

- Connected Limited Device Configuration (CLDC)

- Connected Device Configuration (CDC).

A Java ME profile sits on top of a configuration to complete the JRE by adding more high-level APIs, thereby preparing a device for a specific device category. Currently, there are two common Java ME profiles:

- Mobile Information Device Profile 1.0 (MIDP1)

- Mobile Information Device Profile 2.0 (MIDP2).

---

[1] See the CLDC specification at ***http://jcp.org/aboutJava/communityprocess/final/jsr139.***

Optional packages add functionality to the Java ME platform by offering standard APIs for various technologies such as advanced multimedia, 3D graphics, file system access, web services and much more.

A widely adopted standard is the combination of MIDP and CLDC to provide a complete Java platform for mobile phones and similar devices. MIDP applications are known as MIDlets. With MIDP 2.0 the specification becomes rich enough to support a wide variety of sophisticated applications.

12.2   Requirements of the Java ME Subsystem

Java is an important application environment for mobile phones, supporting a multiplatform third-party market for downloadable and installable programs (including games, utilities and media players), a standard environment for enterprise developers seeking to extend information systems to mobile phone users within organizations and a standard platform for mobile phone services, as well as a branding and personalization mechanism (through custom applications) for network operators and others in the phone retail chain.

Platform independence is an important part of Java's philosophy that applies as much to the MIDP context as to desktop Java implementations. The Symbian OS implementation of Java ME provides a standard environment for installing and running MIDlets, with access to the underlying operating system services through the supported MIDP 2.0 APIs, which include a number of optional APIs – for example, Mobile Media, 3D Graphics and File GCF – as well as the Java Technology for the Wireless Industry (JTWI) standard, which aims to standardize MIDP support for mobile phones, and the Unified Emulator Interface (UEI), a development tools standardization initiative.

12.3   Design Goals for the Java ME Subsystem

Symbian OS provides a rich and powerful host for the Java ME implementation, but also poses some great challenges. These are a result of the architectural differences between Symbian OS and the Java platform and the differences between Symbian OS components and the MIDP APIs which are built on top of the Symbian APIs.

- There are some specific mismatches between Symbian OS and Java models; in particular their threading models are incompatible.

- Symbian OS has its own native application model (for C++ applications), with which the MIDP lifecycle must be integrated if MIDlets are to have a seamless, native application lifecycle.

- Symbian OS has a rich set of native controls, localization abstractions, custom dialogs, input mechanisms, and so on, all of which are expected to be customized for look and feel by the providers of the variant user interface with which the operating system is integrated on any particular Symbian OS device. Therefore, any look-and-feel dependent aspects of the MIDP implementation must be customized for the different variant user interfaces.

- MIDlets should, as far as possible, look and feel like native applications. For Symbian OS, this poses a particular challenge since application look and feel ultimately depends on the variant user interface which is running on a given Symbian OS device.

- Symbian OS component APIs have a different design from MIDP APIs, which requires an elaborate internal architecture to bind the functionality between those two orthogonal class hierarchies.

- The Java ME subsystem as a whole can be swapped out and replaced by another. Symbian's licensees are also at liberty to pick and choose which of the JSRs they use and which they do not.

As well as overcoming these complications, the Java ME design must meet some basic design goals:

- Support near-native behavior of MIDlets to enable seamless switching between MIDlets and native applications

- Provide the fullest possible implementation (MIDP specifies both required and optional features) to ensure that the Java ME implementation on Symbian OS is highly competitive

- Enable a 'common platform' by providing a solid core system on all Symbian smartphones

- Provide customizability as an essential part of Symbian OS. Licensees may choose to vary the degree to which they support Java on a particular device by extending or limiting the default Symbian Java ME implementation, including removing or replacing it entirely.

12.4   Evolution of Java on Symbian OS

Symbian's Java support has evolved with Java itself from the earliest days of JDK 1.1.4, which had an AWT-based user interface implementation, to the PersonalJava and JavaPhone implementations of Symbian OS v6. The first MIDP 1.0 implementation on Symbian OS v7.0 (subsequently backported to Symbian OS v6.1) was followed by the arrival of MIDP 2.0 in Symbian OS v7.0s.

Through Symbian OS v8 and v9, Java delivery evolved to a rich Java ME platform supporting a significant number of optional MIDP APIs and enhanced by the latest version of Sun's CLDC 1.1 VM, which uses HotSpot technology.

Although recently the emphasis has shifted towards Symbian's licensees as the ultimate 'owners' of their own Java ME platforms, with Symbian focusing on the core of its delivery, Java has strengthened its position as an important element in any enterprise strategy, where rapid development and rollout of bespoke local applications is an important driver in the choice of devices. It is also an important enabler for rapid development and easy deployment of small but high-value applications, including games and custom applications and services from the mobile phone manufacturer and network operator, all of which help to consolidate the position of Symbian OS in the market.

Java provides a powerful model for acquiring mass business market share and building a technical platform. One of the important, original underlying goals for Symbian OS was to enable Java as a true platform for developing Symbian OS applications, not simply as a lightweight platform for running relatively trivial generic applications.

Symbian (or Psion as it then was) was quick to recognize the significance of Java. The success of its own OPL language[2] and the emergence of a strong third-party developer community was an important element in the success of Psion's Series 3 products. The follow-on 32-bit system was conceived from the beginning as an open development platform for third-party application developers. At the same time, the complexity of its native C++ development environment and its unsuitability for some kinds of application development was recognized early, as was the need for a rapid development alternative. OPL support was present from the beginning, following which a Visual Basic porting project was started. However, attention moved to Java as a more powerful solution for third-party developers.

Sun's first release of Java appeared in early 1995 and the more stable JDK 1.0.2 release appeared in 1996. In early 1997, a few months before the first release of what was still known as EPOC32, Psion started its Java port based on JDK 1.1.2. 'Hello World' first ran in July 1997 and the first graphical application ran in October of that year. By the summer of 1998, graphical applications were running on an upgraded version of the port, based on JDK 1.1.4, and in August 1998 certification was granted by Sun. The first Java run-time system for EPOC32 was released in May 1999.[3] Demonstrations of Java applications running on early Symbian

---

[2] OPL lives on as a rapid development language for Symbian OS, having been released under an open source license by Symbian in 2003. It can be found on the web at **www.allaboutopl.com/wiki/OPLWikiHome**. See also [Spence 2005].

[3] The full history of these early implementations can be found in earlier Symbian programming books, including [Tasker 2000] and [Allin et al. 2001].

OS smartphones caused quite a stir at the Symbian developer conference in June 1999.[4]

By that time, Psion had become Symbian and EPOC was evolving towards the first release of Symbian OS. However, porting Java had not been particularly easy. For one thing, Sun had at first insisted on a binary-only license for the VM. Since the VM assumed an ANSI C/POSIX platform, a basic C Standard Library implementation was needed to interface it to Symbian's native C++ APIs. Fundamental differences in the thread model between Java and Symbian OS posed further problems. And at that time, since the port was still based on 'full-sized' Java, graphics were AWT-based. On a small platform, it was simply not possible to extract acceptable performance from AWT.

Even before its first Java release for EPOC had shipped, Symbian had started work, early in 1999, on a port of PersonalJava, a newly defined, scaled-down Java specification targeting smaller devices. In July 2000, PersonalJava together with an implementation of the JavaPhone API was released as part of Symbian OS v6, appearing for the first time in a phone product later that year in the Nokia 9210 Communicator. JavaPhone opened key Symbian OS APIs to Java applications, giving basic access to the underlying address book, communications services and telephony APIs.

PersonalJava was a first attempt to define a Java environment suitable for constrained devices such as PDAs and mobile phones, and a forerunner of what eventually became Java ME. By the time of the first Symbian OS v7 release, the Java ME specification had been defined in terms of MIDP/CLDC. MIDP 1.0 was included for the first time in Symbian OS v7 and subsequently back-ported to earlier releases, appearing in the Nokia 7650, based on Symbian OS v6.1. However, the PersonalJava and JavaPhone combination was still offered as an option to licensees.

Symbian OS has tracked the evolution of the MIDP specification (indeed, as a member of the MIDP Expert Group, it has played an active role in shaping it) through subsequent releases of the operating system, supporting MIDP 2.0 for the first time in Symbian OS v7.0s, and extending its support in Symbian OS v8 and v9 with the addition of important optional packages, as well as improving the compatibility, interoperability and completeness of Java ME technology by providing support for the JTWI standard and increasing developer productivity by supporting on-device debugging and standard integration with Java IDEs through the Unified Emulator Interface (UEI).

MIDP 2.0 is a significant enhancement of the original MIDP specification. In particular, it supports 'push' applications, in other words, MIDP applications that are launched in response to 'external' events (i.e. events

---

[4] The Java team demonstrated a Rubik Cube application, running in color on a Psion netBook, to an enthusiastic audience.

originating outside the application's own process), for example, alarms, incoming messages or other network events.

At the same time, the CLDC specification that defines the execution environment has evolved (again with Symbian participating as a CLDC Expert Group member). In practical terms, the most significant change is the move in Symbian OS v7.0s from the KVM to CLDC-HI 1.0, with HotSpot VM technology, and in Symbian OS v8 to CLDC-HI 1.1. The current CLDC1.1 configuration is likely to be the final configuration from Symbian, although licensees may continue to evolve their own extensions and track future evolution of the MIDP profile.

12.5   Architecture

In the Symbian OS system model (see Figure 12.1), Java ME is shown as a self-contained block spanning the UI Framework and Application Services layers, which emphasizes the external view of Java ME as an application platform.

The system model idealizes the representation of Java ME. As conventionally represented (see Figure 12.2), MIDP/CLDC forms a software stack that provides the execution environment for MIDlets; the CLDC configuration consists of the VM itself and the basic Java languages libraries; the MIDP Profile defines the frameworks for application support expected from the device class and the various packages (both required and optional) providing the application APIs.

Figure 12.1   Java ME in the system model

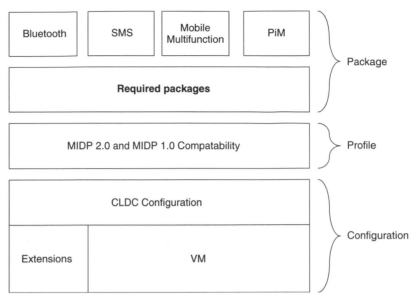

Figure 12.2   High-level Java ME architecture

From an internal architectural perspective, however, the Java ME implementation hooks deeply into the supporting layers of the operating system. Complex system interactions are required between the two, and indeed requirements originating from the successive Java implementations have had significant impact on the evolution of some fundamental features of the operating system, including the native Symbian OS application model, inter-process communication (IPC), and the thread and process models. In particular, the different threading models of Java and Symbian OS posed particularly thorny issues for the early Java implementations, which in turn influenced later design decisions.

At a high level, the architecture (see Figure 12.3) can be broken down as follows:

- the SystemAMS server, which is responsible for the lifecycle of MIDlets and VM processes, and static and dynamic resource management for Java applications
- the SystemAMS plug-ins for licensee customizability (i.e. customized security policy)
- the SystemAMS extension plug-ins that extend internal AMS frameworks
- the VM executable, which includes the VM, MIDP APIs and frameworks
- the VM plug-ins for licensee customizability (i.e. graphics customization)

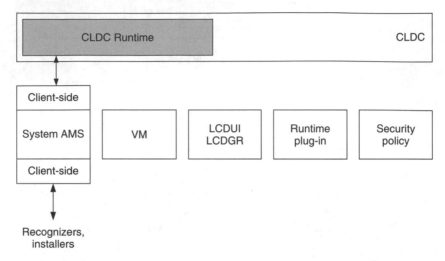

Figure 12.3   Internal architectural view of Java ME on Symbian OS

Java ME Applications Management Software

The System Application Management Software (SystemAMS)) operates as a managing agent between Symbian OS and the Java ME run-time (see Figure 12.4) for all stages of the MIDlet lifecycle from installation, launching and stopping MIDlets, controlling execution and launching VM processes as required. SystemAMS also manages static and dynamic push connections and alarms registered by MIDlets, which is the basis for the MIDP 2.0 'push' support.

SystemAMS is implemented as a server which is run from device boot time. From an application perspective, the MIDlet can initiate some state changes itself and notify the MIDP run time, which eventually notifies SystemAMS of those state changes by invoking the appropriate methods.

From a system perspective, SystemAMS provides client-side interfaces used by the VM and the Java installer to support MIDlet recognition, installation and launch; manages resources such as registered push connections and Symbian OS alarm notifications; and is responsible for managing the MIDP policy-security model.

The CLDC Configuration Layer

An important feature of Symbian's CLDC implementation is its support for various VMs. For example, in earlier versions of Symbian OS, the KVM was used and later replaced by the CLDC 1.0 VM which was eventually replaced by the CLDC 1.1 VM. As VMs may change, an abstraction layer is required between the VM and the various CLDC and MIDP

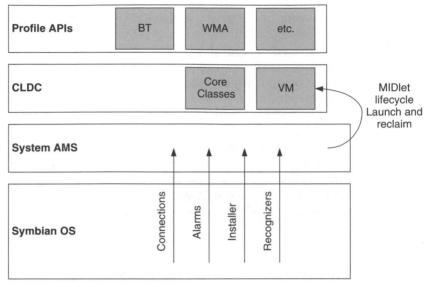

Figure 12.4 SystemAMS Architecture

libraries to avoid dependencies between them and any particular VM. The abstraction layer also interfaces Java event-handling with the native event-handling model.

CLDC-HI is designed as a high-performance JVM and run-time environment for resource-constrained small devices, in particular mobile phones and communicator-style devices, and is optimized for performance, footprint, and efficient resource management, with a specialized Just In Time (JIT) compiler for the ARM processor architecture. Sun claims an order of magnitude improvement over the performance of the older KVM, for example.

The MIDP Profile Layer

LCDUI

MIDP is specifically targeted at small, mobile devices. It includes the LCDUI specification, which is optimized for this device class. LCDUI defines both the UI event model and the standard GUI widgets available to MIDlets (lists, forms, textboxes, and so on). LCDUI therefore requires integration both with the native UI Framework and Application Architecture.

In order to make MIDlets (as far as possible) indistinguishable from native applications, LCDUI uses the native widget set as peers for the Java widgets. A MIDlet runs as a single native application owning its own window group, listed in and launchable from the task list (if the user interface variant has a task list) and integrated with the save notifier framework, power events, and foreground/background notification. Input methods are also inherited from the native widget set so that all native functionality is available to MIDlets, for example, mechanisms such as T9, handwriting recognition and non-keyboard character set input (e.g. Chinese), which are all based on the Front End Processor (FEP) framework. This also harnesses the native locale support, for example for bi-directional text and capitalization.

The LCDUI implementation consists of a framework that implements the core user interface functionality and provides the high-level interface between Java ME LCDUI APIs and the concrete user interface platform implementation areas that are implemented in separate graphics plug-ins (which licensees customize to provide integration with the graphics system of their specific user interface platform).

GCF

The MIDP Generic Connection Framework (GCF) provides the generic mechanism for creating a connection from a URI and enables a wide variety of connections including networking connections such as HTTP and HTTPS, sockets and server sockets, secure sockets and datagrams, as well as support for 'push' connections and on-device mechanisms for local file and directory access (which is known as 'File GCF').

The MIDP GCF design maps the Java class interfaces to the underlying Symbian OS communications models and provides core communications functionality for MIDlets including:

- opening, closing, and disposing of connections

- opening Java streams for appropriate types

- a server connection pattern for types capable of receiving incoming connections.

Symbian's Java ME implementation enables all the relevant MIDP 2.0 GCF protocols and its framework is intended to be used by extensions that provide support for future protocols. In particular, it supports push connections, using the SystemAMS dynamic and static resource management for managing the registered connections.

Mobile Multimedia

Mobile Multimedia implements access to the multimedia support provided in the underlying Symbian OS, enabling MIDlets to play and record audio and video data from a wide range of inputs using a range of possible mechanisms, including streaming. The design follows a framework-plug-in architecture:

- the framework provides the high-level interface to MIDP Multimedia functionality

- the reference DLL contains all dependencies on the underlying native multimedia services and can be customized.

PIM and RMS

PIM support is provided for accessing native Symbian OS contacts (i.e. phone book or address book) and agenda (i.e. calendar) entries, including Event and ToDo classes.

Record Management System (RMS) support, which enables MIDlets to store persistent data, is implemented over the native Symbian OS DBMS APIs, using the DBMS in client–server mode and thus enabling database-like functionality including transaction integrity and sharing between multiple clients for Java applications.

Security

SystemAMS and the MIDP run-time are responsible for supporting the MIDP security model, through static (at installation time) and dynamic (at run time) checking of permissions, which provides for trusted and untrusted MIDlets, and for protection based on security domains. Licensees implement specific security policies by customizing the MIDP security plug-ins.

12.6   Component Collections

The system model divides the Java ME block into a number of separate collections (see Figure 12.5), broadly layered to reflect the conventional layering of the Java ME software stack.

The foundation is provided by the core Java class implementations and the CLDC-HI VM, together with the low-level plug-ins that integrate the MIDP frameworks with Symbian OS. The MIDP profile and packages collections are layered over this foundation.

MIDP 2.0 Profile Collection

This collection (see Figure 12.6) implements the Java ME MIDP 2.0 Profile.

- The MIDP MIDlet component implements the MIDlet lifecycle, which defines how MIDlets are started, paused and destroyed and how they interact with the host environment.

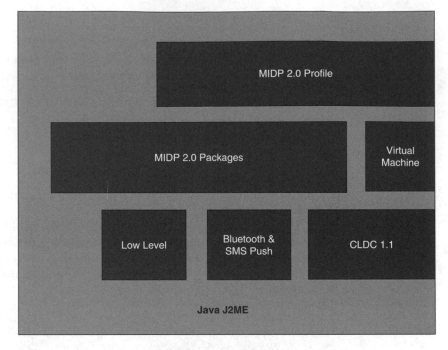

Figure 12.5   Java ME Block

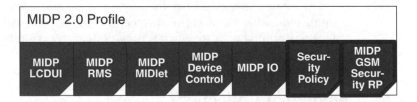

Figure 12.6   MIDP 2.0 Profile components

- The LCDUI component is specifically designed with small LCD screens in mind and provides compact, device-independent controls that can respond to user input ranging from keyboards to phone keys to touch screens. MIDP graphics APIs are implemented in terms of generic native controls which acquire platform-specific look and feel through the UI Application Framework implementation, which is customized by the UI variant.

- The RMS component provides MIDP persistence APIs. RMS is implemented internally over Symbian OS native DBMS, using the DBMS in client–server mode.

- The I/O component provides MIDP high-level input–output APIs, including networking support and HTTP connections.

Table 12.1 MIDP 2.0 Profile Components

Component Name	Development Name
MIDP MIDlet	MIDP2
MIDP LCDUI	JAVAX.MICROEDITION.LCDUI
MIDP RMS	JAVAX.MICROEDITION.RMS
MIDP IO	JAVAX.MICROEDITION.IO
MIDP Device Control	MIDP2
Security Policy Reference Plug-in	MIDP2SECURITY
MIDP GSM Security Recommended Policy	MIDP2SECURITYRP

- The Device Control component provides an interface for implementations of MIDP device control APIs, for example controlling device vibration.

- The Security Policy Reference Plug-in provides a reference implementation of Java security policy, implemented as a replaceable plug-in.

- The GSM Security Recommended Policy component adds specific protection domains to the security model (for example 'manufacturer', 'operator', 'third-party' and 'untrusted'). Licensees should customize and provide their own concrete implementation plug-in to be used by the framework.

MIDP 2.0 Packages Collection

The Java ME MIDP 2.0 packages components (see Figure 12.7) extend the MIDP 2.0 Profile implementation with additional APIs.

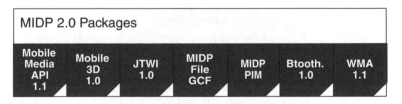

Figure 12.7 MIDP 2.0 Packages components

Table 12.2 MIDP 2.0 Packages Components

Component Name	Development Name
Mobile Media API 1.1	MMAPI11
Mobile 3D 1.0	M3GIO
JTWI 1.0	Java ME9.12
MIDP File GCF	GCF
MIDP PIM	MIDP2
Bluetooth 1.0	BLUETOOTH
WMA 1.1	WMA

- The Mobile Media API 1.1 component comprises Mobile Media APIs (JSR-135).

- The Mobile 3D 1.0 component comprises 3D-graphics APIs for scalable, small-footprint devices (JSR-184).

- The JTWI component implements the JTWI specification, which improves the compatibility, interoperability and completeness of Java ME technology implementations in mobile phones by reducing API fragmentation and raising the bar of functionality to specify a common set of APIs and standards such as MIDP 2.0 and including optional APIs (WMA 1.1 and MMAPI 1.1).

- The MIDP PIM and File GCF components support MIDP personal-information-management (PIM) and file-connection APIs (JSR-075), enabling reading and writing of event, contact, and to-do items, and file system access. File system access is implemented through the GCF communications framework, which generalizes framework support for HTTP, IP and socket-based connections.

- The Bluetooth component implements two optional MIDP2.0 Bluetooth 1.0 (JSR-082) packages: the core Bluetooth API and the Object Exchange (OBEX) API.

- The Wireless Messaging API (WMA) provides platform-independent access to wireless communication resources, enables send and receive of SMS messaging, including SMS push, on GSM, CDMA and other networks supporting asynchronous messaging protocols.

Both Bluetooth 1.0 and WMA 1.1 add 'push' capability to the support for MIDlets, allowing a MIDlet to respond either statically (at install time)

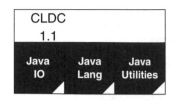

Figure 12.8 CLDC 1.1 components

or dynamically (programmatically) to an incoming WMA or Bluetooth connection (i.e. to a 'message'). Both implementations are integrated into the GCF communications framework.

CLDC 1.1 Collection

This component implements CLDC 1.1 Java class libraries (JSR-118). CLDC 1.1 specifies the core subset of the Java language required to support MIDlets. The language libraries define basic types and objects, including Byte, Integer, Object and Thread; the I/O libraries define the data-stream-based input–output APIs, as well as APIs for reading and writing bytes and basic Java types; the utilities library supplies basic utility classes, including Date and Time, and collection classes including Hashtable, Stack and Vector (see Figure 12.8).

Table 12.3 CLDC 1.1 Components

Component Name	Development Name
Java Lang	JAVA.LANG
Java IO	JAVA.IO
Java Utilities	JAVA.UTIL

Virtual Machine Collection

The CLDC-HI 1.1 component is the Sun CLDC HotSpot Implementation VM, which is part of the CLDC 1.1 specification (JSR-139). The HotSpot

Figure 12.9 Virtual Machine components

Table 12.4 Virtual Machine Components

Component Name	Development Name
CLDC HI 1.1	CLDCHI

VM applies a variety of advanced performance-optimization techniques to deliver a highly competitive execution environment for Java applications. See Figure 12.9.

Low-Level Plug-ins Collection

These plug-ins allow customization of the CLDC run-time framework and bind LCDUI to the underlying graphics system. See Figure 12.10.

Table 12.5 Low-Level Plug-ins

Component Name	Development Name
Runtime Plug-in	MIDP2RUNTIME
LCDUI Plug-in	LCDUIB

- The Runtime plug-in component is the MIDP 2.0 run-time plug-in module. It can be customized by the licensee.

- The LCDUI plug-in component consists of low-level graphics APIs with direct screen access, implemented as a plug-in that is replaced with an alternative implementation.

Bluetooth and SMS Push Collection

These plug-ins bind the Bluetooth 1.0 and WMA 1.1 packages to the underlying system. See Figure 12.11.

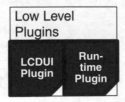

Figure 12.10   Low-Level Plug-ins

Figure 12.11 Bluetooth and SMS Push components

Table 12.6 Bluetooth and SMS Push Components

Component Name	Development Name
Bluetooth 1.0 Push Plug-in	BLUETOOTH
WMA 1.1. Push Plug-in	WMA

13

Notes on the Evolution of Symbian OS

13.1   The State of the Art

Symbian OS reached market for the first time towards the end of 2000, following on from the Psion EPOC32 releases. The last release of EPOC32 was Release 5; the first release of Symbian OS was Symbian OS v6.0.

The most recent release is Symbian OS v9, but Symbian OS v8 remains very much an active platform on which new products are still being developed and brought to market. Phones based on earlier releases still ship in their millions, even though Symbian's licensees have moved up to the latest releases for new products. Indeed until relatively recently, phones based on Symbian OS v6.1 continued to ship in quantity, particularly in Japan. Since then, Japanese licensees have led the way in adopting the new real-time kernel architecture, and have brought to market new 3G phones based on Symbian OS v8.1b and are likely to follow with phones based on Symbian OS v9.

In other markets, licensees are shipping new phones based both on Symbian OS v9 (platform security, new real-time kernel) and v8 (original kernel architecture). Symbian OS v7 (Sony Ericsson P910, Motorola A1000 and FOMA M1000) remains a volume-selling release and phones based on Symbian OS v6.1 (N-Gage QD) are still shipping.

At the time of writing, in late 2006, phones are shipping on all releases from Symbian OS v6.1 to Symbian OS v9.1, although new product pipelines from licensees are based on Symbian OS v8 and v9.

13.2   Summary of Symbian OS v6 Releases

Symbian OS v6 was the immediate result of a major and long-running project, working with Nokia as lead licensee, to re-engineer and re-architect Symbian OS from its EPOC32 baseline (ER5U). EPOC32 did support some phone and messaging functions, for example 'two-box' telephony solutions in which an EPOC-based PDA could use a GSM mobile phone as a dialup modem, as well as driving it directly to send SMS messages and synchronize with the SIM phone book. However, EPOC remained substantially PDA-centric. Even more importantly, its Eikon GUI was not suitable for phones.

Among the most significant changes in Symbian OS v6, therefore, was the refactoring of Uikon to support multiple user interface implementations, so called 'variant UIs', and the more general re-architecting of phone-centric functionality to suit a true phone operating system. The Symbian OS v6 system architecture was based on a component-based release model and representation.

Symbian OS v6.0

Symbian OS v6.0 was the common platform for what were branded as the Crystal and Quartz reference designs, in keeping with Symbian's DFRD strategy.

One Crystal-based product, the Nokia 9210 Communicator, was brought to market. No Quartz devices reached market (although devices from Ericsson and Sanyo were demonstrated, including at 3GSM in Cannes).

The system architecture of Symbian OS v6.0 was based on a components representation inherited from the original Psion EPOC32 binary-component release model.

Symbian OS v6.1

Symbian OS v6.1 (announced in March 2001 at CeBIT) was the original platform for the Nokia S60 UI (which began life as the Pearl DFRD and launched as Series 60) The first Symbian OS v6.1/S60 phone (arguably, the first Symbian OS phone as opposed to PDA–phone hybrid) was the Nokia 7650. Other Symbian OS v6.1/S60 phones were brought to market by Panasonic, Sendo and Siemens.

Symbian OS v6.1 was very much an addition to the Symbian OS v6.0 baseline. No functionality was deprecated (although there were one or two significant reworkings); some functionality was added (around 150 new classes and other types). In almost all cases, changes were both binary and source compatible. Significant changes included:

- UI Framework and UI Toolkit changes and related changes to FEP and Text Formatting

- Application Services changes including new Contacts Model file format, new Chinese calendar and locale support, including Character Encoding Conversion updates, and improved VCard and VCalendar support

- OS Services changes including major Bluetooth revision (full Bluetooth 1.1 compliance), Infrared IrObex over Bluetooth, screen driver split out from BitGDI, 256-color-mode palette support added to Font & Bitmap Server, Free Type enhancements, Multimedia Server streaming added, new WAP Push and messaging functionality, telephony support for GSM/GPRS and SIM Toolkit, and Comms Database improvements.

13.3   Summary of Symbian OS v7 Releases

Symbian OS v7 was significant as the platform for the first UIQ phones. It also provided Symbian with its first real taste of the problems of fragmentation, with the divergence of Symbian OS v7.0 from Symbian OS v6.1 threatening the common platform model for UIQ and S60, subsequently corrected by the Symbian OS v7.0s release.

Symbian OS v7.0 was the platform for the Sony Ericsson P800 family; Symbian OS v7.0s was the platform for Nokia phones beginning with the 6600 and remained the mainstream platform for Nokia phones until the 6630 3G phone was released.

Symbian OS v7.0

Symbian OS v7.0, announced in February 2002 at 3GSM, was the platform for the UIQ UI (an evolution of the Quartz DFRD). The first Symbian OS v7 UIQ phone was the Sony Ericsson P800 (announced in Q1 2002 and released in Q4 2002).

Symbian OS v7.0 was, at a functional level, substantially backwards-compatible with Symbian OS v6.1, however there were numerous compatibility breaks, as well as significant new functionality added and significant restructuring of the UI Framework components to improve the separation between the framework support and the overlaying user interface. The TechView reference user interface components were also introduced (although TechView never became a complete reference user interface implementation).

Symbian OS v 7.0 also significantly reworked the source tree, introducing a subsystem-based release model and representation. Subsequently, Symbian OS v7.0 became the baseline for the architecture representation based on the system model, which has become Symbian's standard architectural representation for releases from Symbian OS v7.0 forwards.

Among the most significant changes from Symbian OS v6.1 were:

- Application support
  - The Time/World application refactored into Alarm Server and World Server

- ○ New Help file format; Agenda and Contact format changes (for vCard and vCal)

- ○ System agent updated to support two-box system.

- ○ Improvements to text handling and text views (formatting) support for user interface

- ○ Microsoft Word and Excel converter support

- UI and Graphics

  - ○ Further refactoring of Uikon to support user interface separation and pluggable Look-And-Feel

  - ○ New standard controls including animation in menus, menu-highlighting options, variable-height list items, customizable text wrapping, line breaks, automatic hyphenation (editable windows), improved error handling, a generic dialog server, flip support, and many other minor enhancements

  - ○ Graphics changes for anti-aliasing, key click and long keypress support, direct screen access and 2D hardware-acceleration support, hardware bitmaps, font name aliases, polygon fill, bitmap scaling, fading

  - ○ Front End Processor Base optimizations

  - ○ New TechView test user interface

- Multimedia

  - ○ Media Server updates to support audio and video streaming and graphics acceleration

  - ○ Improved audio codec support

- Comms and Telephony

  - ○ Telephony re-architecture introduced Multimode ETel (in place of the Basic, Advanced and GPRS extensions), enabling CDMA support and performance improvements

  - ○ New 3rd-party telephony API, non-third party APIs restricted

  - ○ New SIM Toolkit phone applications support

  - ○ New reference and test TSY implementations

- Networking

  - ○ Dual v4/v6 IP stack introduced, networking support for packet telephony (GPRS) and IPSec, including integration with new Multimode ETel

  - ○ Socket Server changes relating to IPv6, internet sockets, and secure sockets

- ○ Improved emulator Ethernet support
- Bluetooth and short link: Simplified HCI implementation to assist porting
- WAP and browsing
  - ○ WAP stack withdrawn (reliance on licensee implementations), WAP messaging implementation refactored
  - ○ HTTP Client API added
  - ○ WSP Adaptation Layer and Protocol Handler APIs
  - ○ Opera-specific web-browser support component added
- Messaging
  - ○ Support introduced for multimedia messaging (MMS) including SMIL message content markup. SMS messaging re-architected including support for Multimode (non-GSM) phone messaging refactored from ETel
  - ○ Smaller fixes in internet mail, fax client and scheduled send
- Cryptography: support for x.509 parsing and ASM.1 library added
- Connectivity: support for SyncML connectivity protocol added
- MIDP JAVA ME introduced with fully compliant support for MIDlets on Symbian OS; PersonalJava enhancements but JNI compatibility maintained
- System Libraries
  - ○ ECom (also known as 'Magic') Plug-in Framework introduced implementing new plug-in architecture
  - ○ StringPool API factored out of Uikon and re-engineered
  - ○ Support for non ROM-based localization added
  - ○ Support for Shift-JIS (Japanese) character set added
- Kernel, Base Porting, and Build Tools
  - ○ Emulator target build system migrated to Metrowerks CodeWarrior from Microsoft Visual C++
  - ○ Added XScale processor support
  - ○ New kits delivery model
  - ○ New Backup and Shutdown server, USB support, MMC card support, power management improvements, performance improvements (speed and ROM footprint).

Symbian OS v7.0s

Symbian OS v7.0s, announced in April 2003 at Symbian's developer event, repaired the fragmentation resulting from incompatibilities between Symbian OS v7.0 and v6.1 and the scale of the Symbian OS v7.0 changes, which threatened to create permanent divergence between S60-based product lines from Nokia, and its licensees, and UIQ-based products (for example, from Sony Ericsson and Motorola). The Symbian OS v7.0s system architecture was based on a subsystem release model and representation, retrospectively updated to the architecture representation based on the system model.

13.4   Summary of Symbian OS v8 Releases

Announced at 3GSM in February 2004 and reaching the market in phones such as the Nokia 6630, the Symbian OS v8 release was a significant increment on Symbian OS v7. In particular, it marked the first appearance of the real-time kernel.

In large part, the feature set is common to both Symbian OS v8.0 and Symbian OS v8.1, except that Symbian OS v8.1 offers the option of the new kernel.

The main new functionality in Symbian OS v8 includes the following (by no means an exhaustive list):

- CDMA telephony support
- Multimedia Framework replacing Media Server
- new connectivity, data synchronization and device management services architectures
- new WAP stack architecture and implementation
- OpenGL ES vector graphics support
- new implementation of Certificate and Key Management
- new App Installer architecture (preparing the way for Symbian OS v9 Platform Security)
- new content-access and content-handling frameworks, supporting policy-based content management, that is, DRM
- new JAVA ME JSRS
- USB-device class support
- MMS support, including parsing of SMIL markup, and support for OBEX over Bluetooth and Infrared

- Improved VCard and VCal conversion
- New XML parsing framework

Symbian OS v8.0

Originally Symbian OS v8.0 was envisaged as the release which would introduce the EKA2 kernel option for the first time. However, in the event, only one Symbian OS v8.0 release was made, based on the original EKA1 Symbian OS kernel.

Symbian OS v8.0 marked a substantial increment on Symbian OS v7, with new features spanning most layers of the system:

- Communications and telephony changes including the new communications framework based on the Comms Root Server and MBufs, first stage of CDMA telephony support, and Quality of Service (QoS) for GPRS
- Bluetooth and short-link changes introducing new USB class support providing control for USB devices and Bluetooth stack changes to support new Java ME JSRs
- New WAP short-stack WAP Messaging API, providing a limited functionality WAP stack and message API
- OpenGL ES Framework introduced, as well as multi-client access to screen, keyboard, and pointer or digitizer for GUI applications
- New Multimedia Framework replacing Media Server, new ECam camera API, Image Conversion Library and codec plug-ins, and low-level Media Device Framework providing low-level MIDI, video, speech recognition, and audio hardware-acceleration APIs
- New implementations of Certificate and Key Management and secure application installation
- Content-access and content-handling frameworks to support DRM content
- New connectivity, data synchronization and device management architectures
- Java ME new JSRs including Mobile 3D 1.0 (JSR-184), Bluetooth 1.0 (JSR082), Mobile Media API 1.1 (JSR135) and JTWI 1.0 (JSR185).
- New VCard and VCal conversion support and new character encodings
- New XML parsing framework, including XML Parsing Framework, WBXML Parser for WAP Binary XML, WBXML XML Parser Plugin
- Messaging support for OBEX over Bluetooth and MMS messaging

Symbian OS v8.1a

Symbian OS v8.1a is the Symbian OS v8.1 variant built on the original EKA1 kernel, but otherwise sharing the same features as Symbian OS v8.1b:

- ETel CDMA telephony extensions introduced support for CDMA networks, including CDMA SMS and WAP messaging support, and CDMA, Multimode, and SIM TSY reference plug-ins
- Graphics support for multiple simultaneous display, multiple display sizes and multiple display orientation
- SyncML device management
- Java ME upgrade to CLDC 1.1 from 1.0
- New Comms Database compatibility.

Symbian OS v8.1b

Symbian OS v8.1b is the real-time EKA2 kernel release of Symbian OS v8.1. This is the release, therefore, in which the EKA2 kernel becomes available for the first time.

In Japan, Symbian OS v8.1b has been the platform for a wave of 3G DoCoMo FOMA phones (from Fujitsu, Mitsubishi and Sharp), in particular with rich music and multimedia capabilities.

13.5   Summary of Symbian OS v9 Releases

Symbian OS v9 is the platform for the latest 'generation 3.x' UIQ and Nokia user interfaces, starting with UIQ 3 and S60 3rd Edition. It is also the release which introduces the new Platform Security architecture and completes the transition to the new real-time kernel architecture (first introduced as an option in Symbian OS v8.1; in Symbian OS v9, the original EKA1 kernel is retired).

Symbian OS v9 is the release with which Symbian has set its sights on the high-volume, mid-range market and the Symbian OS v9 releases to date take incremental steps to improve the fit of Symbian OS with that market, in particular in terms of performance and scalability (improving critical, basic performance and providing a cleaner architecture for porting to new hardware, support for single-core phone designs to reduce bill of materials (BOM) cost, improved peripherals support, and other porting and tools improvements to help time-to-market and reduce development cost). In keeping with these volume goals, it is also the platform on which Symbian has made its 'compatibility promise', the promise of API stability for all releases from Symbian OS v9 on.

Compared with Symbian OS v8, the headline changes are the retiring of the EKA1 kernel architecture, in favor of the new, real-time kernel, and the introduction of platform security, providing a trusted application model.

From a developer perspective, since the new kernel maintains user-side API compatibility, the kernel changes have little application-level impact. In contrast, platform-wide security has significant impact for all developers, introducing a security-capability model to protect system APIs and data caging for all application data.

The Symbian Signed signing and certification program grants capabilities (required to access protected APIs) to applications. Java ME MIDlets are also integrated into the security model. A free certification process is provided to freeware and shareware developers. Experience to date suggests that, for third-party developers, the more general certification requirements for robust and safe handling of extreme conditions (such as out-of-memory) are as much of an issue as the immediate requirements of the security model.

In other respects, Symbian OS v9 is very much an incremental update to Symbian OS v8. There is little radical architectural change, but a significant amount of re-engineering and improvement (in particular to improve performance, with boot time and RAM usage the key target areas, along with a set of user-oriented critical use-cases, for example addition and deletion of multiple contacts).

Symbian OS v9 also makes the transition from GCC to the ARM RVCT compiler, supporting the ARM ABI versions 1 and 2 and therefore promising tools interoperability for ABI-compliant tools (including compilers).

Future Symbian OS v9 releases are likely to continue the focus on performance, including high-performance graphics and support for continued user interface evolution, improved suitability for the mid-range, evolution of the build toolchain, and backwards compatibility, while introducing headline new technologies (likely candidates include location-based services).

Symbian OS v9.0

There is no productized Symbian OS v9.0 release, which instead served as a baseline release for the integration of Platform Security on top of the EKA2 kernel. All Symbian OS v9 'new features' (as opposed to platform changes) therefore appear in subsequent releases, from Symbian OS v9.1.

Symbian OS v9.1

The first phones based on Symbian OS v9.1 (which was announced in February 2005) shipped during the first half of 2006. They included the

Nokia N80 3G phone and Sony Ericsson P990, both with Wi-Fi and multimegapixel cameras (3 and 2 megapixels respectively).

For the enterprise market, device-provisioning enhancements and group-scheduling APIs are important additions.

Headline features include:

- Platform Security

- Real-time EKA2 kernel, with support for a range of ARM CPU architectures and memory models

- System Starter, a new policy-based mechanism for server startup

- New Device Management and Client Provisioning services, including a new Client Provisioning Framework

- Broadcast Tuner APIs

- Bluetooth Remote Control Framework

- Improved Timezone support

- Networking enhancements including RTP/RTCP support

- Java ME MIDP 2.0 (JSR 118) and CLDC 1.1 (JSR 139), plus new MIDlet security policy support

- Hardware reference platform incremented to TI OMAP H2.

Symbian OS v9.2

Symbian OS v9.2 continues performance improvements, supports several new locales and provides build-toolchain and porting improvements, while maintaining baseport compatibility, all of which can be seen as steps along the way to an improved mid-range offering.

New features include some important new APIs and technologies (the SIP Framework, for example). The major platform features, however, are those which it shares with Symbian OS v9.1: platform security and the EKA2 kernel architecture.

Compared with Symbian OS v9.1, the key changes are:

- New RTP/RTCP and SIP multimedia protocols, together with underlying networking support and communications architecture evolution to support high data rates

- New Hindi and Vietnamese and improved Japanese language and locale support

- Device management, messaging, email and multimedia enhancements

- Continued improvements in boot time, RAM usage, and range of specific performance use cases
- Hardware reference platform incremented to ARMv6.

Symbian OS v9.3

At the time of writing, the latest release (announced in July 2006) is Symbian OS v9.3. Compared with Symbian OS v9.2, the key changes are:

- additions to SIP protocols and continued evolution of the communications architecture
- support for Wi-Fi wireless networking.

Part 3
Design Case Studies

The case studies presented here provide an in-depth examination of significant turning points in the evolution of Symbian OS or of significant aspects of the wider context in which the operating system continues to evolve. They take an unashamedly historical approach, which I hope helps to capture the flavor of the times, as well as providing some insight into how real systems get made and what forces – often contingent and accidental – shape them.

Each study explores a different aspect of the evolution of Symbian OS: the adoption of object-oriented ideas in designing and creating the system, the choice of C++ as the implementation language and the consequences of that choice; the early decision to implement telephony support and what that meant for Symbian OS; the radical solution to the question of user interface customization and what led up to it and shaped it; the challenges of renewal and evolution, which all software systems face and which all system designers must overcome; and, finally, taking a small step back from the operating system itself to the wider context of its production, the tensions exposed by success and the scaling up from small-scale to large-scale software production and what it implies for the future of the system.

These chapters are exploratory and not always conclusive but, I hope, illuminating nonetheless.

14

The Use of Object-oriented Design in Symbian OS

14.1   Introduction

Symbian OS traces its lineage back to the operating systems that Psion created for a succession of innovative and market-leading handheld devices, from the early Organisers through to the Psion Series 3 and, finally, to the Psion Series 5, for which the first versions of what became Symbian OS were designed. Psion was an early adopter of object-oriented programming techniques and an early adopter of C++, opting to build a commercial, production operating system in C++ significantly ahead of the mainstream.

This case study explores that history, and the consequences. It goes on to survey some of the ways in which object-oriented techniques are used in Symbian OS, for example the adoption of model–view–controller (MVC) as the basis for the application model, the widespread use of frameworks, the active object idiom, and the way object-oriented ideas influenced the design of the kernel itself.

The adoption of object orientation, generally, and of C++, in particular, were radical steps for the company; both posed challenges, both were motivated by some clear expected benefits. Within the company at the time, both decisions were controversial and the risks were enormous. The history and the controversies are instructive and still relevant today – both within and outside Symbian. They give a flavor of the company's particular character at that time, of its openness, its willingness to take risks, its commitment to understanding what the 'right thing' was to do, and then doing it; as well as giving a flavor of what the broader context was for the design of Symbian OS and of some of the influences that shaped it.

14.2   Pioneering the Object Approach in Psion

The operating system that eventually became Symbian OS started life as Psion's 32-bit, 'next generation' operating system, an all-new operating system designed for a new generation of 32-bit hardware which was planned to replace the then current (and highly successful) Series 3 range of 16-bit palmtops. Psion in early 1994 already had a history of experimenting with an object-based design approach and object-oriented programming techniques. The 16-bit operating system for the Series 3 (SIBO), for example, had used object-oriented design ideas and implementation techniques heavily at the application and user-interface levels, with great success. The Series 3 had a powerful but easy-to-use graphical user interface (GUI), unique for a machine of its class. Its user-interface and application architectures were fully event driven, with an object-oriented implementation. Martin Tasker, new to the company, saw object orientation as a perfect match for the problem.

> **Martin Tasker:**
>
> They had understood that objects helped you to do GUIs and systems, and in the SIBO GUI they had the perfect event-driven system.

Object-Flavored C

The big question facing the company as it prepared to start work on the new 32-bit system was how best to build on its object-oriented legacy. An important focus of the debate was which object-oriented language to choose.

> **Peter Jackson:**
>
> The big debate that was going on was, 'Do we write another proprietary object system like we did for SIBO or do we go to C++?' People were saying, 'C++ is the next big thing, everyone will know how to do C++ programming, therefore it's easier to get engineers without special training.' And of course this is complete nonsense, because what you really need to know is not C++, but how to program against a particular object model and the APIs that have been developed for the system you're programming. It's just as big a learning curve as learning some proprietary object framework. But the C++ faction won out.

For the 16-bit system, Psion had evolved a proprietary object-flavored dialect of C, along with tools which pre-processed this 'C with classes' code and generated standard C output, which was then passed into a conventional compile and link stage. Objective-C was the explicit influence.[1]

In fact, this was the second time around for the C++ debate. At the outset of the 16-bit project, the team had evaluated a number of language options including C++, before deciding to develop its own object flavored, C-based solution.

David Wood:

We looked at C++ as one alternative, but we were concerned by a number of things. First, we saw it as being a very large language compared with C and we thought that we would just have no constraints on our design. Second, we saw it as still being immature. There weren't mature compilers for it at that stage, certainly not for the PC which was our development platform. So we took the view that it would be a big, big jump to adopt C++ and we didn't know much about it, whereas we thought we could do a more constrained job ourselves.

Psion's solution was home-grown and decidedly arcane, but on the other hand it did what it was meant to do and it did what was needed. Since the users were all in-house, the fact that it was a proprietary solution was not an issue. It also gave the team complete control of the most important components of the development toolchain. There was, however, rather more to the detail than simply pre-processing pseudo-class definitions into standard C.

Andrew Thoelke:

You had a class description file which basically said, 'This is a class' and it knew what its base class was, it knew what methods you had added, what methods you had overridden, and it knew its data members. That was then processed by a preprocessor, in effect a class compiler, which generated a C header file with lots of #defines in it which defined method numbers and offsets of data and that sort of thing. It also defined data structures for the classes themselves, so you could actually use pointers to the class objects, and then it generated an object file which contained a binary representation of the class structure. These were then all compiled together, so in a given dynamic library in SIBO the header contained initially a lot of information describing all the classes within that dynamic library, and then following that was all the

---

[1] Objective-C (see Chapter 4) is an object-oriented superset of C with a Smalltalk-style (infix) message syntax, which emerged in the early 1980s more or less concurrently with the Cfront versions of early C++, in an independent effort to harness the plain syntax and underlying efficiency of C to an object model. Objective-C is still in use as the native system programming language for Mac OS X, having been inherited by it from the NextStep operating system.

> actual code itself, the actual functions, the methods. And normally you would have written that in plain C.

High-level object-oriented design could therefore be directly implemented using class-like constructs. Underneath, the class (or pseudo-class) method implementations were vanilla C. The class compiler was known officially as the Category Translator.

David Wood:

It looked at the class definition file and generated C output. There were also helper functions in the operating system which enabled indirect function calling to methods in the classes you defined, which was the equivalent of C++ virtual tables. Our implementation didn't use virtual tables in the same way as C++. We tried to implement that in a more efficient way, driven by the desire to save memory.

The importance of saving memory in order to maximize the small ROM and RAM footprint of the hardware was deeply engrained in the Psion culture. Necessarily so, since early Series 3 models were based on either 128 KB or 256 KB RAM with 512 KB ROM, increasing in later models to 1 MB or 2 MB RAM with1 MB ROM, and to 2 MB ROM for the last model in the series.

Another optimization was the replacing of the standard C library with a dedicated library of support functions implemented as thin wrappers around native operating system calls. This kept the dynamic libraries small because, instead of each library including the same several kilobytes of standard library code, they simply made direct calls to operating-system functions. It also improved run-time performance, another important driver.

Peter Jackson:

All in all, it made for some very lean and mean code and it was also a very elegant system. It was precisely targeted to meet the requirements of the object space you would need to have for a PDA.

Throughout the system, assembly code was also extensively used to achieve speed and small code size. Again, this code was written inside an object framework which integrated with the C-based object model.

> **Martin Budden:**
>
> The assembler stuff was written to the same kind of interface, so it was effectively object-oriented assembler, and it fitted in the same framework, at least at the application level.

While object-oriented techniques were used extensively in the higher levels of the system, the lower levels of the system were more conventional. The kernel implementation, for example, and the lower-level system services, such as the file server, were written in conventional C.

> **Geert Bollen:**
>
> It was a layered OO [object-oriented] mechanism, using message dispatch on top of a C system. That was the Psion tradition. Underneath was a classic C-API-based operating system with a small number of system calls exposed as C functions.

This is exactly how it should have been, in David Wood's view. Object orientation was a pragmatic choice, not an ideological one. It was intended to achieve some clearly defined benefits without undue risk either to the system (in terms of size and performance, for example: code bloat was the big fear) or time-to-market (which a wholesale jump to C++ might have jeopardized).

> **David Wood:**
>
> OO came in stages, which is how all large software systems should evolve. One of the very important rules of software, or indeed of anything else in life, is, If you're not sure what you're doing, don't do it on a large scale. I think the phrase is attributable to Tom Gilb. What it means is that you should go and experiment, and you should evolve through iterations. So we introduced objects first of all just for the user interface, and then we realized that we could apply those ideas elsewhere too. So we proceeded through iterations.

C++ Language Choice and Adoption

At the outset of the 32-bit project the team assumed that a straightforward software migration effort would suffice to move the major part of the software base to the new hardware architecture. In fact, the first plan was for automated migration. Since the move up to a 32-bit system involved a switch from Intel to ARM processor architectures, the focus for rework was on the hardware interfaces of the low-level system and on the operating system kernel itself, for example to migrate to new process and thread models and a new MMU architecture.

> **David Wood:**
>
> In 1994, we had at first a plan that we would do a quick project to convert from 16-bit to 32-bit. We assumed we would keep more or less the same syntax and that we would actually write a tool to read in everything and then spit out 32-bit versions.

There were a number of reasons for moving from Intel. The Intel x86 architecture was primarily aimed at desktop computers. It had no real power management features, making it power hungry and giving it much worse MIPS per Watt performance than ARM. Just as importantly, even as early as 1994 Psion had identified mobile phones as the most promising future mass market for handheld devices. ARM had established itself as an ambitious and successful semiconductor player in the mobile device market. In contrast, Intel had little or no mobile device presence.

The plan for automated migration did not last long. The language question quickly came to be seen as an important strategic choice, not just a practical one, in much the way that the processor choice was both pragmatic and strategic. One consideration was the nature of the intended platform so far as third-party applications were concerned. The Series 3 had provided a run-time support system similar to Visual Basic for installable third-party applications, and a thriving hobbyist culture had grown up around it. However, this was far from offering a natively programmable platform (which would imply that the system-level APIs were open to third-party applications as native calls rather than serviced through the run-time language support). Native programmability was emerging as a key requirement for the 32-bit system.

The more general question was how far Psion could afford to ignore the mainstream. Continuing with a home-grown system involved more than software maintenance costs (supporting the toolchain and keeping it fit for purpose). It would incur the larger, negative opportunity cost of locking the 32-bit system out of the mainstream. As ambitions for the new operating system scaled up, with its design life projecting well beyond 2000, moving with the mainstream became an increasingly important consideration.

> **David Wood:**
>
> As we saw the scale and longevity of the 32-bit system we thought, 'Wait, we don't want to be stuck in some custom programming system, we want to go

with the mainstream', and increasingly the mainstream was C++. That's what software engineers were learning at university, that's what they would know about. We thought it would be harder to recruit people if they then had to go and learn this comparatively arcane technology which we had. So we adopted C++ for sociological reasons as well as technical reasons. But the technical reasons were important too. As we studied C++ more over the years we saw there were benefits in it being, quotes, *a better C*, rather than just being an object-oriented C.

While not everyone was convinced by this line of argument, it was certainly clear that the alternative to C++ was an in-house system; that C++ was almost certainly going to be one of the dominant object-oriented languages for the foreseeable future; and that the mainstream would be there if it was anywhere. Dialects such as Objective-C were being relegated to the margins by the momentum which was gathering behind C++. Arguably, even Smalltalk, which had some claims to priority, was losing ground to the C++ juggernaut.[2]

While C++ was still the same language which the company had previously rejected as too big, too complex, and insufficiently constraining, the context had changed. Looked at more positively, C++ not only retained many of the advantages of C as a systems programming language (its ability to get 'close to the metal') it was indeed by design 'a better C'.

Stroustrup characterizes this aspect of the language as including stronger type checking; more convenient and less error-prone memory allocation (operator `new`) and release (operator `delete`); function overloading; the ability to initialize variables with values as well as to assign values; references; and, in general, less reliance on exploiting tricks needed to get the best from the language. (Type casting, for example, was a ubiquitous pattern in C that was only rarely necessary in C++ [Stroustrup 1994, p. 171].[3])

---

[2] A 1995 survey conducted at Object World identified 77% of respondents as C++ users compared with 28% as Smalltalk users [CIO Magazine, ***www.cio.com/archive/110195/object.html***].

[3] It is also worth considering Stroustrup's views on what programming languages are for. A language does two things, as he sees it: it provides a vehicle for specifying actions to be executed and it 'provides a set of concepts for the programmer to use when thinking about what can be done.' [Stroustrup 1993, p. 8] In other words, on the one hand, it provides a model which abstracts a programmable machine (a 'machine model') and, on the other hand, it provides a conceptual model for translating real-world problems into program solutions. Arguably, the history of the evolution of programming languages is the history of the stretching of the gap between the two, so that the conceptual model is increasingly distanced from the underlying machine model, as well as of the increasing abstraction of the machine model away from a literal physical interpretation toward a logical interpretation.

Andrew Thoelke:

I remember hearing about the discussions, but I was too junior in the company really to be directly involved or influence them. But C++ was still quite young at that point. This must have been 1994. It was way off being standardized, probably four or five years before the first pass at standardization. But it clearly had most of the elements of object orientation that had been put into the Series 3 system, and clearly it was going to be standardized, it was going to be much more mainstream, it was important. And the experience of using an object-oriented language and having an event-based application environment was one they didn't want to lose (because the 16-bit system already had that) by migrating to a new hardware architecture and going 32-bit. So it's not surprising they went for it.

Martin Budden, already a veteran at that time, was an early C++ advocate.

Martin Budden:

I was actually advocating C++ fairly early on, but I wasn't the only one, there were other people advocating C++ too. There was a debate about whether to move to C++ or stay with C and, fairly quickly, the C++ argument won. Since Colly was doing the first cuts of the operating system, once he had been convinced, it followed from there.

Geert Bollen arrived at Symbian in the early summer of 1995 with impeccable credentials in object orientation and a significant background with C++, specifically. While others in the company were still learning C++, Bollen already had three to four years heavy-duty implementation experience behind him. By the time he arrived, the decision to go with C++ had already been made and the team was six months into the project.

Geert Bollen:

C++ seemed the obvious choice from my perspective. But I suppose I hadn't really been exposed so much to some of the issues in an operating-system context. But consider, it's only in the last few years that the C++ ABI (Application Binary Interface) has been standardized, so that completely determines the interoperability of different toolsets. That gives you an idea of how young the language was, at least as an industrial tool. All that had been taken for granted in the C world since the '80s; they just weren't issues. In the C++ world they were wide-open problems.

When Bollen joined the project, the lack of standardized low-level tools meant that no code was yet running on ARM hardware, and all development was still emulator-based. The project was still feeling its way, climbing the language-learning curve while throwing its energy into the object-oriented design opportunities of a clean, from-the-ground-up, re-engineered system.

> **Geert Bollen:**
>
> Moving to C++ was a very brave decision. I don't know if it seemed so at the time, but in hindsight it was a very, very brave decision. They had selected C++ to build this system, with some trepidation, as a leap of faith.

Challenges of Switching Languages

Switching incurs costs. There were what David Wood might call the 'sociological' issues, the questions of what tools developers knew, the problems of bringing in new developers familiar with C++ but not with the in-house object-oriented style, while training up the seasoned in-house developers in a new language; and all this, of course, as the design itself evolved fairly freely.

But there were other problems too. Performance problems, perhaps inevitably, quickly became apparent.

> **Howard Price:**
>
> C++ was slow and big in various situations. For example, the OPL VM grew by about six times when it was ported from S3 8086 Assembler to C++. There was some argument, to the point where there was talk of dropping C++ and moving back to the faster, leaner, meaner approach – just sticking to C. But, thankfully, David Wood and Charles Davies were really strongly committed to C++ and object orientation, and they persuaded the rest that this was the way to go.

But the biggest practical headache almost certainly was the lack of a suitable, stable toolchain.

> **Geert Bollen:**
>
> The target tools were not in place when I arrived, so whatever the system was, it was emulator-only. The kernel implementation was an emulator implementation. The on-the-metal version of the kernel was started and delivered after I arrived. Colly Myers assembled a team for that. Before that, he had been a one-man band.

There was no standard toolset capable of targeting both Intel and ARM. Or rather there was, of course, the GNU GCC toolchain, but not one that would work naturally in a Microsoft Windows development environment. Since most development took place on Microsoft Windows for the emulator targets, a productive Microsoft Windows programming environment was essential. Microsoft's compiler and toolchain targeted only Intel processors. Because there was no standard Application Binary Interface (ABI), Intel binaries were in a world apart from the GCC ARM binaries.[4]

In the short term, this meant that bespoke tools had to be written to enable dual-target compile and build using two different compilers, each with its own make and link tools and other utilities, but in a way that integrated reasonably well enough with the Microsoft Windows-based IDE so as not to cripple developer productivity.

The longer-term legacy is that this bespoke system still persists, in essentials unchanged, to this day, a source of low-level discomfort for the external developer community. (Internally, it is as it is and you get used to it. Externally, it is reasonable to ask why you should have to get used to something that causes development pain.)

The underlying problem of course is that phones now, and PDAs then, are essentially embedded devices, and the development process for embedded systems is inherently more complex (typically in arcane and opaque ways) than development for desktop systems. A cross-compilation model is natural and inevitable in an embedded-systems context. For Symbian OS now, as then, initial development is based on an emulator hosted by Microsoft Windows and only later (when the basic implementation is running) moves to a cross-compilation model targeting real hardware, either prototype or production devices or reference boards.

Another lasting legacy of the early tools problems is the nature of the emulator implementation. Essentially, the emulator consists of the user libraries (the base-level services and utilities above the kernel itself) and higher-level kernel services implemented on top of native Microsoft Windows libraries. This is nominally transparent to the higher layers of the system, but key kernel services and hardware-level services are really being provided by the host operating system, in this case Microsoft Windows, to which the low-level calls of the hosted system are mapped. Historically, the problem with this solution is that for anything beneath application-level programming, the true behavior of the code is obscured by the underlying Microsoft Windows implementation. This was not an issue in the early development of the operating system, since the whole point of the emulation was to allow the application developers to get on with work on the application suite, but it rapidly became a problem both for internal and external developers. While the worst problems have

---

[4] Standardization at the binary level is only now appearing across the toolchain with the standardization and adoption of ARM's EABI, enabling interoperation of tools from different vendors.

been fixed for some time, some awkward issues have remained even up to the most recent releases of the operating system. Those challenges can be traced all the way back to those first problems of early adoption of a pre-standard C++.

As the kernel moved onto real hardware and began to exercise the ARM-targeted toolchain, yet more challenges emerged.

David Wood:

We hit various quite significant obstacles along the way, for example we were calling C++ functions from one DLL into another DLL in a way that I don't think had been widely done before, and certainly this required lots of changes in the GCC compiler that we used. So several times, the implementation for ARM as opposed to the emulator got held up because we were using C++ to the limit, with the DLLs, and we needed to get Cygnus (who were at that stage under contract from us, maintaining GCC) to issue numerous fixes. So we were pushing it to the limit.

Again the issue was the immaturity of the C++ development environment and the lack of standardization of the low-level interfaces.

Yet another area in which early adoption proved to be costly was the lack of standardization at the level of the basic language libraries.

David Wood:

There was no sufficiently agreed standard library that covered the things that we wanted. There was no text-handling library that Colly Myers was satisfied would provide either the necessary degree of security or the necessary degree of efficiency and performance.

This is the reason that descriptors were invented as type-safe, memory-safe, string and binary-data container classes. The alternative would have been to use standard C++/C-subset strings, which provided neither protection against overruns nor type-safety.[5] For a system with pretensions to robustness, they were out of the question.

Error handling was another unstandardized area of the C++ language. Again, a solution was invented, specifically aimed at the needs of the class of device that the operating system targeted, based on the notions of 'leaving' functions and a cleanup stack.[6]

---

[5] C-style strings are just arrays of raw data bytes, which can be accessed as either character data or raw binary data, and as bytes or multi-bytes. Indeed, that very versatility is their point, but it makes them both error prone and open to accidental or deliberate abuse.

[6] The cleanup stack is used to hold pointers to automatic (i.e., stack-based and, therefore, function-scoped) objects, so that if the function should fail ('leave') the objects are deleted and not left allocated but inaccessible (which is a memory leak).

Other features of the language were treated with caution.

Andrew Thoelke:

We were very cautious in our adoption of advanced C++ features, partly because not all compilers supported them or did them well, but also because you just don't know whether you're using a feature which has lots of power but has a lot of background overhead to it. And we were trying to ensure that we had some control over what was happening behind the scenes. It's still important today, although compilers are better at implementing all of those odds and ends, and having a good relationship with the compiler vendor makes it easier to have confidence in the compiler being efficient.

Charles Davies:

We were experienced in object-oriented programming. We were battle-hardened in object-oriented programming. But, in all honesty, we were still novices in C++.

Managing C++

Beginners' mistakes are the mistakes you make because – well, because you are beginners. C++ is a powerful language and there are almost certainly cases in which its powerful object-oriented features overrode the better judgment of even seasoned developers.

Charles Davies:

We over-used inheritance, which everybody does when they get introduced to object-oriented programming. But we weren't new to object-oriented and should have known better.

David Wood:

We probably did take some of the inheritance hierarchies too far. That was driven by an understandable wish to avoid duplication of code, but some of the hierarchies as a result became, shall I say, unnaturally clever. That cleverness solved the problem of the moment, but ended up itself being difficult to maintain.

Possibly the problem was that because people felt they knew what they were doing with object orientation, they saw in the language only the opportunity to exploit object orientation more fully, rather than the dangers of over-abstraction and over-complexity. However, overuse of inheritance is symptomatic of confusion between class relationships (e.g. the difference between is-a inheritance relationships and has-a composition relationships) and the misuse of so-called 'implementation' inheritance.[7]

Andrew Thoelke:

C++ is an extremely flexible language because you can ignore all the extras and just use raw C with all of its ability to manipulate hardware and to do whatever you like. But C++ also allows you to be overly excessive with the use of things like inheritance until you get to the point where you lose track of what exactly you are doing anyway.

David Wood:

There's actually a spectrum of reuse. On the one hand, you end up with multiple copies of the same code and that's no good. At the other end of the spectrum, you end up with just a single bit of code which is so convoluted it manages to do everything but no one can understand it. And there's a happy middle ground which nowadays we occupy much more readily.

Certainly there were language features which were considered dangerous, in some cases because they were not well-enough understood in the company (there just was not sufficient grounding in the language for there to be enough experts to make the judgments and provide the design and coding guidelines, short of outright banning), and in other cases because they were not mature, whether in terms of the language specification or of the implementation of the available tools (compiler). Templates, multiple inheritance and namespaces are three examples.

Templates

Templates are a C++ mechanism for writing type-independent code, typically library-like classes that are useful for managing objects of any type, which are then invoked in subsequent code with a parameter of a concrete type, for which the compiler generates (or selects) type-specific code at the point of invocation. They were added to the language particularly to solve the problem of providing useful, generic container classes capable of acting as containers for objects of arbitrary type, in other words, of any type whatsoever that a program might invent (see [Stroustrup 1994, p. 339]). (Ada is another language that provides a template mechanism, though it is different from that of C++.)

---

[7] 'Implementation inheritance' (in which a derived class inherits its base class implementation, that is, the implementation of the base-class methods) contrasts with 'interface inheritance' (in which the derived class inherits its base-class interface but not the implementation of the base class methods); see [Stroustrup 1993, p. 413].

The problem is that naive use of templates causes code bloat. As a template user, each time you instantiate a given template for a given type, the compiler generates a fresh copy of the complete template code for that type. Each time, every time.

> **Martin Tasker:**
>
> All major systems that existed at that time had been written in C, and C++ was a language that people were still getting comfortable with. There was still a lot of opinion that some of the features of C++ were too expensive to use in a ROM-based small device. Templates being one of them.

Careful design of the initial template classes can avoid the problem, but to do so requires expertise.

> **Andrew Thoelke:**
>
> Certainly parameterized programming using templates is something that's at least a degree harder to understand than some other features. It's very easy to bury yourself in complexity that you assume the compiler will get you out of, but you really don't know. The actual side effects in terms of size of code or performance are hard to measure sometimes. It's hard to write very complex template code well.

Especially in the early days of Symbian's adoption of C++, when the language was fluid and the standard libraries (which were almost wholly template-based) were still evolving and had not been fully standardized, templates were regarded in the company as potentially dangerous and therefore treated with extreme suspicion. [8]

> **Martin Tasker:**
>
> The Standard Library is the basis of what people call generic programming. We do not do that at all in Symbian OS. We have a few templated types, but we don't use them aggressively enough to call it generic programming. I don't think there were many people anywhere that understood the Standard Library at the time it was being designed, and we certainly didn't. Collectively in Symbian, or Psion as it then was, we didn't understand it.

---

[8] Even the experts find templates hard. See, for example, Scott Meyers in his introduction to [Alexandrescu 2001].

Instead, an in-house idiom called 'thin templates' was evolved for some generic, container-style classes, based around type-independent, pointer-based concrete classes (which therefore can accept pointers to any type of object), wrapped by a template class that contains no implementation, but just serves to enforce a parameterized interface to force type-safety for the object type for which it has been invoked. (The technique is quite well known in the C++ literature, see [Meyers, p. 191, Item 42].) Beyond this, templates are not used in the Symbian OS library classes. If used internally elsewhere, they are used with extreme caution.

Andrew Thoelke:

It was more about the fact that the compilers didn't do a good job with templates. They could only do very basic things. So we were very careful about how we used them. We didn't do anything very advanced. We were also careful to tell people that templates can lead to code bloat if you use them aggressively. So instead we used templates to give type-safety over essentially type-unsafe container objects. Really we were trying to make sure that we could have more maintainable, better written, higher quality code in the first place, by actually constraining the use of the language and helping the external and internal developers to help themselves.

Multiple Inheritance

Inheritance is the basis for structuring the object relationships which underwrite the collaboration and delegation between objects in (class-based) object-oriented design.[9] Inheritance relates objects logically and provides the mechanism for sharing common behavior.

Some object-oriented languages, and C++ is one of them, allow multiple inheritance. (Eiffel, CLOS and Dylan are other examples; languages which do not allow multiple inheritance include Beta and Smalltalk). Multiple inheritance allows objects to have multiple parents. As with templates, casual use of multiple inheritance can lead to multiple instances of identical code being generated. One of the early rules within Psion was therefore 'no multiple inheritance in C++'.

The idea of multiple inheritance is trivial enough to grasp. For example, a police car inherits from Car, but also has Emergency properties, which fire engines and ambulances also share, even though they are not derived from Car (for example, they may be derived from Truck); multiple inheritance allows emergency vehicles to share Emergency properties, inherited from an Emergency class, while deriving from different vehicle base

---

[9] There are non-class-based systems for example the language Self, based on prototypes (see Chapter 4).

classes.[10] While the principle is simple, the implications for implementation in an object-oriented language go deep.

Martin Tasker:

The multiple inheritance chapter in the Ellis and Stroustrup book is staggeringly difficult! It's mind-bogglingly difficult! So we made a really conscious effort: no multiple inheritance at all. And in fact our solution, Mixins, serves exactly the same purpose as the equivalent in Java, that is, interface classes, and they are really easy for the user.

Andrew Thoelke:

Some of the constraints, like 'avoid multiple inheritance unless the additional base classes are interfaces only', was partly to avoid the Evil Diamond inheritance graph, which you can acquire without always realizing it, and which all the text books said was a Bad Idea, unless you used virtual bases, which we always said, 'No, don't do it', on grounds of performance and footprint and questionable value.

The 'Evil Diamond' pattern is one in which class K derives from both classes B and C, which both derive from A. (K, by multiple inheritance, inherits from both B and C; both B and C inherit, possibly also through multiple inheritance, from A.)

The problem with this pattern is that, because C++ implements inheritance of behavior at compile time not run time, if class A contains concrete method implementations then its code may be and its v-table is compiled into the code for both B and C (because they inherit the behavior) and appears twice in class K.

The immediate problem that outlawing the use of multiple inheritance causes is that no class can present multiple interfaces. An example of the value of interface inheritance is the Observer pattern. Some object has behavior derived through one inheritance hierarchy, say Timer or Alarm, but is also an Observer of some event that triggers its behavior. Not all timers and alarms are Observers, but some certainly would like to be.[11]

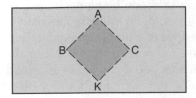

Figure 14.1   The 'Evil Diamond' pattern

---

[10] The example is from [Stroustrup 1993, p. 405].

[11] [Stroustrup 1994, p. 271] cites Stream I/O as an example of the value of composition of interfaces; another example is composing a class from an implementation and an interface.

Java (as a 'better C++') explicitly provides machinery for adding interfaces to objects. But in the absence of that, multiple inheritance is the most natural way to get it, and the only way to get it by derivation. (So-called 'fat' interfaces are an alternative approach, see the discussion on 'Streams, Stores and Persistence' in Section 4.3.)

Eventually a compromise was reached and 'mixin' ('M' or interface) classes were introduced, following the solution which had actually been first adopted for CLOS.[12] Mixins solve the problem of how to get the best of multiple inheritance, for example, so that objects can present multiple interfaces, without getting the worst of multiple inheritance, code duplication and bloat, and over-complex and over-designed class hierarchies.

With the mixin pattern, while only one inheritance path may inherit behavior, in other words inherit from (and, therefore, include the code from) a concrete base class, a class is allowed to inherit from as many M classes as it likes because M classes may only define pure virtual functions. In other words, they define abstract behavior (interfaces) only, not concrete behavior.

A typical use of an M class is to define an Observer class, so that a CCoeControl-derived control (GUI widget) also presents an MObserver-derived interface. Because MObserver is an abstract class (all of its methods are pure virtual), there is no inherited implementation, and therefore no danger of duplicated implementation code (when you define an MObserver-derived class, you must provide all method implementations yourself).

Martin Tasker:

The notion of implementing an interface, which incidentally I think is used very nicely in some of the printer driver classes for instance, seems very natural. So, one pattern for implementing is-a uses mixins; another pattern is that you have an attribute and it turns out that the interface paradigm is actually very natural.

There is a more general and interesting point about the similarities with Java.

Martin Tasker:

We ended up with a lot of the same things that Java did, and we invented them more or less completely independently. We didn't really start looking

---

[12] Mixins were reputedly named for the way of specifying the flavors you wanted at Steve's Ice Cream shop, near MIT: vanilla plus mixins. The original LispMachine implementation was indeed called Flavors.

> at Java until somewhat after our major design decisions were committed. But funnily enough we ended up using multiple inheritance in the same way, we constrained C++ multiple inheritance and ended up using it in the same way as Java uses inheritance interfaces.

Before the mixin solution was adopted, some designs suffered from the injunction against multiple inheritance and had to find other ways to offer multiple interfaces, resulting in convoluted design.

The lesson is that early adoption involves multiple leaps into the unknown. Leaping into the unknown can harm your design.

Namespaces

Namespaces are another example of a useful C++ language feature which was not used because it was not standardized in early C++ implementations. (At least, namespaces were not used early on in Symbian OS.) More recently, some technology areas (networking is an example) have pioneered the use of namespaces, generating some internal debate about what the rules for using namespaces should be.

For example, in designing the networking components, namespaces have proved useful as a mechanism for associating Symbian OS implementations of standards-based behavior (in other words, externally specified APIs) with a standard API, without polluting the global namespace, making it easier for licensees to swap out the Symbian implementation in favor of their own alternative implementations and to do so in principle simply by changing the namespace qualifier.

Freedom Through Object Orientation

One of the more intriguing aspects of the way that object orientation was used inside Psion was to liberate the design process from the need for central oversight and scrutiny, below the level of API definition. While, in some respects, this liberation meant freedom for developers to shoot themselves in the foot with complex features of C++ or just over-use of features (such as over-elaborate use of inheritance), in the wider sense of applying object orientation as an enabling technology to unleash talent to solve a problem while still providing a sufficiently constraining mechanism, it was hugely successful and partly explains how Psion was able to create a fully-fledged modern operating system, all the way up to the application level, in a remarkably short time with a relatively small number of engineers. (In 1997, not long after the first operating system release, Psion Software was still a company with barely 200 employees.)

Martin Tasker:

Something I think Charles Davies did extremely consistently and well, was he basically said, 'Look, you can design an object-oriented system and train all those programmers that you had to recruit because this fantastic system is just too big for your elite team of ten, and you can design the interfaces and give them a sandbox in which to play inside those interfaces, and they can't hurt anybody else by just adding another function onto that library over there', which is what you can do in C, and which is what you are strongly inclined to do in C.

I think Charles, in particular, did this; he paid minute attention to the details of his APIs, he used their explanatory power to motivate his people, and he almost didn't need to look at what they produced in terms of implementation code or test code. He basically said, 'If it meets the requirements of the API and if you feel it's correct then I trust you that it is correct.' Whereas I think that process is harder to do in a non-object-oriented system, because the boundaries are much harder to draw, both for the architect and for the implementer. I think Charles used that to massively good effect. He was outstanding in that.

David Wood was another exponent and practitioner of the principle of freedom and his approach remains to this day: hire talented people, point them at the problem, provide the minimum constraints necessary to bound the solution space, and let them get on with it.

However, this was not what initially attracted him to the benefits of object orientation. Rather, it was the holy grail of reuse. Inheritance as a mechanism for delivering reuse seemed to be a perfect fit for the problem domain of constrained devices.

David Wood:

My first interest in object orientation was in the notion of inheritance, which was that you could reuse somebody else's code but still modify it without having to end up with a complete copy of it. So, the same principle of looking for small code size which runs through many of our early design decisions is there, and we saw that we would otherwise have duplications of code, where there would be a system component whose behavior couldn't be fully parameterized just by data, so you couldn't say, 'well, here are all the flags that will completely govern the behavior of specializations of this object.' We saw that yes, you had to have a way of specializing the code as well as specializing the data. And that is the idea that really struck home. I know exactly the book that I read that made that impact, it was a book by Bertrand Meyer about Eiffel. So that was a very pragmatic consideration, it wasn't at all an ideological consideration.

The object-oriented design approach also seemed to offer something else that was particularly necessary in GUI design: a systematic design approach (abstraction hierarchies) capable of bringing elegance and order to a complex problem domain and empowering designers. Probably the first, large-scale example of a complete object-oriented system (even though it was written in C) was the original Carnegie–Mellon design and implementation of the X-Window system for Unix.[13] Abstraction hierarchies (whether called classes or not) lend themselves elegantly to solving the GUI design problem, while also being open to full-scale extension.

David Wood:

I bought into the wider philosophical view which was that if you followed an object-oriented approach you could handle greater complexity. Because you will have items in your code that map better to concepts in the real world that you're trying to model, or map better to concepts in the user's mind. So you'll have data and code together and you can organize them in hierarchies which correspond to how they are arranged in the real world, or correspond to how they are arranged in the users mind. So, going forwards, that was a very important reason to adopt object-oriented principles. It allowed a single programmer to hold more in his head.

Again, the efficiencies delivered were not just machine efficiencies (more effective, efficient code) but human ('sociological') ones too: more productive, effective and empowered developers.

David Wood:

You always fought against code bloat, and there are two ways to fight it. The first way is to encourage people to practice efficiency in the small, which was to understand the side effects of the code they were writing and to think about every line of code they wrote to make sure it wasn't unnecessarily long and they weren't duplicating code unnecessarily. The second way is to look for efficiency in the large scale as well, which is when you realize that two programs, or more than two, are actually trying to do essentially the same job, so instead of them having independent copies of code which is largely the same, you have one copy of the code, together with small bits in each of the applications where they provide their relevant specializations.

Object orientation provides a principled approach to code reuse. In contrast, nothing is easier or more tempting or more fatal, than reusing code by literally copying and pasting code sections within programs and

---

[13] At the time of writing, there is a good history at ***http://en.wikipedia.org/wiki/X_windows.***

between programs; object orientation is the ultimate principled antidote to 'copy and tweak'.

David Wood:

What we were fighting against was something that's sometimes called 'copy and tweak', when you're writing some code and you notice that something you've already written doesn't quite do the job but it nearly does, so you just copy it wholesale and then tweak it, so you end up with two not quite identical copies of the code and before you know it you end up with dozens of copies of essentially the same code, but it's not quite the same. And that's bad for all kinds of reasons. It's bad because it's inefficient. It's bad because it's very hard to maintain, if you discover that there's a problem with your original algorithm you may manage to fix it in one place and then the same bug remains very probably in all the dozen separate copies that you made in the meanwhile. Some people nowadays say, 'Well, the hardware has advanced so much, the same efficiency considerations don't apply.' And there's certainly an element of truth in that. We needn't work quite so hard to save every single byte. But avoiding duplication is still an important principle, because if you duplicate unnecessarily then you end up with maintenance problems and comprehensibility problems.

14.3 A Thoroughly Object-oriented Operating System

The rest of this chapter looks in more detail at how object orientation is used in practice in Symbian OS, identifying some of the most characteristic examples of object-oriented style and techniques. Symbian OS contains some important object-oriented patterns. What follows is not an exhaustive list, but it captures the larger scale object-oriented patterns in the system.

- Frameworks and plug-ins are a good example of the power of polymorphism and are ubiquitous in Symbian OS. As a pattern, frameworks operate at the level of the static architecture of the system (what parts are in the system, in other words) although they also have an interesting run-time, dynamic aspect.

- Active objects are thoroughly object-oriented and are probably best thought of as a design and programming idiom for avoiding the complexity of multithreadedness, while enabling asynchronous activities to be spun out of a single thread and spun back in.

- Descriptors are good examples of object-oriented encapsulation, best thought of as an idiom for achieving type-safe and memory-safe string handling.

- Cleanup is another example of good object-oriented design applied to solve the error recovery and propagation problem and is probably best thought of as a programming idiom (although it also touches the built-in typing model).

- Streams and Stores (the persistence model, in other words) is pure object-oriented design, that underlies a programming idiom for externalizing and internalizing data through an abstract, non-file-based abstraction.

- The User Library is a good example of simple encapsulation-style object-oriented design used to package some system services.

On the other hand, there are also some designs in the system which are more conventional, non-object-oriented patterns.

- The client–server model is not an object-oriented model (unlike frameworks, for example), even if it is given an object-oriented flavor by modeling the client–server relations with objects.

- The file system, beneath the client–server and framework architecture and the object-oriented interfaces provided by the stream–store persistence model, is in fact a conventional, non-object-oriented, files-on-disk system. (Compare for example with the thoroughly object-oriented 'object soup' storage model made famous by Apple's Newton and the similar database approach later adopted by Palm.)

- Critical aspects of the kernel design, for example the scheduler, are conventional in design, with hand-optimized assembler implementations for speed.

It is worth making the observation that the device for which the operating system was originally designed, the Psion Series 5, was quite a conventional device in terms of its technologies. So while the use of the technologies was innovative, the technologies themselves were conventional: a small, ROM-based device with screen and keyboard using conventional, removable (Compact Flash) data cards for external storage. To take the file system as an example, the choice of a FAT filing system was conventional but eminently reasonable. And in fact, if users wanted to use their removable cards to swap data with other devices, then FAT was the necessary choice, for removable media at least. Deciding for a conventional file system simply reflected the realities of data exchange for a handheld device in a world dominated by PC consumer computing.

This argument in favor of a conventional file system beneath the object-oriented wrapping is one that Geert Bollen, for example, has heard before.

> **Geert Bollen:**
>
> That's very much the argument you would have expected to hear from Colly Myers. But, it doesn't follow from the argument that the internal storage representation should be FAT, though that's quite a valid design choice.

The point is really that, while the system indeed is thoroughly object-oriented, adoption of object orientation was motivated by pragmatism and, in some cases, that same pragmatism led to more conventional, non-object-oriented design choices in the system.

Frameworks

A framework is a component that defines an abstract interface which it does not implement, but for which instead it provides a runtime loading and management mechanism for locating and loading external plug-in components that provide concrete implementations for its interface.

Framework–plug-in is a classic pattern therefore for separating interfaces from implementation, which is one of the driving design principles of object orientation (inherited, as it were, from the notion of the abstract data type, an important influence in the emergence of object-oriented languages). The framework–plug-in pattern is therefore classically good object orientation. It is also a natural pattern for enabling extensibility.

But frameworks are more than that. As [Beck 1999, p. 258] puts it, 'Design is hard'! A framework is, in effect, reusable design because it expresses a part of a system as a set of abstract, cooperating classes, scopes the behavior of those classes by defining their interfaces, and implements the interface between the framework and the underlying system; in other words, it wires the design into the system. At that point, the design work is complete. What remains is implementation, and implementation (the actual instantiation and also, if you like, the interpretation of the behavior scoped by the interfaces) is left to others including third parties.[14] Frameworks are an important design choice, because a framework design strongly determines how we should expect to work with some part of the system. In the object-oriented literature, frameworks go back to ideas of Deutsch and Johnson in the late 1980s.[15]

Of course, patterns are also 'designs for designs' and, arguably, so too are class libraries, but frameworks go further than patterns, because they

---

[14] Or as Martin Budden put it to me, 'Frameworks are for people who just can't make up their darned minds!'

[15] My references here are taken from [Beck 1999, p. 257].

are actually coded ([Rising 1998, p. 375]), and they go further than class libraries, because they are already pre-wired into the system.[16]

There is an objection: frameworks are complex and hard to write and need 'refining by fire' (see [Rising 1998, p. 184]). In many cases there is an additional, important dimension. Where frameworks expose interfaces on two sides, an 'up-side' application-level interface and a 'down-side' hardware-level interface (which is a common pattern in Symbian OS – the frameworks in the communications and multimedia areas are good examples), the question of the 'thickness' of the API, (how much code there is between the up-side and down-side interfaces) can become critical for retaining control of the design. The thinner the API, the more exposed it is to the opposing and even hostile forces of its up-side and down-side clients and the easier it is for them to subvert the original design intention.[17]

Frameworks can also be complex to use. Because they constrain the user, they must be understood well to be used well. Frameworks are a pervasive design choice in Symbian OS, because they are a particularly effective mechanism for enabling customization and extension at a deep level; the operating system implements an overall design in some particular area but licensees are still able to contribute highly customized behavior. Frameworks therefore can be found in all layers of the system, from top to bottom. Indeed, in some senses, the different ways that frameworks are used in the different layers may be the dominant design characteristic of those layers.

Model–View–Controller

Model–view–controller, or MVC, is a well-known design pattern which applies the idea of frameworks to the design of an application, 'the earliest framework that was recognized to be a framework', indeed, according to Johnson [Rising 1998, p. 376].

The high-level design of an application in Symbian OS really resides in this pattern, which provides a general model for the classes that an application needs, what their basic collaborations are, and how they are achieved.

The simple version of MVC, and the headline from the point of view of Symbian OS, is the separation of the application model (the data and the APIs which operate on it) from the application logic (the way the user uses the application and its data), not just as a coding rule, but as a well-founded design principle.

---

[16] As Johnson puts it, the difference between patterns and frameworks is the difference between reusing knowledge and reusing code. Frameworks reuse code and, thus, enable more immediate reuse than patterns [Rising 1998, p. 382]. Patterns are a degree of abstraction further out from frameworks.

[17] See Chapter 16 for some examples of the 'thin framework' problem.

Howard Price:

Model–view–controller goes way back to SIBO. It probably was Charles Davies and David Wood who introduced it. But, right from the beginning, that's always been the design. Right from the beginning, there was a decision that you really should separate these three things.

Active Objects and the Event Model

The event model is an important part of the wider application framework, but it is itself a part of the system which makes rich use of object-oriented principles. Its key principle is to abstract the event loop using the active object pattern. Again, the driving point here is to fix the design in the framework and so simplify the way that applications interact with it through the framework.

Martin Tasker:

Think of it this way. Firstly, the fundamental requirements of doing a GUI and the fundamental requirements of event handling motivate object orientation, and polymorphism in particular. So you have an event and you have a control, you send that event to that control or you get that event from that control. But then of course you need concrete types underneath that, you need concrete actions for concrete controls and concrete events. The point is that the fundamentals of event-orientated programs in a GUI context take you in a very short sequence of steps to object orientation.

At the heart of the message dispatch system, active objects are used to encapsulate and serialize (i.e., make sequential) incoming events. In Symbian OS, this is probably the object-oriented pattern which goes back the furthest. The alternative, of course, is the conventional 'big switch' statement style of classic Microsoft Windows.

Active objects provide a simple, natural, serialized, alternative.

Martin Tasker:

Colly Myers was right, active objects are a dream. For people who know that they are dealing with event-handling programs, they're an absolute dream. In an event-handling system, active objects are a really natural way of handling things and they are so much easier for programmers to work with than pretty much all of the alternatives actually, or any alternative that I had seen.

The key advantage of active objects is that they provide a simple model for maintaining a single flow of execution through the main program logic of an application, while still enabling a responsive, event-based implementation. They offer a simple alternative for what would otherwise typically require a complex, multithreaded approach.

> **Martin Tasker:**
>
> With multithreaded programs you have to do locks, so you always have to ask yourself, 'Am I handling these locks correctly?' And what eventually happens in those systems is that you get some conservatism, so you eventually end up with two or three global system locks which only your kind of super-elite programmers ever touch, and then you get a couple of local locks which are more or less private business, and you don't quite have to be a genius to do stuff with those locks on the basis that you only penalize your own code, so you're probably going to get it right eventually. But it's much more complex and it's much less elegant.

There are times, of course, when the active-object model is inferior to a more raw-thread-based model.

> **Andrew Thoelke:**
>
> Of course there are times when the active scheduler gets in the way, but for many things active objects are a better model. It's actually a better model to have cooperative active objects running in a single thread than to go multithreaded. But there are certain tasks for which multithreaded programming is definitely more useful. Certainly when you want to be able to do finer time slicing or you don't want to have to manually break up a task into discrete chunks. Sometimes it has been hard initially to integrate that sort of software with Symbian OS. In the early days, a client–server session was inherently tied to a thread. The client was the thread and that was engrained in the whole framework and model. And there were even cases where that was designed into the server. The server required that the thread and client were synonymous. That has changed now, explicitly to enable multithreaded clients. So you have the choice.

The Framework for Frameworks

An important change in Symbian OS v7.0 was the introduction of the ECOM 'framework for framework plug-ins', the Plug-in Framework component.

> **Andrew Thoelke:**
>
> The plug-in pattern was pervasive in the OS. Everything has plug-ins: the kernel does, the file server does, the window server does, and in all cases the actual

mechanism for tagging and identifying and locating a suitable plug-in that matched your interface protocol and statically matched your actual specific requirements right now, was very ad hoc, except for the fact that you generally used a UID to match the server protocol interface. But the way you located them, whether you searched for them, whether the loader searched for them, it was all very unpredictable.

Worse still, new variations were continually being created, of varying quality. Standardizing framework–plug-in design by introducing a framework for framework plug-ins not only brought more discipline to bear on the way plug-ins were searched for, found and loaded, it allowed the best design (the plug-in system created for the web and Internet browser) to become the template.

The basic principle adopted is the abstract factory design pattern.

Andrew Thoelke:

Our design is a fairly obvious pattern as soon as you see it. You can see that actually I can apply this anywhere somebody wants an abstract factory. You put that abstract factory something in a DLL, and then when you request that DLL you have it conform to your interface. And the ECOM plug-in framework just provides a standard way of doing all of that.

As it happened, standardizing the plug-in mechanism fit in well with the new requirements of the platform security architecture. (Unchecked loading of externally written plug-ins into system frameworks, or applications for that matter, is a potential security risk.) And that in turn helped to enforce the standardization, because it could be pushed through as part of the effort to implement the system-wide, pervasive changes required by the security architecture.

Andrew Thoelke:

With platform security, it was highly desirable to get away from everybody searching for their executable content, which is what's really going on when you're looking for plug-ins to load into your framework. And because we already had put ECOM there and it was an established part of the system, we could enforce it as the standard mechanism and police plug-in loading at that single point. So we had a way to force the migration to ECOM through. After PlatSec, there is no way for your framework to find out what plug-ins are available by enumerating the actual executables in the system because it's not possible to do that any more. So you either use ECOM, because it's got its own registration mechanism with resource files, which regulates add-in plug-ins, or you have your own registration mechanism where something which gets

> installed not only installs the executable, it installs something else which tells you that there's an executable plug-in that you can load, so you control what plug-ins can be added.

Streams, Stores and Persistence

Symbian OS, possibly surprisingly for an object-oriented system, supports a conventional FAT file system. At the same time it provides a native, object-oriented persistence framework that, for example, enables applications to externalize and internalize state without explicitly invoking file system APIs. It also provides a native database storage model, which is extensively used by applications such as Agenda and Contacts. Geert Bollen was the original architect of the persistent-storage frameworks.

> **Geert Bollen:**
>
> At my previous company I had implemented a persistent object database over RDBMS in C++. Actually, I had done that several times. So I was brought in as a C++ and design expert, and I was given the persistent-storage job. I had a team of two, me and Andrew Thoelke.
>
> **Andrew Thoelke:**
>
> It was only 18 months after I started and EPOC 32 was really kicking off in a big way. So I started working with Geert Bollen shortly after they had the major discussion about 'Is it going to be a database? Is it going to be a file system? Is it going to be some object soup? Are we going to buy it in? Are we going to write it ourselves?'

To start with they designed and implemented a DBMS API based on the DBMS API in the 16-bit system (which had originally been written by Colly Myers).

> **Geert Bollen:**
>
> The design and implementation is that of a classic, by-the-book RDBMS light, because it was always designed, for example, not to have joins. It was light.

The design layered a stream serialization interface over a dictionary-style persistent store over a (conventional FAT-style) file system.

> **Andrew Thoelke:**
>
> Initially they were going to say, 'Right, well pretty much every file's going to

be a database' and that was refined a bit later on, so it turned out that every file should be a database but they were going to go with a file system, and you had different sorts of databases for different sorts of files. And then the ideas got refined somewhat to having two separate layers, having a store architecture, and the notion of Streams and Stores which is slightly more basic than a full database, and then you have the database sitting on top of that for applications that explicitly want that kind of data storage.

Geert Bollen:

DBMS was always intended to be the heart of application storage. The Psion Series 5, recall, was to ship with a suite of built-in applications, including Agenda, which was a diary, meetings, and to-do list application, as well as office style apps. DBMS was intended to provide storage for them all.

For more conventionally file-based applications, such as the spreadsheet and the word processor in the original Psion Series 5 application suite, a more document-centric Store interface was available.

Later, the decision was taken to design a less abstract storage framework with object serialization at the heart of it. DBMS is a natural storage model for a database application, such as Data, but, for document-based applications, the requirement was somewhat different.

Geert Bollen:

We needed a Store for document-based applications which load up their application data when they launch and write it out when they suspend.

The question was then how this fit with the needs of the more transaction-based applications (such as the Data application, which explicitly used a database format, or the Agenda application, which implicitly used a database format); in other words, applications which didn't load their data into working memory, but had a store on permanent storage and conducted transactions against the store.

Geert Bollen:

The requirement was to build a framework which combined the needs of database operations with the simplicity of object serialization. That led to a transaction-based API for Store and we provided that as a fat API-style framework.

It is a 'fat API' because its single interface had to encompass the multiple interfaces it needed to support the multiple different kinds of objects. In other words, Store exposed a single, 'maximal' interface which was only partially implemented by any one of the concrete objects which implemented the Store interface and which were available at run time. In other words, the concrete objects which were available at run time each implemented a subset of the complete interface. The 'fat API' approach was adopted as an alternative to multiple inheritance, against which there was a strict injunction. Since at that time the 'mixin' solution had not yet been adopted, there was no other way to expose multiple interfaces.

Just as the database layers over the generic Stream−Store architecture, so it in turn layers over the more or less conventional filing system.

The goal of the Stream−Store design was to define a generic persistent storage mechanism suitable for any application type and robust enough to guarantee bullet-proof data safety in a model in which 'the user never saves'. Data safety was required no matter what the user might do or fail to do, including pulling out a removable media card while an application was using the data stored on it or even pulling out the batteries while applications were active.

Peter Jackson:

The end-user requirement is that you don't want corrupt data still around, you want a transaction-oriented filing system at some level so that if something goes wrong in the middle of what you're doing you don't have to do some expensive repair process. You might have lost a couple of transactions because that's when it went wrong, but the transactions that have already happened are safe. And the whole Stream−Store technology gives you that kind of layer. And for that reason I would say it's a lot better than just having a raw I/O system.

Andrew Thoelke:

I think for Streams, in particular, the actual class design went through something like four or five iterations, because we were trying to deal with the many ways you might want to use Streams and chain them together and run them back to back without resulting in a system that passes one byte around between different classes, which gives you performance problems.

Store is a complex design problem because what's required is really not a single solution at all, but multiple orthogonal solutions, serving the fundamentally different needs of different applications working with data in essentially different ways, that is, document-based versus transaction-based.

Andrew Thoelke:

DBMS is just a template API for a database, whereas Store is an extensible

> Steams and Store architecture where you can have different types of back end to it.

No design allows an application to continue writing to or reading from a media card that the user has removed from the device. However, different design choices provide different levels of protection against such unexpected failures. A conventional file system is the least robust solution. From a developer perspective, on the other hand, file-based semantics are so widespread and so engrained as to seem like second nature. In contrast, object-based serialization using externalize–internalize and store–restore semantics imposes a new learning curve at what should be a very basic level of programming, saving and restoring data. Inevitably too, if developers don't understand a model, they use it wrongly (or, indeed, even deliberately subvert it). In some respects then, the persistence architecture is another example of a Symbian OS idiom which has been described as a barrier to new developers, but which is firmly rooted in the original design requirements for the system (unrivalled robustness for a device class which is quite different from the conventional desktop device).

Object Orientation in the Kernel

The Application Architecture framework, the Control Environment hierarchy (CONE), the Graphics Device Interface framework and Store all make aggressive use of object-oriented and C++ techniques including interface inheritance, polymorphism, templates (in the Symbian OS 'thin template' style) and so on.

At first sight, most of these techniques are absent in the design and implementation of the kernel (whether the EKA1 or EKA2 architecture).

> **Andrew Thoelke:**
>
> If you look at the features of C++ that get used in the kernel, you don't see very much in the way of templates and you don't see very much in the way of derivation and you don't see very much in the way of virtual functions and overrides and frameworks, so it's a simpler use of object-oriented design. But it's still object-oriented design.

In fact, the kernel design has some interesting object-oriented features and they are sufficiently fundamental to persist in the design of EKA2.

Geert Bollen:

There is something interesting. There is a key insight of Colly Myers's which is very interesting, and it's an interesting use of OO concepts. A lot of what an OS does is very close to the metal, which is not an area that lends itself to standard OO-programming mechanisms. But Colly's key insight was that at one level the kernel is a model of an executing system. That's what the kernel's job is, parceling out the hardware resources of the underlying machine, and of course managing the computation, which is what the central processor is there to do. So what else really is the kernel? The kernel is really just a dynamic model of the computation which is in progress, of the thread of execution and the processes which exist and the way they are locked and communicating. You can make an OO model of that. You can implement that model using a bunch of objects. And essentially that's what Colly did. That was a very interesting use of OO concepts to design a kernel.

The basic kernel objects derive from this approach.

Andrew Thoelke:

Any kernel is going to have a control block for a process, but rather than just saying it's a control block, in Symbian OS it's actually an object with responsibilities to manage that process and the things that belong to that process. So it's not just a data structure that says, 'This is the process control block associated with this dynamic collection of threads and memory space.' It's an object and it has the responsibilities for that management within the kernel.

The kernel takes a different approach to the use of object orientation than is taken in the UI or Application Services layers, for example. Nonetheless the 'object-oriented clouds' are there.

Andrew Thoelke:

In the kernel in some sense you can say there are clouds, because you've got encapsulation, and you've got objects. Perhaps it depends on whether you say, 'Well, if you don't really use advanced inheritance and virtual functions then you aren't really using OO', right? So OO is derivation, using virtual functions, polymorphism and whatever, but a key part of it is about encapsulation and roles and responsibilities and the way the different objects in your system interact with each other using defined methods or messages, rather than by just invasively tinkering with each other's data structures. That's OO too. You can see that in the kernel.

Object orientation is not just a way of making a design tidier. It enforces discipline, not just on the design of a piece of software, but also on the way that users must go about designing their own use of the software.

> **Andrew Thoelke:**
>
> From the point of view of modeling a system and having clearly defined components which interact in well-defined ways, the kernel is really quite object-oriented. Whether EKA2 is as rigorously so, it probably is at the same level although for the sake of performance occasionally we don't always try and protect data. But from the point of view of saying, 'Are we utilizing virtual functions and polymorphism substantially?' – well the kernel has got one or two examples of it, the key one really is the device-driver framework, where you have a base class which represents the basic device driver and device driver-channel, and then a real device driver which implements a concrete device-driver object and device-driver channel. But, on the whole, that isn't really used in the rest of the kernel, for the good reason that the rest of the kernel is a closed system, and there is not a great deal of value to be added by pursuing that kind of design extensively.

There are some other choices that Colly Myers made which are interesting.

> **Geert Bollen:**
>
> Colly Myers decided to use polymorphism for implementing the differences between the target machine and emulator implementations. For example, `DThread` is the abstract base class and you have a derived emulator thread class and a derived on-the-metal thread class. I'm not so convinced about that.

Bollen's objection is that polymorphism in this case is the wrong concept to apply.

> **Geert Bollen:**
>
> I'm not convinced, because it's using a dynamic binding mechanism, that is, polymorphism, to represent a static property of the system, that is, whether you are an emulator build of the system or an on-the-metal build of the system. You don't need to do that at run time. In fact, you really don't want to do that at run time. That could as well be a compile-time or link-time binding.

Using a complex mechanism where a simple one would do (as Bollen says, 'A typedef would do it: `#ifdef baretothemetalthread...`') seems to break a basic design principle, though no doubt there was a reason for the decision when it was made. But there is a deeper point, echoing one previously made.

Geert Bollen:

Modeling the computation, making that literal and the operations that you do is still an interesting choice. And it points out something else which was going on, the desire in the architects to constrain the design choices, to have a limited number of patterns in the system. This is an OO system, so let's use OO mechanisms. There's some possible impact on runtime? So?

15

Just Add Phone

15.1 Introduction

Mobile phones are uniquely complex devices: more complex than PDAs; more complex by far than PCs. This case study looks back to the critical early points in the evolution of Symbian OS into a fit-for-phones operating system. It looks at why phones are different, what particular challenges they pose and the impact of those challenges in shaping Symbian OS.

It is often said that Symbian OS was 'built for mobile phones' and the claim, while true, does not remind engineers of the roots of the operating system. Its predecessor, EPOC, was conceived and implemented first as a mobile operating system for PDAs, even though, by the time it first shipped, mobile telephony had been identified as a critical market in which the operating system would win business and work was well in hand – driven by phone licensee collaborations – to put Symbian OS inside phone handsets.

As this case study shows, the shift in emphasis from PDAs to phones involved some very real challenges. Understanding what those challenges were helps us to build a deeper understanding of what Symbian OS is today.

15.2 Anatomy of a Phone

'Mobile phones', says David Wood, 'are just irreducibly complex.' He should know. As one of the five founding directors when Symbian was created,[1] Wood took responsibility for the Technical Consulting arm of the company – with a mission to support licensees, old and new, to create 'great phones' on Symbian OS. Technical Consulting has always been at the forefront of the process of securing licensees' use of the system and following through to shipped products.

---

[1] The others were Bill Batchelor, Colly Myers, Stephen Randall and Stephen Williams.

> **David Wood:**
>
> There's no getting away from the fact, it's not Symbian OS that makes smart-phones complex. Smartphones are complex simply because of the enormous number of different technologies that are contained in every single smartphone.

Compared with PCs or even PDAs, phones pack an astonishing number of different technologies into a tiny package. The things which in a PC are peripheral become integral in any pocket device; screen, keyboard, speakers, microphone and soundcard are packed in with CPU core and memory and permanent storage. But phones, of course, go further. There is the phone radio hardware itself and possibly multiple other radio interfaces, such as Bluetooth, and their associated software stacks, full networking, and a complete multimedia system. Megapixel cameras with true optics have arrived (Zeiss lenses, for example, and optical zoom) as, of course, has stereo sound with real-time compression and decompression for stereo playback. Integrated high-definition TV is the most recent arrival, hard on the heels of Wi-Fi, with Wi-Max waiting in the wings. Add in the power-management technologies needed to deliver long battery life while fueling this impressive array of technologies with ever-increasing processor speeds yet also avoiding the device becoming (literally) hot in the user's pocket; the PC, in comparison, starts to look trivial.

Looking back, the hardware architecture of the early devices for which Symbian OS was originally designed now looks remarkably simple too. A hardware schematic of the Psion Series 5[2] shows little more than an ARM core connected via a data bus to ROM, DRAM, and removable media-card memory, with direct connections to the remaining hardware: an audio codec for microphone input and speaker output, RS-232 and infrared UARTs, LCD screen and digitizer overlay (via an analog-to-digital converter) and keyboard (via parallel I/O pins). A power supply unit drives the system from two standard AA batteries and a wristwatch-style flat cell backup battery. Interestingly, the Series 5 hardware was itself more complex than that of some competitors, for example, Apple's Newton or Motorola's Magic Cap devices [Wolf 2001, p. 548].

15.3   The Phone Operating System

A phone operating system must manage complexity at a number of levels.

- It must manage the sheer hardware complexity of a converged device but, more specifically, it must manage a double platform: a highly specialized, data-centric (including voice data[3]) radio hardware device on

---

[2] [Furber 2000, p. 366] shows just such a schematic.
[3] Voice-centric GSM/2G has morphed into data-centric 3G.

the one hand and a multimedia-ready, networked, application-centric device on the other.

- It must cope with the sheer software complexity and, more specifically, a double software stack: specialized communications and data-centric protocol stacks and real-time channels on the one hand and a GUI-based, friendly, application-rich consumer system on the other.

- It must deliver the performance expectations of a general consumer market (toasters 'just work' and so phones had better 'just work': when they stop working, they are thrown away), quite different from the expectations of users of desktop computers, PDAs or gadgets.

- It must conform to the usability expectations of a general consumer market (no one expects to read the manual for a toaster; why should they read the manual for a phone?), again quite different from specialist users.

- It must stay fit and keep up to date with rapidly evolving technologies, a rapidly evolving network-services (operator-services) infrastructure and evolving open standards (often multiple, competing global standards).

Depending on who you talk to, 'phonification' means different things but, in principle, it means all of the above. 'Being a phone' is different from being merely a small, pocketable, mobile device.

Architectural Impacts

The impacts on the system architecture of the move from PDAs to phones are worth examining.

- The most obvious impacts are on the Kernel Services and Hardware Interface layer. The Kernel itself must either support the real-time needs of the baseband or it must be made amenable to an alternative solution. It must also support licensee ASIC or ASSP custom chip packages, which may mean supporting alternative memory models and memory-hardware architectures. It must support phone-specific device peripherals, including screens and keypads and possibly dedicated hardware such as a phone flip.

- In the Base Services layer, the File Server must support multiple file-system architectures and media types (both NOR and NAND flash, for example, and probably also hard drives), providing specialist services such as wear-leveling (for NAND) and demand paging.

- In the OS Services layer, the Comms Services architecture must support telephony protocols and integrate with networking support; graphics and multimedia services must support phone use cases (such as camera-phones and music-player-phones); connectivity requirements are probably also different between PDAs and phones.

- The Application Services layer is significantly affected too, with the different phone use cases for applications and new services requiring support.

- Finally, the UI Framework must provide support for a dedicated phone user interface.

There are almost certainly other system-wide impacts:

- system performance characteristics are likely to be quite different for a phone and a PDA

- greater modularity in the system may be necessary to enable the different product cycle for phones (a more platformized model, in other words) and a full product matrix

- adaptation almost certainly becomes more complex, with a much greater range of technologies needing integration (web browsers, viewers for content such as Flash and PDF, and Office file viewers, font technologies, etc.) and plumbing to close hardware support

- service assumptions turn into architecture headaches; phone operators are simply not used to open network models (third-party software availability, for example) and a system-wide security model becomes a requirement.

Supporting the Baseband

The mobile phone 'baseband' is the software signaling stack that accepts a data stream (e.g., the output from a digital signal processor which has been fed voice input from a microphone) at the top layer and emits a stream of encoded data frames at the bottom layer onto a data channel to radio–air-interface hardware and vice versa; accepting encoded data frames at the bottom and emitting a data stream at the top (if it is voice data, it is fed to a DAC for conversion to audio output).

The signaling stack sits on a software layer of some kind, traditionally a real-time operating system (RTOS), which in turn drives the baseband processor, typically a general-purpose CPU such as an ARM or StrongARM dedicated to the RTOS and the stack that it hosts. The radio hardware is likely to vary between phone makers, from custom silicon to standard bought-in parts, and the data channel might be old-fashioned serial or, more recently, USB.

Non-voice data on 2G and 2.5G and all data on 3G is packetized (into TCP/IP packets). The DSP/DAC steps are omitted but otherwise the packets follow the same path through the signaling stack and are tunneled as GSM/2.5G/3G frames.

The basic hardware design choices for a Symbian phone revolve around whether to use one processor or two and, if two, how to connect them. The two-CPU option treats the baseband, or phone side, as a completely separate hardware subsystem and interfaces it to an application subsystem with its own application processor (an ARM or StrongARM CPU, for example), running Symbian OS and applications. The single-CPU option creates a single hardware system, and shares it between an RTOS and Symbian OS or (enabled by the Symbian real-time EKA2 kernel) uses Symbian OS exclusively to host both the baseband and the application stack on a single CPU.

While to some extent how licensees architect their phones around Symbian OS and the design choices they make are opaque (and often jealously guarded), the consequences clearly impact the operating system design and the assumptions it makes about the environment in which it runs.

The hardware design options are as shown in Figure 15.1.

A: Two processors connected by a fast serial bus: CPU A is the baseband processor and hosts an RTOS, which in turn hosts the baseband stack. CPU B is the application processor and hosts Symbian OS, on top of which is layered a bespoke or licensed user interface that hosts applications.

B: A custom package with two CPUs and shared memory at the register level. CPU A is the baseband processor and hosts an RTOS, which in turn hosts the baseband stack. CPU B is the application processor and hosts Symbian OS, on top of which is layered a bespoke or licensed user interface that hosts applications.

C: A single processor hosts both the RTOS and Symbian OS. RTOS runs the baseband stack and Symbian OS runs the user-side processes; the two operating systems have a mutual agreement to share the CPU, RAM, device drivers, and other system resources.

D: A single processor hosts Symbian OS with real-time kernel EKA2. Symbian OS, abstracted by a custom 'personality layer', runs the baseband stack, device drivers and user-side application processes.

But telephony is not just a matter of getting raw data to the baseband. The baseband needs to be under application control, which means that there must be application interfaces to the phone side from the application side. On a typical Symbian phone, the phone application is simple and can therefore be relatively hard-wired to the phone side, but phone

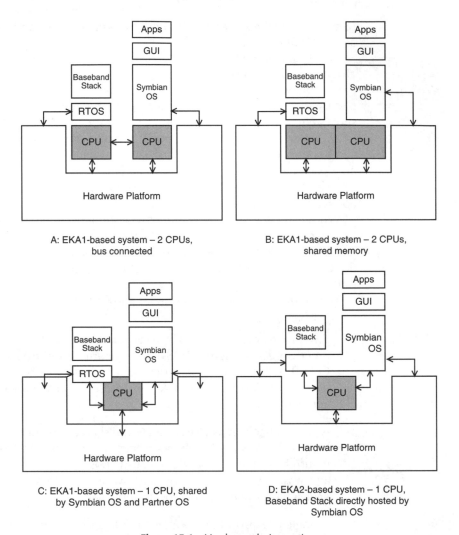

Figure 15.1   Hardware design options

books, call logs, messaging (SMS, MMS, email, and so on) require access to phone protocols and data services (such as networking and Web browsing) need to control the phone side in a modem-like fashion.

While the telephony application that the user sees is relatively simple, the underlying engine which sits beneath it is quite complex. It needs to handle a number of cases such as ensuring that emergency calls are always possible, even in low-memory conditions, and it must interoperate with hardware accessories such as headsets, as well as specific call-handling and over-the-air (OTA) settings protocols (e.g., call-handling and SIM toolkit functions).

Supporting New Hardware

The real-time kernel (EKA2) is still valuable even in designs that retain a separate, dedicated RTOS or partner operating system. Whether or not it hosts the baseband, EKA2 has advantages on the application side too.

Ian Hutton:

Hosting the telephony stack directly on Symbian OS requires real-time capability – and this is an issue. But the other argument for EKA2 is that it will allow mobile-phone manufacturers to integrate more multimedia hardware. The increased complexity of the hardware puts demands on the operating system that are increasingly hard to sustain, just with the level of interrupts and so on, and that's where EKA2 makes the difference. So enabling the integration of more and more hardware without compromising performance, which is what we are witnessing with phones, is the real bonus.

The other aspect of new hardware – cameras, audio-codecs, high resolution displays and multiple displays, multiple radio interfaces and new memory formats such as NAND flash, to give the most obvious examples – is that it needs drivers and, increasingly (and especially for more exotic hardware), the drivers are likely to be proprietary rather than supplied by the operating system. Easy integration of third-party and partner drivers becomes a significant matter. EKA2 has been designed with these needs in mind.

Supporting Services

Networks are driven by services and services are supported by phones. As voice services become increasingly commoditized, existing non-voice services such as messaging, browsing and new services (alternative network access through Wi-Fi, broadcast TV, presence and navigation) become more important.

Supporting Features

There is no good definition of what features make a smartphone into a smartphone, a mid-range phone into a mid-range phone or a low-end phone into a low-end phone.[4] Typically the measure is Bill of Materials (BOM) cost; as manufacturing techniques improve and Moore's law[5]

---

[4] In [Lindholm *et al*. 2003, p. 172], 'smartphone' is still an emerging category defined by its balance of phone, personal productivity, imaging and gaming features.

[5] Moore's Law, derived from an article by Gordon Moore of Intel, originally published in Electronics Magazine in 1965, is popularly formulated as predicting a doubling of computing power every 18 months.

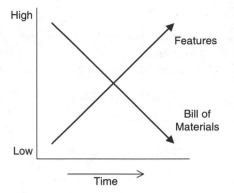

In 2007, best guesses are that entry-level phones will target BoM costs of less than $20, while high-end phones based on Symbian OS and competitors (Linux, Microsoft) will target costs of less than $100. Feature phones, the so-called 'mid-tier', are likely to be aiming at less than $50.

Figure 15.2 BOM costs fall but feature pressure rises

continues to hold, BOM costs fall (See Figure 15.2). As volumes continue to push inexorably upwards, driving marginal costs down, so the cost of a given feature set drops inexorably.

Meanwhile feature pressure (the demand to pack more and more functionality into phones) exerts a degree of counter pressure, driving ROM/RAM peripheral hardware requirements up. (For example, the camera-phone has evolved into the multi-camera-phone; still pictures have become video sequences; the ringtone phone has become a music-playing phone and has subsequently evolved into a direct competitor to MP3 players.) It's not clear whether that means that the BOM costs are not falling as fast as they might and therefore the line between high-end and mid-range is not falling as quickly as it might or whether it just means that we are all migrating to the high end.

Supporting the User Model

Phones are branded goods, fashion items, consumer appliances (see [Lindholm *et al.* 2003, Introduction]), and a host of other unlikely things that drive user expectations for how they behave, how easy they are to use and what they do. Phones have hard keys and soft keys; some have keyboards, some have pens, some accept voice commands; all may raise issues about handedness and screen orientation. Users expect 'natural' interaction models, associate interaction styles with brands, and fuel both performance pressure and feature pressure (where the performance and features of mid-range phones are increased to match high-end phones).

Supporting the Market

Phone manufacturers typically want rapid product cycles as part of their drive towards increased volume of sales. Feature pressure (again with a helping hand from Moore's law and the economics of volume) drives a rapid technology-proliferation cycle. The pace of the drive towards increased volume of sales continually quickens and the cycle times for technology moving from research labs to products speed up.

The result is a demand for continuously greater agility from suppliers, including the operating system supplier, as well as continuously greater predictability, fuelled by the relatively long product lead time coupled with a short product cycle and lifetimes.

There is additional pressure on a system when it becomes a platform (being a supplier is easier: there is only one customer to please).

1997: The State of the Art

It is worth recalling what a mobile phone was in 1997, the year the Psion Series 5 was launched.

Mobile phones had been mass-market products since perhaps 1994–5, depending on which geographical area you look at, with the UK's penetration of a little over 20% about average for Europe, excluding Scandinavia (which was more than twice that). Penetration in the USA was a little ahead of the UK, Canada a little behind, and in the rest of the Americas almost non-existent, as it was in China [Haikio 2002, pp. 157–9].

Worldwide, mobile phone sales were just under 108 million units (in 2005, just under 800 million units were sold).[6] Motorola was dominant with 23% market share, Nokia was a little behind with 19%, and Ericsson was a little more behind with just under 15%. Nokia, indeed, seemed to have stumbled, issuing profit warnings in both 1995 and 1996. Otherwise, Vodafone in the UK had a little over two million customers (15.5 million in 2005) and had just introduced a pay-as-you-go service. The WAP Forum had just been created, with the first WAP phones two years or so away.

Motorola's phone of the year was the SlimLite (see Figure 15.3a), with a 4×16 character monochrome display and a 100-entry phonebook memory. Among Nokia's hot phones was the 3110 (see Figure 15.3b), in which a new, easy-to-use 'one key' (the Navi-key) user interface debuted. The design context was still dominated by users' propensity to use anything that looked like a dedicated 'Call' key to try to get a dialing tone before keying in a number. One of Navi-key's goals was to help

---

[6] The statistics and product specifications in this and the following paragraphs come from public sources. Some useful URLs include
- *www.gartner.com/press_releases/asset_132473_11.html*
- *www.gartner.com/5_about/press_room/pr19990208a.html*
- *http://en.wikipedia.org/wiki/List_of_mobile_network_operators*
- *www.gsmarena.com/motorola_slimlite-78.php*
- *http://en.wikipedia.org/Nokia_7610*
- *www.paconsulting.com/news/by_pa/1997/by_pa_19970115.htm*

(a)   (b)

Figure 15.3   a. Motorola SlimLite and b. Nokia 3110 phones

users learn to 'punch in the number before trying to contact the network' [Lindholm *et al.* 2003, p. 75].

Typical screens that year were 84×48 pixels, monochrome, giving 3 + 2 character lines (3 lines of 'user' text and 2 status lines). A typical 1997 Nokia (based on their DCT3 hardware platform) had 400 physical parts ([Lindholm *et al.* 2003]). DCT4 arrived in 2001 and halved the number of parts to 200.

Mobile network data services were largely unused except for SMS, which the analysts were still describing as underused. But the two-box PDA–phone solution, using a GSM phone as an infrared or serial cable modem, was seen as the coming thing for enabling email and Internet access and fax transmission from PDAs via phones.

The still somewhat revolutionary alternative could be glimpsed, though, in the Nokia 9000 (Figure 15.4), Nokia's first generation Communicator and the first converged PDA–phone device, introduced in 1996.

Figure 15.4   Nokia 9000

Convergence

Convergence may or may not have been inevitable but, despite its clunky physical form factor and monochrome display (640×200 pixels), the Nokia 9000 had put it on the cards. (It was a big, exciting, and very secret project inside Nokia; so much so, according to [Lindholm *et al.* 2003, p. 74], that no-one wanted to work on basic phones, which suddenly looked 'trivial' in comparison. Lindholm's Navi-key project had a hard time competing with it for resources.)[7]

The two-box solution, pairing a data-centric handheld such as the Psion Series 5 with a communications-centric GSM mobile, still seemed to be where the market was leading. Convergence, on the other hand, seemed to suggest putting a phone into a Series 5 or, more to the point, putting the Series 5 operating system into a phone. At some point, it became the inevitable next step.

Putting a Phone into the Series 5

The challenge which Symbian's engineers faced was to take an operating system that was designed and created with an almost obsessive attention to the details of a particular product context and evolve it to suit a different product just as well. It does not seem far to travel, after all, from the (ARM-based, pocket-sized, 1/2 VGA screen, clamshell-case, keyboard-centric, battery-powered) Psion Series 5 PDA to the (ARM-based, pocket-sized, sub-1/2 VGA screen, clamshell-case, keyboard-centric, battery-powered) Nokia 9210 Communicator. And it seems not that much further from the Communicator to an even more phone-like and less PDA-like device, in fact to a mainstream (if high-end) phone.

The Symbian OS mantra 'built for mobile phones from the ground up' doesn't quite tell the complete technical story. The operating system which shipped in the summer of 1997 in the Psion Series 5 knew exactly what device it was built for: a clamshell, AA-battery-powered, always-on PDA, with a keyboard, a touch screen and a couple of serial ports. It was optimized for mobile devices, but it also required several evolutionary steps to properly address mobile phones specifically.

> **Peter Jackson:**
>
> I don't know the point at which we really got to grips with the idea that we were making an operating system for a phone. Because when we started we weren't, we didn't. We were making an operating system for a Series 5, not even for a generic PDA but for a Series 5, and for other things like it that we might invent.

---

[7] Lindholm was later the architect of the Series 60 user interface and arguably the driver behind its platformization.

15.4   Telephony

Telephony services in Symbian OS are organized around the ETel server and framework, which is at the heart of the application-side interaction with the phone baseband or, as it was originally conceived, any modem at all.

As Andy Cloke remembers it, development work on ETel started even before the Series 5 had shipped. See Figure 15.5.

Andy Cloke:

We started doing ETel when Roger Nolan[8] was running the Comms group. I don't think it was entirely clear that we were going to do the next Nokia Communicator. At least, it wasn't clear to me at that point. There was still quite a focus on PDAs. Certainly, the thought that we wouldn't be doing PDAs hadn't become clear. So it could have been PDAs in a variety of forms: the modem could have been built in; it could have been a plug on; it could have been wirelessly linked. It was quite different from the way we are today.

The device that came to market was the Nokia 9210 Communicator, the direct descendant of the Nokia 9000. But at least two other phone-based projects ran more or less concurrently with the Communicator

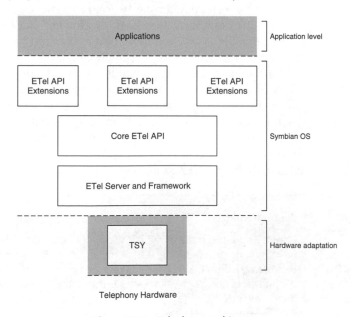

Figure 15.5   Telephony architecture

---

[8] Roger Nolan had previously been a member of Colly Myers's kernel team.

project, one for a Philips phone 'companion', which did not come to market, and one for the Ericsson R380. (See Chapter 2 for more about the background to these early projects.)

The starting point for the design was not really a converged device at all. While assumptions had moved beyond being simply an operating system for the Psion Series 5, the thinking was closer to a 'super PDA'.

Andy Cloke:

We were thinking about the possibility that you might have multiple modems that may appear and disappear. So the concept of PCMCIA cards that could be inserted and removed dynamically was still very prevalent. So that was the motivation for having multiple phones: you might have an internal modem and a PCMCIA modem and another one accessible over Bluetooth, for example.

ETel first appeared in the codeline at version 001 in July 1997 and by the end of the year was providing basic 'Hayes' control of a GSM phone (as modem) over a serial connection. As it happens, the test phone was none other than a Nokia 9000. If there had been uncertainty about whether the Nokia Communicator project was the target, it was resolved by the end of the year and ETel was explicitly delivering into the Nokia Communicator project.

Andy Cloke:

We established the ETel API first and then, in parallel as I remember, we started developing an AT-command-based TSY, and we didn't quite realize how big a job that was. And in parallel with doing that, we started talking to Nokia, to the Communicator development team. I remember going to talk to them and we thought we were going to get completely toasted when we presented our ideas because we were fairly new to it, and clearly these people knew everything about telephony and we didn't. I can remember coming back on the plane after the meeting in Tampere and thinking 'Wow!' Because we had expected to get roasted alive and that hadn't happened, so we just thought, 'How great we are, we've managed to get it right first time!' It wasn't until later that we realized how easy they'd been on us and how much we had to learn. In the end, we got strong design steers from Nokia and Ericsson.

In 1998, the first GSM extensions began to appear; the fax server was integrated from a standalone component into the ETel framework (as a framework extension). By the end of 1998, the code had reached a degree of stability as an alpha-release component of what was by then the 'EPOC Release 5' platform, aimed at the new Psion Series 5MX, a souped up palmtop intended to have full phone and networking capability (in two-box mode).

In January 1999, ETel branched for the Ericsson R380 and by May it had become a component of the ER5u baseline, the so-called 'wide' or Unicode build of the operating system, and part of the main codeline. By then, there were multiple licensees taking the component.

Andy Cloke:

As you can imagine with companies working together but competing in the same market, there were quite a few political considerations; the opportunities for getting the different parties into a room together were fairly small, and the amount that they would discuss in the room together was also fairly small. You have to understand that we hadn't shipped a phone at this point, the Ericsson R380 hadn't shipped and the Nokia Communicator (9210) hadn't shipped either. So it was difficult for us. In terms of the TSY development, both Ericsson and Nokia were saying, 'Don't worry about that, we'll do the TSY'. We all thought that this was a good thing, little knowing that with us creating the API and them creating the TSY that plugs into the bottom of it, we were creating an integration nightmare. If we'd done the TSY design more in parallel it would have been better. We would have avoided some pain later.

ETel Design Goals and Architecture

ETel's design drivers are clear enough. It was designed to support multiple clients and multiple dynamically loaded TSY modules with the goal of enabling a PDA to access either a built-in or external modem – indeed, multiple modems at any time – typically to enable data calls (for example, SMS messaging and fax) and Internet access (for example, email and Web browsing). From the beginning, there were also some clever phone-specific extras, designed for the case in which the 'modem' being accessed was in fact a GSM mobile phone. For example, SIM toolkit functions allowed synchronization of on-phone SIM-stored and memory-stored phonebook entries with the PDA contacts application.

The design took the existing serial communications architecture as its starting point (Figure 15.6). The design goal was to provide an abstract model for controlling phones from Symbian OS. Phone hardware was understood in classical Symbian terms as a resource to be shared by multiple applications with serialized access; in other words, the server model applied. Analogously with CSY serial communications modules, TSY telephony modules were defined as the abstractions for actual hardware and a similar framework architecture to the serial communications system was adopted. Depending on what hardware was available to the system (and what application was requested) the TSY would either interface directly to the hardware (the baseband–built-in modem case) or access the hardware through the serial communications system (for example, a TSY sending AT commands to a true Hayes modem or to a GSM phone presenting an AT interface, over infrared or a cable serial connection).

The ETel architecture closely followed the tried and tested architecture of the Comms services. A classic Symbian OS client–server interface shares access to telephony services and hardware between multiple clients. A framework architecture provides for a core API which is extensible in two directions by plug-ins; horizontally, ETel extensions add richer functionality to the basic core set with extensions for fax, packet data for GPRS and 3G, the Multimode extensions for CDMA2000, and SIM toolkit extensions; vertically, plug-in TSY adapter modules, modeled on the Comms CSY modules, that are loaded on demand interface the abstracted ETel APIs to the actual hardware available.

ETel therefore interacted closely with the serial communications system and, in fact, did so through the generic mechanism of the Socket Server (requesting a serial socket connection, for example). Looked at from the other direction, ETel equally became a socket provider, providing telephony sockets to serial or networking components.

The communications design analogy was an obvious starting point, especially because the design was proven and had shown itself to be both extensible and flexible. In effect, the whole communications architecture was elaborated horizontally so that the telephony system was created as

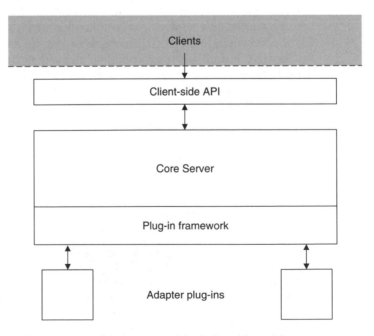

Figure 15.6 ETel design mirrored the design of the serial comms server

a peer of the networking and serial communications systems. The serial system was 'first among equals', primarily because, in the use cases that drove the design, serial communications via modem always provided the physical access to the network (phone or Ethernet).

With hindsight, however, the PDA or two-box use case dominated over the built-in phone use case. ETel started with a design goal which it met admirably, but which was rapidly overtaken by the change in context. Symbian OS in a phone has different telephony requirements to Symbian OS in a modem-connected PDA.

Andy Cloke:

Our early design assumptions were not really true any more; modem wasn't the primary use case. And certainly the primary use case that we're dealing with now is a phone that has a single baseband, and consequently ETel has morphed into a baseband abstraction.

The more complex question is whether ETel's design supports the needs of an abstracted baseband or simply provides an API for application access to the baseband.

In design terms, ETel is a classically good example of object-oriented abstraction. Its three key concepts are phones, lines and calls. Clients request an `RPhone` session, from within which an `RLine` subsession can be opened and a further `RCall` subsession can be created.

The TSY instantiates corresponding derived classes from the framework: `CPhoneBase`, `CLineBase` and `CCallBase`. In turn, these objects create AT commands, instantiated as objects derived from `CATBase`, which are then sequenced through the TSY's command sequencer, which controls a communications port requested through a serial communications session owned by the TSY. The communications port in question provides a direct (docking-style), infrared or serial-cable connection to an actual phone, which responds to the AT commands.

Andy Cloke:

I wish in hindsight I had spent more time looking at the ISDN specs, standard wireline phone specs, and the GSM standards. I think `Phone` and `Call` are still quite valid abstractions, but I think `Line` was a bit of a waste of our time. It was there originally because, on a single GSM phone in the GSM specs, you can have a different telephone number for voice calls, data calls and fax calls.

Fax, indeed, turned out to be troublesome.

> **Andy Cloke:**
>
> Fax over GSM is tricky because of timing issues. You have to spoof packets on the base-station side to stop the fax machine timing out because the transmission delays are too long over GSM. Well, and then it all gets more complex because of the alternate numbers. But anyway, I think I would have dropped the `Line` abstraction.

Another, subtle, design assumption that proved a problem comes back again to assumptions about the nature of the modem that is providing access to the phone network.

> **Andy Cloke:**
>
> That's something else that ETel was predicated on, that the modem would become completely commoditized, and while it is slowly moving in that direction, with the advent of 3G and all the different services on there and all the different ways that packet data works and some of the services work, that is not true yet. And coming you have HSDPA and HSUPA – High-Speed Downlink and High-Speed Uplink Packet Access – which will be the next service that will be required. It's already being referred to as 3.5G. So these sorts of forces stop the modem software – I'm using these words, modem and baseband, interchangeably – they stop that software becoming a commodity because it just moves on and on, it's moving so fast. How many non-proprietary 3G signaling stacks are there out there that have any kind of market credibility? I think the answer is as small as three or four, ones where you're going to buy off the shelf, so it's certainly not yet a commodity item.

The Problem With ETel

> **Charles Davies:**
>
> We did a telephony API, whereas what we needed was a distributed computing solution with a baseband. Or to put it another way, it was not so much a telephony API we needed as a subsystem that does more than telephony, but of which telephony is one application.

The more general case of supporting data communications, including abstracting the baseband, is the direction in which Charles Davies is heading. With hindsight, the requirement was not to provide abstract control for a phone, because outside the two-box context that is not what devices require. What devices really require are high speed data connections between the application side and the phone baseband. As Davies puts it, something much more like what he calls 'a distributed computing solution with a baseband.'

> **Charles Davies:**
>
> What that really means is that you supply a reliable by-value RPC mechanism between the two sides which is independent of the application you are doing it from, and then telephony would be one of the applications.

On the question of the importance of telephony, Charles Davies is even more radical.

> **Charles Davies:**
>
> You can debate how important a telephony API really is anyway. Many other APIs are much more important from the point of view of the operating system, in the sense that connections to packet data are more important, because telephony is only needed by one application, although it certainly is an important application. But it's all the rest of it that allows a phone to be built.

Andy Cloke takes a more sanguine line.

> **Andy Cloke:**
>
> With hindsight we would have preferred to establish a narrower API upon which Symbian would have built its services, in other words, data calls, establishment of PDP contexts and secondary contexts, transmission of SMSs, reception of SMSs, these sorts of things. It's very important that we have an API which works well there, a downward, hardware-adaptation-interface-style API.

The Danger of the Thin API

For Andy Cloke, the real lesson here is a more general one about API design in the case where both the application-level API clients and the hardware-level adapters are extrinsic to the operating system, and the operating system has the role of defining a thin, stable API against which partners can develop their value-adding components. In this case, two big issues arise: how thin can a thin API be and still be sustainable? And, how do you manage the design and development process to ensure that the API is viable and the right one?

Andy Cloke:

ETel today is a baseband abstraction layer and suffers because it tries to provide an abstraction for the whole of the baseband. Not just modem functionality, but also the ability to make calls, put calls on hold, do multiparty calls and all that stuff, and it also contains all the supplementary services, call forwarding, notifications that somebody else has put you on hold, so all of this kind of Layer Three GSM signaling, all these notifications need to come through. As well as that, it also has to be able to transmit SMS messages, which are actually a reasonably complex beast when you dig down because there are various different levels of ACK-ing that occur in the SMS protocol stack, and then there is USSD too, the unstructured service data.

Symbian does not build the telephony application above ETel, and we do not own the TSY below ETel. Where you have code like that which is so thin between two parties, between the top-side API users and the bottom-side API users, especially when they're very broad, well it's almost just worth getting out of the way and letting the users sort it out between them. Actually an API produced by the creator of a telephony engine might well deliver the best result.

It is possible, of course, to define, create and successfully manage 'thin' frameworks. There are successful examples in Symbian OS, as well as industry examples.

Andy Cloke:

A good example is Direct X, although it's in a slightly different area. You have the games developers on one side and the graphics drivers creators on the other side. It's a tricky path to walk, creating that kind of thing. You have to have everybody bought into the fact that you need it, and then create a forum with a number of graphics card creators and a number of games developers, and you need to mediate between them, which can be quite hard. Similarly, this is a hard thing to do in the telephony area.

What makes it difficult in the case of ETel is the very specific dynamic between Symbian, as platform company, and its licensees, as phone manufacturers. Because licensees already have their own solutions in this area – it is, after all, their very particular expertise – they do not necessarily see this as a place in which the operating system either can or should try to add value. From the Symbian perspective, however, giving up its telephony offering would reduce the value of the platform for potential new licensees lacking existing investments in telephony. Symbian stands in the centre of these conflicting positions.

Andy Cloke:

Some licensees do not see this as something they require. They would like to talk straight to the silicon, direct to the baseband. They don't want anything in the way. So they just want to know what is the minimal TSY that they need to create in order to support Symbian OS on top of it, and the rest of it they bypass. They just talk straight to the baseband.

But that's not the position of all licensees. For example, it is a matter of public record that Nokia does not license its own telephony application (its crown jewels, or part of them) to competitors. It is important therefore for the viability of the S60 platform that its licensees should be able to integrate their own (or a third-party) telephony application. For these licensees, an application interface is critical and it is also critical that it is supplied as part of the operating system, so that it can be standardized and controlled, since without a Symbian OS API between the application above and the TSY below, there would be no standard for the interface between the two, or at any rate no controlled standard.

This implies a clash of interests between Symbian, those customers who are in the business of making complete telephony solutions themselves and simply want the rest of the operating-system functionality, and those who may not have a telephony application at all and want the operating system they buy to offer a complete solution.

This is, of course, not a design or engineering problem; it's a business problem and there is no engineering solution to it. What is the lesson?

Andy Cloke:

You should do an incremental design and you should also make sure that you have an interested community both above and below the interface and that you involve them, that you validate the interface by continuing effort on it through the development process of the clients and the plug-ins on both sides, really to prevent yourself being circumvented wholly by those people. Especially when you have the same company building both the client and the plug-in. Because they will build assumptions into their client code which are fulfilled by their plug-in code, but which are completely absent from the specifications. And of course they will extend it further than you would ever expect.

15.5   Messaging: It's Different on a Phone

Messaging is another area which proved tricky in the transition to being a full-on phone-focused operating system. The Psion Series 5 was specifically intended to provide integrated email and, in particular, to be 'Internet ready', offering standard, Internet-based mail solutions, as well as access to other Internet services. Support for Internet-based email was therefore an important design driver.

Keith de Mendonca joined Psion in 1994 in one of the early, small expansions of the company which came with the success of the Series 3 and the start of the preparations for the Series 5 project.

Keith de Mendonca:

I started with SDK work as a grounding, and I was working with Colly Myers on the Psion remote comms protocol SDK for the Psion Series 3a. Interestingly enough, I remember getting a fax of appreciation from a small company in America that I didn't really know, a small company called Palm Computing that it turned out we were in communication with about them perhaps writing connectivity software.

At around that time there was talk of cash-rich Psion buying Palm. But in the end Psion and Palm went their own ways. Meanwhile, after a stint writing applications for some of the later Series 3 machines, de Mendonca moved to the new Messaging team.

Keith de Mendonca:

I think at the time it was the biggest team that Psion had ever had working on a single application.

Its first task had been to create a messaging application for the Series 3.

Keith de Mendonca:

On the Series 3 the focus had been on working just with corporate mail. It was very much before the real Internet explosion, so it was just bespoke. It worked with ccMail and I think one or two other mail clients. It was only later that there was a decision to go Internet, where the open standards were.

For the Series 5, the team started from scratch with a new, modular and flexible messaging-client application which was designed to provide a seamless, unified, single-point-of-access interface to email and other message types.

> **Keith de Mendonca:**
>
> The intention for the Series 5 was very much that open standards were the future for us and the company. This was the most flexible way of actually interacting with servers and hence getting the best market share. So all the work really was to do POP3 and SMTP and IMAP4, which we started doing for the first time. In addition to that it obviously did SMS and there was a fax component as well, so you could send and receive faxes. So that was all in the messaging application and the design of the application was informed by discussions that we had been having at the time with Nokia, because obviously in the background there had been lots of work thinking about what was to be in the Communicator.

The Nokia Communicator project was already having an influence on the design. The ambition level was high.

> **Keith de Mendonca:**
>
> That was probably representative of the Nokia Communicator requirements generally, that what everybody shipped in the end was much less than what we intended to do when the project was scoped at the very beginning. I think the idea about what this next generation of communicator should do was beautiful and visionary, but the vision was very much greater than the reality of what both companies could produce in the time available.

Messaging Design Goals and Architecture

The key design drivers were to support open standards, with a flexible and extensible solution. The core of the design was a message storage framework for all messages, regardless of type, with plug-in protocol modules for particular message types. This was a modular solution, based on sound object-oriented principles of abstraction from a generic notion of message to the individual message types, designed for openness and extensibility.

> **Keith de Mendonca:**
>
> You had one message store but you could actually plug in different modules that allowed you to add different messaging protocols as and when they came up and even add those after market. So you could literally just download some SIS file and install some components, and suddenly you had fax or IMAP4 support, for instance. So the message architecture was designed from the very beginning to be very modular and flexible and that is the architecture which we still have today.

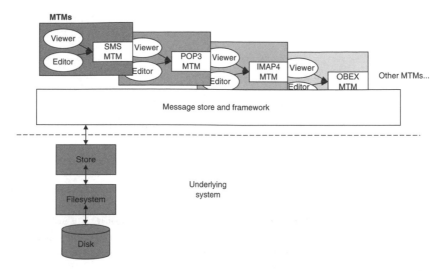

Figure 15.7   Messaging architecture

Following the general principle of Symbian OS design, access to the message store is through a server, which offers a client-side API. The plug-in modules (Message Type Modules or MTMs) are loaded dynamically by the framework on demand, based on the type of the message, and handle everything from the bitmap which is displayed in the inbox to indicate the message type and status, to creating, copying, moving, deleting, parsing and editing contents of the message type. See Figure 15.7.

Keith de Mendonca:

It also provided a back door for any special commands, which was literally just a special ID and any arguments that you chose to send to it in case those generic ones didn't actually represent what you were trying to do with a particular message type. The framework also allowed a simple API for another client of that server, another application, to send content as a message and choose the type dynamically based on the plug-ins which happened to be in the framework at that time, the Send-As API.

While the design met the stated requirements, the result was a large and complex component with complex APIs. In at least one area, the flexibility of the application had an immediate pay-off. With the messaging application still not ready to ship on the final Series 5 launch date, the messaging architecture made it possible to ship a complete messaging package as a user-installable after-market upgrade. It duly shipped some months later.

That flexibility proved just as useful for the first release of Nokia's 9210 Communicator. With the product pushing the limits of its ROM budget, messaging saved the day by shipping some features not in ROM but as optional installable packages on a companion CD.

Keith de Mendonca:

When we ran out of ROM space on the Communicator they decided to put the fax MTMs on the CD instead of built into the actual machine. It allowed them some flexibility in the ROM. Likewise, IMAP4 was originally delivered as a SIS file because it was finished a little later than the messaging application, which kind of proved that the architecture worked, you really could deliver things after market.

The flexible architecture and, in particular, the support for after-market delivery of MTMs by third parties and partners, which enabled messaging capabilities to be extended for any future message types, had been among Nokia's headline design requirement.

Keith de Mendonca:

It was responding to a customer-level requirement as far as we were concerned. But I'm not really sure to what extent that flexibility was really needed and you could argue that it was over-engineered or the requirement was a bit too heavy-duty, for the actual reality of what the application required. There was a vision of a new community of programmers writing corporate email plug-ins and suchlike, but the downside of such an extremely flexible architecture is that, unfortunately, often it can be quite complicated. This was quite a barrier for programmers and there's not really been a large market for those kind of MTMs to this day. But of course then you are left with your architecture, which you need to maintain and continue. A lean development model might have delivered the best value first, and then grown it as and when the appetite of both companies grew and we understood better what the market really wanted.

Perhaps one of the more puzzling facts of the phone market to date is that, despite the apparently huge popularity of products such as the Blackberry, email on phones has not yet proved a core function. In general, messaging – other than SMS – seems to have made little impact on users to date.

Keith de Mendonca:

The Communicator was very advanced but also much of the same code went into the Ericsson R380, which was much smaller, but that was a really visionary product as well. Everybody in the messaging team has picked up their mail on their phones ever since that day whenever it was, 1999 or thereabouts, yet that's only becoming a relatively recent phenomenon for other people buying phones. So it was extraordinary power that there was inside the actual product right from those earliest days.

What is the lesson? That making a phone operating system is not just about putting cool technology in the box. The harder formula seems to be – the right technology at the right time in the right product for the right market.

Meanwhile, flexibility implies complexity and the complexity of the messaging architecture has, arguably, had its downside.

Keith de Mendonca:

If I had my way again we would have presented less complex APIs for messaging. This complexity is one reason for the small number of people actually innovating using messaging. For instance, it's quite a big job to port your own corporate email solution that you might already have running on WinCE or some other device, to then package those into the MTMs, understand that paradigm and make it work on a Symbian phone. So it's been a barrier to that.

But also performance became an issue, primarily with email, due to some design decisions that were made early on. We also have been held back, perhaps, by feeling very constrained by APIs and data structure storage methods. That looks ridiculous looking back now. Of course we are now very closely restricted by such considerations, because we sell so many phones a month, but I felt very restricted about changing those things in Symbian OS v6.0 or between v6.0 and v6.1, when we would have survived making changes. But there was this compatibility promise very strongly enshrined in Psion and Symbian in those times, that you couldn't break binary compatibility, so it was quite difficult to make substantial changes.

The Universal Inbox

There are two other lessons from the messaging experience, one about designing for the wrong use case and one about performance.

Keith de Mendonca:

Although we knew our primary goal was to produce mobile phones and the phone would always be there, there was still quite a lot of work, say in the Symbian OS v6.1 project, for two-box support, even though the particular product never came to fruition. That probably resulted in the two-box concept not dying so quickly, although the company was focusing 100% on its new environment.

One particular issue concerns the design of the message folders.

Keith de Mendonca:

When we designed messaging we had a concept of 'the universal inbox' [See Figure 15.8], so it didn't matter if it was SMS or emails or whatever, they would all appear in the same message store and appear physically in the same inbox view. That's exactly what we used to have on the Series 5, where, of course, it was a two-box solution. You put your phone next to it or connected a modem to it and you downloaded all your email, that was the concept, and it all appeared in the inbox. And if you synchronized your SMSs, they would appear there too.

But from the beginning we also started toying with an idea of offline operations, a remote mailbox view. The point was that because you were a small device and you didn't have that much memory, you didn't want to download everything, because you didn't know how big those things were going to be. Quite frankly you might not want to have them. So you would synchronize the headers, that's all that appeared in the inbox, and then if you were interested in anything, then you would populate the individual items and fill them up with their content. The view was then represented as an external view, those messages didn't appear as just headers inside the inbox, they were actually shown as a separate independent view representing to the user that they were looking somewhere else, and in fact they were inside my very nice tree-structured message store browsing the complete structure of the message contents, which was very expensive in performance, not at all like just looking at the headers.

Now, in effect, the view on a phone is actually that offline view, certainly it's the way that Nokia displays them, it's the view of your mailbox. You no longer just synchronize the headers, on a phone you go straight into browsing and opening and reading. So the concept of the universal inbox immediately dies, because in the new model that we followed you had a remote view of each of your email accounts and those messages never appeared in your Inbox.

The difference between browsing the simple, inbox view of the message headers and browsing the elaborate, tree-structured representation of the complete contents of messages has a significant performance impact, one which, moreover, is tricky to resolve because the design is being used in an unintended way.

There are other impacts too, which ripple down throughout the whole design. For example, the basic assumptions underlying the design of the messaging APIs are the wrong ones for some of the most common phone use cases.

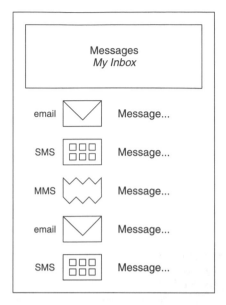 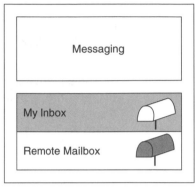

Figure 15.8   The universal Inbox

Keith de Mendonca:

Consider the generic API for the MTMs, which includes Copy and Move, for example. Actually, you never copy or move anything to your inbox on our phones. Those methods on that generic API for all of those MTMs are probably never used nowadays, because the API doesn't match the common usage, so the methods are redundant. Should we have predicted that? But take a different example, along comes MMS and suddenly our original assumptions about the universal inbox start to make sense again, because now you can have a mixture of SMS and MMS messages in the inbox. It's become universal again! But it's only when that MMS technology reared its head that you actually had something else to go in your universal inbox, because remember, emails are always set off alone in their own view anyway. But can we claim that our design was right after all?

It is not so much a point about user interface design, because the underlying conceptual design is interpreted by the user interface in a way that suits the user interface paradigm.

Keith de Mendonca:

Of course it's up to the UI how it actually presents those things. On UIQ phones, it doesn't mix them all together in the inbox. It shows you different message types and it 'pretends' they're all in the same general location.

> Conceptually they're stored in a different structure underneath these remote services.

The underlying conceptual design does, however, have a strong impact on how the system APIs below it, and which support it, are used. In the case of the original design of messaging, the supporting APIs are misused.

Misusing Store

Keith de Mendonca:

The conceptual model for the message store is a logical tree structure with the root at the top. Underneath the root there are the folders that you know about, the Inbox, Outbox, Drafts and Sent, also some representations of other things that are on the device; those are all seen as local services. It also has the concept of remote services, so your email at work technically would be the SMTP email server at work, or there's the fax machine you are trying to send to or the SMS service centre that you're sending your message to. Normally those things are invisible, but they still exist physically and are represented inside the message store.

Symbian OS supports folders, and it's very flexible – you can create any number of folders and put any number of messages in folders. Logically that's just a tree stored in RAM, in a RAM index effectively, but also it's written onto disk in case the battery fails, and it uses the Stream interfaces which are part of the Store in Symbian OS. It was designed in the same way that we designed most things, to be robust and never to lose data if you lost power. Therefore there's a lot of writing to disk, and as we know and have learnt, the Store components themselves, because they are so careful in making sure that you can never lose any data, are also quite expensive in disk writing.

This wasn't a problem when we wrote the message architecture, because it was designed for the Psion Series 5 which had a RAM disk and didn't have any performance problems. However, it started to create some problems if you put your message store on a removable CF card, which was the first-generation flash technology.

The team made minor modifications to their implementation when performance problems began to manifest themselves on real hardware. The fact that development was in the first instance carried out under emulation didn't help matters and the inevitable desire of licensees to keep prototype hardware under wraps didn't help either.

Keith de Mendonca:

A lot of last minute changes were made to try to improve the performance, including for the Symbian OS v6.0 release which went into the Communicator. One change was that we decided we couldn't afford to use Store for those small files, and we actually created our own store without changing the API. So from the public interface it would seem as if Store classes still exist, but we just wrote a binary file format which was faster, with some loss of robustness if the battery failed.

Also, email messages were described in the filing system not as a single message containing all the data for an email, but as a tree of objects representing exactly the MIME structure of that email. This meant you had an awful lot of files or objects for representing one message. This was true especially for HTML-style emails which are the standard way of sending emails now, because they not only send you the HTML version which has the complete content associated with it, but they also send you a simple text version as well, and of course we stored that as well. You may have 8 or 10 individual files representing an email, and if you have 8 to 10 files to write then obviously it will be slow. That's a basic challenge that we have even today, the performance of downloading emails, where we still have to create all those physical objects.

As in any other areas of the operating system, performance issues are continually reviewed and implementations tuned. But no one likes to have been wrong-footed by designing for a use case which changes because the overall device design point changes, as happens, for example, in the move from PDA to phone assumptions.

Keith de Mendonca:

Obviously, you would like to start again knowing where a lot of the faults are. But we are still constrained, we can't break this binary compatibility which we have offered so long, because the cost to customers and others of changing the message applications is quite expensive if we make any changes. So we work within the boundaries of our compatibility requirements and we're always making improvements.

Another compounding problem was the unfortunate interaction of the Store design with the behavior of Flash memory systems. Having been designed in the context of a conventional, fast, random-access, RAM-disk-based file system, the design turned out to be weak in the context of sequential-access, fast-reading but slow-writing flash hardware. Store, designed to offer a transaction-style, robust, append-based file system does not have a natural concept of updating already written data. Streams are appended sequentially to the physical medium and the header index is updated to point to them.

In the worst possible use case for a flash-based system, the design of Store ensures that it only ever grows and, in the worst possible use case, it even grows to delete things, depending for its space performance (so that it does not outgrow available memory resources) on a regular compaction cycle. The reason, of course, is data integrity, and that most basic, fundamental guiding principle of the operating system: 'Thou shalt not lose the user's data'.

To some extent, the problems with the early version of the messaging framework highlight a more subtle problem.

Martin Tasker:

The ground rules that we gave ourselves when we were making the PDA are not the same as the ground rules in the mobile phone space. We told ourselves that the ground rules are that you should assume you've got very little memory, and you should assume that the process is going to run forever, but also to assume that you're not in an embedded OS context where you can give a fixed partition to everything.

The thing is that in the mobile phone world, you do have system restarts and memory management is more complicated. If you look at the phone from a high-level perspective, don't just ask about the details of programming. Instead, ask about the dynamics of the number of applications and the number of services and how often it is turned on and off and how you allocate memory when you're sharing it between the programs, and look at the data storage and the RAM. The answers to those questions are different from those that we originally looked at in the PDA context.

It is not that the problem is one of greater complexity; it is different complexity.

For Ian Hutton, the distance between the early phone projects such as the Nokia 9210 and the Ericsson R380 and the phone projects that Symbian is involved in today, is the best demonstration that the hard lessons have been absorbed and the big step has already been made.

Ian Hutton:

I think the fact that we really are now beginning to see very, very ordinary phones, not just the incredible phones but ordinary phones with Symbian OS inside, phones that will still do anything you want, but are very similar to ordinary mid-range phones. In a way, that's the biggest step.

16

One Size Does Not Fit All: The Radical User Interface Solution

16.1 Introduction

Symbian OS is an inherently GUI-centric operating system which ships without a GUI of its own.[1] Moreover it targets a market (wireless devices generally and mobile phones in particular) in which the user interface is a critical competitive element. Delivering an operating system without a user interface is a radical software solution to a business dilemma: how to create a common, multi-vendor platform for phones, while at the same time not merely supporting but positively driving vendor differentiation.

This case study traces the evolution of Symbian's user interface architecture and strategy. User interface issues are particularly interesting because they highlight many of the unique problems of the mobile phone market. It is significant that Symbian's user interface architecture and user interface strategy have both undergone some quite radical transformations since the company was first created.

Although in computing terms the specification of a typical phone based on Symbian OS in mid-2006 is equivalent to that of a mid-range PC of the mid-1990s and exceeds that of a mid-range PC of the 1980s by an order of magnitude,[2] nonetheless a mobile phone is only incidentally a computing device. It is simply not thought of by users as being 'a computer' (What

---

[1] I use the terms GUI and UI almost interchangeably in this chapter; and use 'GUI' when I want to emphasize the graphical aspect in particular.

[2] 80386-based PCs clocked at a maximum of 33 MHz in 1985; 80486-based PCs reached 100 MHz by 1994; Nokia N series phones have reached 300 MHz in mid-2006. IBM's top of the range PS/2 machine, the 1987 Model 80 (with a list price of more than $10 000 – without adjusting for inflation!) featured a 16 MHz processor, 16 MB RAM, and 140 MB hard drive. The Nokia 5500 shipped with 64 MB RAM and supported removeable cards up to 1 GB.

is it then? It's a phone, stupid!). As I have argued in previous chapters phones are 'just different' from PCs and PDAs.

Phones have evolved rapidly from the clunky business tool of the 1980s to the seemingly essential 'upscale accessory' of 2006. In particular they have become consumer goods – 'and we cannot expect the user to configure them'.[3] But high-end phones have also become application platforms and this, in particular, has been the ground staked out by Symbian, which has deliberately set out to create an open platform for third-party software development. Where mobile phones began as the 'functionally direct replacements of their wired forebears', they have evolved a long way beyond those beginnings, boosted by the rich variety of available applications and by dizzying competition between vendors to outdo each other with new software and hardware features. They have become quite independent platforms for personal communication in modes both new (picture messaging, text messaging, and phoning home from the train, plane, street or shop), nearly new (email on the go) and old (plain old voice from home or office); for personal broadcasting (mobile blogging, web-sharable photo albums); and personal entertainment (games, MP3 players, pocket web browsers, pocket TVs). If they ever were just functional replacements for fixed-line phones, they certainly are no longer. With astonishing speed, mobile phones have become 'platforms for entertainment and commerce and tools for information management and media consumption'; everything in other words from business tools to games players to shopping tools to fashion accessories.

Differentiation: The Big Idea

The goal of differentiation is to avoid selling on price alone. Operators look for ways to apply their own branding to phones and to add value that will not be available from competing operators. On occasion they strike exclusive licensing deals or negotiate periods of exclusivity with vendors for particular phones. At the very least, most operators demand a minimum degree of customization of phones from vendors, for example operator-specific packaging, stenciling of the operator name or branded logo onto phones, and inclusion of custom operator applications or support for dedicated operator services on phones (for example, O2's Homezone and Vodafone live!).

This is not quite the same as vendor differentiation, which is the most visible form of differentiation. For phone vendors, differentiation – of their phones from those of their competitors – encompasses everything from design philosophy and style through reliability and build quality to ease of use and, of course, technologies and features.

---

[3] The quotes in this section are from Keinonen [Lindholm *et al.* 2003, p. 4] and Kiljander and Jarnstrom [Lindholm *et al.* 2003, p. 15].

Naturally all players in the phone market are seeking competitive advantage, but there is something else at work: avoiding commoditization. For both operators and phone vendors, commoditization is an ever-present threat and is one part of the complex dynamic which drives the exponential pace of technological advance and feature growth. 'Commoditization' means the cheap and ubiquitous reproduction of what was once expensive and unique. Commoditization reduces margins because it reduces competitive advantage to price competition. Commoditization naturally centers on hard technologies and features – 'do more, go faster'. Softer product qualities – 'do better, be easier' – are therefore critical points of defense against commoditization.

In the battle to maintain differentiation, the user interface has become a key competitive element for phone vendors. Indeed for phone vendors, design philosophy and style ('cool' is an important selling point for phones) and properties such as usability have become almost as important as features. Since these are all properties which touch or are touched by the user interface, the phone UI has turned out to be a key business competitive edge that drives market segmentation and, as a result, brand leadership.

As Christian Lindholm, the creator of the original S60 user interface puts it, the UI 'is one of the key elements in the fight for customers', creating pull from both end-users and networks, an essential 'competitive asset in the race for market dominance.'[4]

What's in a User Interface?

The user interface is where the phone software and the specific hardware features of a particular phone meet to enable the user to access the features of the device. In large part, 'usability' reduces to questions about the user interface and the myriad decisions made by its designers and implementers about such things as one-handed versus two-handed use, pen versus keyboard input, feature richness versus interface simplicity, file-centered versus task-oriented application design, and so on. At one level of detail down, these become decisions about such details as screen color schemes and fonts, menu structure and sequencing, how hard and soft keys interact with on-screen items, and so on.

At first glance, user-interface style may seem to be a slender thread to hang market aspirations on and a tenuous driver for market share or even dominance, but of course it is not the only driver. But the critical contribution it makes has to some extent become a reality for all consumer products (think of the iPod without its click-wheel), and it has certainly become a reality for phones.

---

[4] Kiljander and Jarnstrom [Lindholm *et al.* 2003, p. 15].

The Multiple GUI Operating System

To succeed in this complex context, Symbian's solution is to support multiple UIs and to engage with others to create those UIs. Licensees either buy in the platform pre-integrated and tested from a Symbian OS UI vendor such as S60 or UIQ Technology or they buy in the operating system 'headless' and develop their own GUI.

In architecture terms (see Figure 16.1), the topmost layer has been removed from Symbian OS. The operating system does not provide the UI layer; instead it provides the infrastructure (frameworks and primitives) for UI creation. Beneath the UI, the common frameworks and the underlying services of the operating system itself ensure substantial platform compatibility at the application-engine (i.e. application-logic) level. Indeed applications can be targeted at the different available user interfaces by customizing the application UI without changing the application engine. A Symbian licensee either creates a bespoke user interface from the frameworks (although this is not a trivial engineering effort, it is the option chosen by DoCoMo in Japan for its FOMA platform, for example) or buys in a UI from a specialist vendor (both UIQ Technology AB, which was spun out of Symbian as a separate company, and Nokia license UIs and supply a pre-integrated platform to phone vendors).

Symbian OS on a Sony Ericsson phone or a Motorola phone with the UIQ GUI – the P910 or P990, say, or the Motorola A1000 – looks and feels very different to Symbian OS on a Fujitsu or Mitsubishi FOMA phone or to Symbian OS on a Nokia, Samsung or LG phone with an S60 GUI. And yet the underlying operating system is the same and the very same applications can run on all of these different phones, sharing identical source code at the application-logic and data-model levels. While Symbian OS is tightly integrated to the GUI which runs on top of it, by way of the UI and application frameworks, it is designed to be GUI-neutral. For application developers and from the application

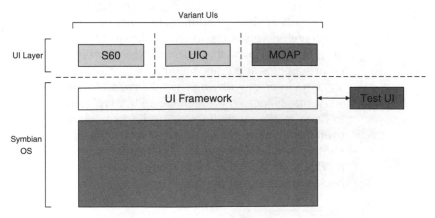

Figure 16.1   Symbian OS user interface architecture

perspective, although Symbian OS applications are intrinsically GUI in nature, applications are only loosely coupled to a particular GUI variant, since the application model and the application-event loop are enshrined in the operating system itself and in the UI Framework support, and not in the variant user interface.

Symbian's solution contrasts with that of other operating systems targeting mobile phones. Windows Mobile, a Windows CE derivative, is architecturally a monolithic UI, like its desktop parent. All phones that use it, from whichever phone vendor, share the common Windows interface and its Microsoft signature branding. Whatever the other opportunities for vendor differentiation, phone vendors supplying Windows Mobile phones are, to all intents and purposes, Microsoft OEMs. Windows Mobile is certainly a platform operating system, but what has so far made it unattractive to many phone vendors is the fact that it is Microsoft's platform.

The situation with Linux is somewhat different. To date, the Linux phones which have shipped (in large numbers in Japan and China) are not Linux platforms because they are not natively programmable. Instead, they are closed phones exposing only limited Java APIs. Further undermining the platform potential of Linux phones is the proprietary nature of the Linux distributions on which the phones are based, with the phone vendor typically directly owning the distribution. To some extent new 'open' user interface toolkits supplied by independent vendors for Linux-based phones (such as the Qtopia QT user interface toolkit from Trolltech) may open platform potential to Linux phones if they become widely adopted. However, it is something of an irony that the question of openness still remains a significant challenge for Linux on phones. Qualcomm's Brew platform, which so far has been limited to CDMA markets but which is now migrating to GSM, offers device vendors a choice between creating their own bespoke UI, or adopting Qualcomm's customizable uiOne user interface.

The early decision that Symbian made was that the phone market would reward UI diversity based on a common underlying platform. Thus far, certainly, the market has found Microsoft-style homogenization resistible.[5] Linux, fast emerging as possibly the strongest challenger to Symbian OS, has not yet solved the platform challenge. So far Symbian OS remains the only operating system which is open to third-party developers across multiple phone vendors, across multiple operators, and in all geographies and all markets (GSM, CDMA, 3G), while also offering vendors strong opportunities to differentiate.

However, the market is still barely in its infancy. The currently competing platforms – most visibly Symbian OS, Windows Mobile and Linux – have until recently been restricted to the high-end, less than 10% segment of the overall market. Symbian's strategy – of 'leaner, faster,

---

[5] To the surprise of some. But since homogenization in the phone context equals commoditization, it is perhaps not so surprising.

fitter' – is based on a drive towards the mid-range, boosted by natural momentum from increasing market acceptance plus a little help from Moore's law (which continues to drive hardware specifications up and BOM costs down, so narrowing the gap between the middle of the market and the top). Meanwhile, the incumbent operating systems in the mid-range are proprietary, for example Nokia's NOS/Series 40 combination of operating-system and UI or the operating system which drives Sony Ericsson phones. These can only be considered 'platform' operating systems in the sense that they support third-party programming in Java. It is in the mid-range that the next rounds of the battle for common operating-system platforms will probably be fought.

Looking further ahead, there are also possible competitive threats in the user interface area from alternative technologies which potentially offer UI alternatives either to the existing incumbents or other embedded operating system alternatives. Examples include building embedded device user interfaces directly in Macromedia Flash, more commonly known as a web-content display technology but proving itself as a UI-simulation tool and jumping the gap to become a potential user interface technology. Similar disruptive approaches include possible 'declarative' user interfaces based on XML-defined interface-description languages. It is not yet clear whether, or how, these technologies will eventually play in the market.

16.2   Background to the Eikon GUI

Symbian OS originates from the desire to create the perfect usable operating system for small handheld devices.

The original GUI – on the Psion Series 5 – was known as Eikon. Eikon itself, at least in the design sense, was an evolution of a previous generation of GUIs written for Psion's earlier 8-bit and 16-bit machines, although in concrete terms it was all new work, a second – or third or fourth, depending on where you start counting – attempt at mastering the GUI.

David Wood:

There were several UIs for the 16-bit software. In a way, it prefigures what's happened to the 32-bit software. For the MC400 which was a full laptop-sized device with a large screen, we designed what we called WIMP, standing for Windows, Icons, Menus, Pointer. And then for the handheld version we created something called HWIM: 'H' for handheld; we took the 'P' out because there was no pointing device. And then we did a new version called XWIM when we increased the screen resolution and it was actually binary compatible with HWIM, which again prefigures something that's happening nowadays, because the applications that ran on the original Series 3 also had to run on

> later versions of the Series 3. So they ran in a compatibility mode with each pixel being doubled up.

The basic ideas about how a UI should work on a pocket-sized machine had been well worked through before any code was written for the Series 5. The iterative approach, however, was quite deeply a part of the Psion culture and scaled well to the size of the company at the time.

David Wood:

HCIL was the first version of the UI for the Series 5 and that was also quite WIMP-ish. For example, you used to tab your way around dialogs. However, after a while we decided that was too complex for most handheld users, so we abandoned having two-dimensional dialogs, we had one-dimensional dialogs which you navigated using the up and down keys. And that change led to the creation of Eikon, actually through two phases. After a while we decided we needed to refactor. After the first implementation we split it into two. We separated out the CONE control hierarchy as something that would be common for all user interfaces on the handheld devices because we viewed Eikon as just the particular interface for the Series 5. And CONE persists to this day. No doubt, it has evolved quite a lot but essential features of CONE are the same as in the 1996 version when CONE was first created.

From the beginning, even though the project was very specifically targeted at the Series 5, design decisions were taken with the later platformization of the operating system in mind, assuming a family of possibly different devices.

David Wood:

Splitting out CONE is a good example of the need to refactor. The need to refactor is an important principle that comes through architecting any large scale system.

A so-called 'Nokie' variant was created and the CONE classes were separated out. Eikon was re-engineered around this separation and the changes were then integrated back into a new version of Eikon.

David Wood:

'Nokie' is Eikon backwards. It's just the kind of black humor of development teams.

Eikon, in its time, was a complete concrete implementation of the Psion Series 5 GUI. Architecturally, it fitted into a complete UI framework comprising:

- Eikon: widget classes with look-and-feel policy and custom behavior

- CONE: generic control classes with no look-and-feel policy and generic behavior

- Application Architecture: the application model, broadly MVC-based,[6] although the detail of the implementation depends on the licensee GUI (UIQ3, for example, delegates most command handling to views).

This is a classically good object-oriented design, with two independent class hierarchies (Eikon and CONE) and an underlying framework exposing some generic behavior (common to all GUIs) directly to applications, with GUI-specific behavior brokered through Eikon.

16.3   Eikon Design Point

Windows, then as now, was the dominant consumer-oriented GUI when Eikon was first being designed, although both Macintosh and even AmigaOS (the Commodore Amiga operating system which included the Workbench GUI) had strong followings. (The line from Microsoft was then and still is 'Windows Everywhere!', from handhelds to data center servers.[7]) In the small device space, although Windows CE ran on some HP handhelds, there were other potential UI models too: Newton, for example, which was a strong influence later on the Palm UIs and PenPoint from GO! Corporation was an innovative and interesting, but commercially unsuccessful, UI. But as the adoption of MVC suggests, there were also other explicitly object-oriented influences.

David Wood:

Windows was one of our reference points, but we were aware of other user interfaces too. We saw what people were trying to do with Taligent for example, which was a combined effort between IBM and Apple that failed in the end but, like many failures, there were lessons to be learned from it. So we read avidly the books produced by people working on Taligent and we looked at that as a reference. Charles Davies was familiar with X-Windows, the Unix windowing system and that influenced us a bit also, though to be fair it more influenced the design of the window server than the UI.

---

[6] The model–view–controller pattern and framework is discussed in Chapter 14.
[7] See the comments in [Petzold 1992, p. 4], for example.

One critical design point for Eikon was usability – simplicity, natural-ness and fitness for form factor of the GUI and the applications which shipped with it. Robustness and an intuitive user experience were the key principles.

> **Geert Bollen:**
>
> There was a logic which dictated that ultimately it is end-user benefit that is important, and therefore any abstraction at all can be broken for the sake of delivering something which delights the end user. And there was another important value, almost a law, 'Thou shalt not lose the user's data'.

Although Psion always practiced a strongly decentralized design regime with almost complete autonomy for the separate teams (sometimes to a fault), a lot of care went into ensuring that the shipped applications were consistent and well-designed from a user perspective. There was a UI Board, for example, which vetted the designs for all the application UIs.

> **Howard Price:**
>
> Bill Batchelor and Nick Healey would run the UI board and you'd go in and show them your design. There was a lot of attention to detail and pretty quickly they would be counting key presses, going down into the details of the look and feel. There was a lot of counting of key presses!

Another critical design driver was ease of application development.

> **David Wood:**
>
> One common principle behind the creation of Eikon was to make the job of writing applications easier and require less code, so that the applications would get more things for free provided they conformed to the framework. So that was a common theme: put the complexity in the system code and allow the applications to get the rich UI without having to do lots of detailed programming of that themselves.

The point perhaps needs some elaboration, because Symbian OS is sometimes perceived as being hard to write applications for. As discussed in Chapter 14, frameworks are a powerful object-oriented concept and the Eikon framework approach did indeed deliver a lot of power to applications. However, the Eikon programming model is considerably more sophisticated than a simple procedural one. Application developers have also recoiled from the complexities of the embedded systems-like toolchain (develop under emulation, then cross-compile for what is, in effect, an embedded hardware target), as well as from the discipline required to develop robust software for devices on which a software error in an application freezing or crashing the system is simply unthinkable. For some, the development experience is a little too far from the desktop experience to be comfortable at first.

At the time, though, the big issue seemed somewhat different. Eikon was evolved together with the system software beneath it and the application suite above it. This approach continued the Psion tradition.

David Wood:

Eikon was created in an incremental process. As more applications were written I looked to see what the problems were that the applications had to solve so when people created a toolbar, for example, then we thought, 'Well, actually toolbars should be in the Eikon framework', and then people did more complex things with toolbars and we thought, 'Well, actually other applications would like to take advantage of this as well'. So systematically things that started their lives in applications moved into Eikon. And that made things a bit tough for the application developers because they had to rewrite things as we went along, but it did mean that with only a small amount of code in the applications, very rich user interfaces could be achieved.

The goal was to maximize the power of the framework to enable graphical applications to be developed with minimal new code and therefore to get the most from the small hardware footprint.

Of course, this didn't make things easy for the application teams working on the built-in applications. Peter Jackson recalls the frustration from the other side, compounded by the time-to-market pressures from the project.

Peter Jackson:

It was changing under your feet the whole time, and you knew that if you put effort into working on something that was based on that framework you were going to have to change it again.

Iteration was the natural model in the company at the time, and there was a long history of evolving from one product to the next. Evolving the Eikon GUI for the phone projects which were starting up in the wake of the Series 5 launch was in some ways just more of the same. Adapting the Eikon GUI to the very different form factor of each different phone was clearly a critical task and Eikon had been designed with adaptation in mind.

The reality, however, was that there was no clear understanding of how that should be done. Martin Budden moved from the application team to become technical lead on one of the very first phone projects, the Philips phone 'companion' (see Chapter 2).

> **Martin Budden:**
>
> We were still a way from forming a portability strategy for how we would develop and deliver the operating system for all these different manufacturers, so we were learning as we went along.

The Philips phone was the first adaptation of the Eikon GUI. The approach was straightforward and pragmatic, a straightforward branching of the components that needed to change. The Philips project developed a bespoke UI and a dedicated application suite with a small team, in not much time.

> **Martin Budden:**
>
> We did the UI but we also did the applications in there: we did the messaging, the contacts, and all those kind of things. To customize Eikon, we essentially rewrote the drawing code and the code for things that we needed to draw differently and we wrote new UIs for the new applications as well, using the underlying engines without change.

Murray Read joined the project in its early days, initially working for Origin, a software consultancy part-owned by Philips and later acquired by Symbian, but at that time supplying specialist software engineers to Psion.

> **Murray Read:**
>
> The UI design Philips wanted was quite different from the Series 5 design. It shared some similarities. I think we had a task list and the major parts of the UI were still there; there was a menu application; there was Window Server and CONE; and they were all there doing the same basic things. But when it came to the UI library itself, we had a much simpler set of controls to work with; one type of button, no keyboard, so we had to make the system work with the on-screen keyboard only, although it had handwriting recognition as well.

When the Philips project ended (disappointingly, without a product coming to market) most of the team moved on to start up the project for the Ericsson R380 and applied the same approach. Meanwhile the first project with Nokia, for the new Nokia Communicator was in full swing (see Chapter 2).

Martin Budden:

The model of doing a bespoke UI was there, and then we did another bespoke UI for the Ericsson R380. The other big project at that time, going on in parallel, was the Nokia Communicator, and again, that involved doing a new UI to Nokia's specification. We swiftly recognized that there was a fundamental conflict between these UIs and it became clear that if we did a UI for every single phone that wasn't going to be sustainable.

The basic problem was clear enough.

Martin Budden:

We weren't thinking about generic problems. We were dealing with specific problems for specific projects that came up.

Each project was in effect a customization project, which created a complete, custom variant of the operating system as required for each device, from the base port to the operating-system services to the bespoke application suite to the UI, including modifications (at whatever level of the system) needed to support bespoke hardware such as dedicated phone keys or the R380 phone flip.

For all these challenges, the first phone projects were genuinely transformational for the company.

Ian Hutton:

The early projects were in fact extremely visionary, not just the Nokia Communicator, but also the Ericsson R380. The Ericsson R380 was a really advanced phone. The fact that it didn't sell in huge numbers, nor the Communicators for that matter, is largely irrelevant. There were really huge advances in design, both in what a phone should do, and how you could do those things on a phone.

Key features of the Ericsson R380 – the flip, for example, and its interaction with the screen modes, flip-closed and flip-open landscape – were good enough to be picked up by the next generations of phones and are still central to the design of phones such as the Sony Ericsson P900 series.

> **Ian Hutton:**
>
> The Ericsson R380 was very much an innovative design. It may have ultimately disappointed in terms of sales, but it did come up with a lot of good ideas and it solved quite a lot of problems in terms of turning what was then EPOC Release 5 into fully-fledged phone platform.

The project also contributed key technology back into the operating system. The View Server for example was originally developed by the Ericsson R380 team but has evolved into a key feature of the UI framework architecture.

> **Ian Hutton:**
>
> Initially it was used to solve a slightly different problem in the Ericsson R380, which was the flip mode. So it was initially written for flip open and, in fact, not just for the flip, it was for landscape mode too. It was written by Symbian's Licensee Technical Consulting (LTC) team for the customer project, and then the larger software engineering group was presented with it by the UI architects. Consequently, it was adopted back into the core architecture.

Accepting licensee (or LTC) changes back into the operating system baseline and evolving them forwards thereafter as part of the platform has been a central principle of the licensing model, right from the beginning. Making the mechanism work has not always been so easy, however.

> **Ian Hutton:**
>
> The issue is how to manage migration of changes initiated by projects to support project-specific and even device-specific requirements which then may have more general applicability and value, and migrate them back into the software base to avoid the base branching. So there's really a continuous process of splitting and some fragmentation, and then reunification – bringing it back together.

In part, the problem came from organizational tensions, the support for licensee projects being provided by a different engineering organization than the main operating-system engineering teams. Over time, the company has become adept at managing the process but there are still key hotspots around defect triage, change requests, engineering changes, and product requirements and the inevitable balancing of priorities between them.

16.4   The Device Family Strategy

Platformization of the operating system had always been a goal, whether or not it was a licensed platform or simply a Psion internal one. The licensing strategy was a logical next step. However, it is reasonable to argue that Symbian was not well prepared for the reality of running multiple licensee projects either in the practical sense of the straightforward logistics involved – it was highly difficult to resource and put immense strain on the company – or in the more narrow technical sense of how to manage the codebase-development practicalities.

Ian Hutton thinks that the Ericsson R380 project was significant in at least one other respect.

> **Ian Hutton:**
>
> Pretty much through the whole of that project it was difficult for LTC to work with the software-engineering organization. It was at the end of that project that the DFRD strategy was written.

For all the problems of branching and the resulting wrangles to migrate changes back into the main codeline, there was an underlying strategy emerging. It was eventually announced at CeBIT in Hanover in February 2000 as the 'reference design' strategy, based on so-called Device Family Reference Designs (DFRDs). As well as announcing a joint Motorola–Psion device, Symbian was showing off Ericsson 'mediaphone' prototypes based on what it called the Quartz DFRD – quarter-VGA, pen-based, PDA-style tablet devices with built-in phones and Bluetooth (enabling the use of remote headsets).

DFRDs emerged out of the need to resolve the problem of multiple, incompatible UIs. And it provided direction for Symbian and its engineering practice.

> **Martin Budden:**
>
> The problem was that it was not possible to come to agreement for a Symbian-based UI that was suitable for all parties. So this was when the idea of resolving all these conflicts with the DFRD approach emerged.

The engineering teams were at least in a position to abstract from the day-to-day problems of competing and conflicting projects and agree some higher design principles.

Martin Budden:

The idea was to have families of UIs. There would be one family which was based on the Nokia Communicator and the Series 5; there would be a family that was based on the Ericsson R380; and at about this time Quartz was started up for Ericsson, which was for the quarter-VGA tablet form factor.

The actual reference design specifications were based around a combination of screen size and orientation and input method. To an extent, the design points were really just rationalizations of the known preferences of the different licensees and were closely modeled on actual products that were already in development in licensee collaboration projects.

The DFRD strategy helped Symbian recognize that the phone world was complex and took a step forward from simply solving problems to systematizing the solutions.

The DFRD model set out to provide enough flexibility to support licensees across a wide spectrum from those, such as Nokia or DoCoMo, who had both the resources and the desire to create their own bespoke UIs, through a middle band of licensees who while not looking for a complete off-the-shelf solution preferred to work closely with a UI supplier like UIQ Technology than make the considerable investment to create their own UIs from scratch, to the smallest licensees who were looking for a near-complete solution and a fast product-development cycle. In effect, it reflected the standard tiering of phone vendors used by industry analysts:

- Tier One licensees create complete devices end-to-end: typically they expect to buy in Symbian OS and create or license bespoke UIs; they create lead products for new Symbian OS releases.

- Tier Two licensees create hardware platforms from standard parts: typically they expect to buy in Symbian OS with a pre-integrated UI; they create follow-on products from proven Symbian OS releases.

- Tier Three licensees focus on custom packaging of external phone design: typically they expect to buy in complete ('80%') hardware reference designs (from silicon vendors such as Intel and TI), with Symbian OS and their chosen UI pre-integrated onto hardware.

Bob Dewolf joined Symbian in early 1999 when Symbian acquired the Origin software consultancy. He had been working on embedded software for pagers and fixed-line phones.

Bob Dewolf:

I had just done a Farsi pager for Philips, and a smaller pager written in 16 KB of assembler, so that was quite the opposite of smartphones. But it wasn't a bad match because you had the embedded aspect from both pagers and telephone work and I had UI experience.

Symbian's DFRD strategy was newly in place. As a veteran of the early evolution from Eikon to what became Series 60, he knows as much as anyone, and probably an awful lot more, about the problems of crafting a phone GUI.

Bob Dewolf:

In that period between April and July 1999, we were in meetings with licensees, primarily in London, and they would get up and talk about their various plans for screen sizes and other key GUI design features. Those meetings were very, very important for Symbian to define a DRFD which fitted those users.

The first two DFRDs, known as Crystal and Quartz, were relatively straightforward to define since each in effect abstracted the actual properties of the particular target devices of projects which were well advanced. Each, in other words, had a single, clear customer. Crystal defined a keyboard-based device in the style of the Nokia Communicator, applicable to the Nokia 9210 and the Psion Series 5. Quartz defined a pen-based, tablet-style device applicable to prototypes that Ericsson (as well as other licensees) publicly demonstrated not long after. Between Crystal and Quartz, however, there was a missing form factor – that of a more or less conventional phone or, at any rate, a high-end device that was recognizably a phone in the sense in which the Ericsson R380 clearly was. The Sapphire DFRD was defined to meet the need of the third category.

Bob Dewolf:

The categories were QWERTY-keyboard-oriented, which was Crystal; pen-oriented, which was Quartz; and telephone keypad, which came to be called Sapphire.

Even then, the licensees who wanted to pursue designs in the Sapphire category had conflicting design philosophies.

Bob Dewolf:

One licensee would not be pen-oriented, while another didn't want to do anything without a pen. One licensee was very Java-oriented; another licensee wasn't. We talked about Blue Sapphire and Red Sapphire at that point, and I remember people trying to figure out how we'd have polymorphism. David Wood then produced some extremely interesting and abstract work about data and layout separation, and about abstract specification of components. I remember thinking, 'Yes, this is the way to do it', and saying, 'we shouldn't talk about those screen sizes, that's the last possible thing we want to talk about, we should be talking about the abstract definition of control sets!'

However, at the end of summer 1999, Sapphire was still blocked.

Bob Dewolf:

Then Nokia said they had a full design concept, and simply wanted to start developing. So one day about four people from Nokia arrived and we decided to see what happened. Originally, they were working from Sapphire documents, but not long after the name was changed to Pearl.

Out of the ashes of Sapphire, the DFRD that emerged was Pearl. By defining a new DFRD, the unresolved problems of Sapphire were circumvented. It seemed clear to those close to it that Pearl was strongly driven by Nokia.

Bob Dewolf:

Nokia had very strong time constraints. I remember sitting in meetings where we cut up responsibility for various types of things like soft keys and notes and queries and list boxes to various people and asked when they could get that done.

A lot of the main decisions for the architecture were made during that period. In fact it's surprising how many of the things we still have are based on decisions made in that period, things like doing all our multitap and key tries, like internationalization in the FEP (Front End Processor); using `CEikDialog` as the base class for notes and queries and forms, which you use in contacts; using the listbox-based classes, using the view architecture, and that was being pushed very strongly because of the Ericsson R380's success with the view architecture.

Some of the core architectural decisions were made at that point. It was a very interesting period. At the same time, the UI team itself was breaking Eikon down into Uikon, so they were refactoring and separating things out, moving things around, and we were trying to understand the customization

> rules. Otherwise, at that point we were trying to stick within modifying C++ files in Eikstd and not touching Uikon at all. So we didn't do something new for notes, we didn't do something new for queries; they were still pop-ups. We were using the FEP architecture. We were very much trying to be a DFRD.

The DFRD approach required the separation of core GUI base classes that were free of a look-and-feel policy; the separation, in other words, of those classes that had no intrinsic look and feel and were therefore common to all DFRDs from the policy-laden, look-and-feel-bearing derived widget classes (which implemented the look and feel of a given GUI). The core classes would be part of the framework; each DFRD GUI implementation would create its own concrete derived classes.

The UI team in the software engineering organization therefore created a design proposal for a refactored Eikon to be called Uikon and the work was planned into Symbian OS v6.0, the first release which would support the DFRD model.

As David Wood sees it, this was very much a natural continuation of the evolution of Eikon, just as the earlier factoring out of the CONE control classes had been.

> **David Wood:**
>
> Just as we split out the control hierarchy and refactored Eikon, so later on people have sought to take other common elements into Uikon.

The original idea was really quite simple, based on a simple implementation strategy.

> **Murray Read:**
>
> The original plan was that everybody would just customize Uikon through what was called UikLaf – Uikon Look-and-feel. This was the DFRD strategy. Uikon provided the core UI, and Eikstd and any bespoke UI libraries were the customizable bits that you could modify. UikLaf provided an interface which you would use to modify the look and feel of Uikon itself, and it was a simple, flat, monolithic interface which would allow you to tweak various parameters inside Uikon.
>
> But in practice, it just wasn't rich enough to give you the customization you needed to implement a UI like S60, for example.

Martin Budden is one of those who still feel that the Uikon architecture and the design choices that were made to support UI separation and customization were not always the best choices.

Martin Budden:

Uikon was an attempt to resolve the conflict between Eikon, which was the Psion Series 5 UI, and the UI developed for the Nokia Communicator. Essentially the idea was to add a further layer of abstraction that would allow both of those to sit on top of the same underlying UI. In effect, we ended up dropping Eikon, so that was how the conflict was resolved. Then Quartz used Uikon, but the so-called look-and-feel layers were just reimplementations; so the fact that they were in a separate layer didn't make any savings and you might as well have just reimplemented elsewhere.

What we could have done was standardize the APIs, rather than trying to make a split; standardize the topside and bottomside APIs of the UI framework so that you could replace it and still be compatible. That way you can run any application on any UI. You could slot in a different UI framework, instead of having a customizable UI framework and trying to separate out the customizable bits, which is problematic because there are no clear separations as to what is customizable and what is not. If you instead standardize the APIs, then you could just write a new framework to those APIs and things would just run.

Budden's views are based on his experiences during the Quartz project, for which he became the first technical lead, commuting between London and Ronneby.

Martin Budden:

The experience of doing Quartz highlighted that there was a lot of code that was not easily separable. For example, the messaging code had UIs in the engine layer, which meant that to change the UI we had to redo a lot of the engine code as well. There were lots of considerations that made it difficult to separate out the different bits. We rewrote a lot of the code, but without necessarily preserving the APIs.

The root of the problem was the UI fragmentation which dated back to the early days of the Nokia Communicator and Ericsson R380 projects. The lesson, Budden thinks, is a simple one.

Martin Budden:

In retrospect, we didn't do enough to support all those various UIs to ensure that they were following the same rules and were compatible.

Some degree of fragmentation was probably inevitable. Certainly, the risks of fragmentation were well understood on all sides and attempts to avoid the worst consequences have been significant drivers of the UI evolution strategy. For example, the history of the Nokia–Symbian UI split shows a repeated pattern of branching and reconciliation of Nokia and Symbian UI APIs; S60 1.0 branched Symbian OS v6.0 significantly, leading eventually to the 'unbranching' Symbian OS v7.0s release, which restored API compatibility across the UI variants. Similarly, Symbian OS v9 has involved the unbranching of some UIQ APIs and will probably continue to include unbranching for both UIQ and S60.

Arguably, however, success is justification enough for the strategy that eventually evolved. With hindsight, different design choices could have been made; but getting devices to market is ultimately the test of success.

> **Murray Read:**
>
> My strategy for UI development has always been to make it work. I know people have this romantic idea that you can take one UI and turn it into another with a few tweaks here and there, but in practice to do that you've got to design that into the core UI in the first place. And that's not where Symbian started from. So my strategy has always been to just make it work anyway, write the new controls that you've got to write, adapt the existing code, branch things if you have to. And it's worked, I think S60 is the proof of that; we have a working UI in S60. And if we'd stuck to the strategy and followed the rules, I don't think S60 would ever have happened.

16.5   Quartz

The Quartz UI was developed at Symbian's Ronneby site in Sweden, which originally had been Ericsson's Mobile Applications Lab, a development lab working with Microsoft's Windows CE operating system. Ian Hutton moved to Ronneby to replace Martin Budden as technical lead on Quartz.

> **Ian Hutton:**
>
> Certainly with Quartz, I think the biggest challenge wasn't so much making that UI from the basis of EPOC Release 5, it was the question of working out what you want to do and then going out and finding something in Eikon to do it. The bigger challenge was really looking at the wider possibilities of the design, which was very challenging. A lot of that was focused around usability, what are people going to do with this UI. A lot of work went into that and working out what our underlying design needed to be to inform the UI design.

The view architecture inherited from the Ericsson R380 project, which had been migrated back into the Uikon framework, became a central feature of the Quartz application model, elaborated in the UI layer into a mechanism named direct navigational link (DNL).

> **Ian Hutton:**
>
> That whole view architecture, which is now in Uikon, was very different from what was being done by Nokia and it's very clever. The design of the views and using DNL was completely new – there's a whole new area of architecture there – and this behavior had to be defined and designed. That was one of the bigger challenges with Quartz.

Ian Hutton describes the DNL idea this way:

> In the phone application, you've got your recent calls, you've got the callers' names, now you want to see the details of this person, so when you tap you switch to the contacts view, and now you can see that person's details, their phone numbers, email address, and so on. In the contacts view, when you tap on the phone number you switch to the phone application and it dials. So you've got some basic UI items and you want to make use of that in another application, and the underlying mechanism which switches you between those views is DNL which is quite a powerful part of the UIQ user experience.

In contrast, to do the same thing in the original Eikon GUI would require the sequence: go to Contacts; copy phone number; task away; find the Phone Application; paste phone number into the Phone Application; tap Dial.

16.6   Pearl

The Pearl team, meanwhile, was working flat out to meet Nokia's timescales.

> **Bob Dewolf:**
>
> The plan was to get the project up and running, and then, about six months later, not exactly freeze the APIs but exert more control over them. Then we would change the layout code so that it was resizable and more generic and customizable, putting in customization layers, and that would become the Nokia product. The DFRD would continue with that codebase and those APIs. But the focus at that point was definitely getting rapid development of their concept. However, there was increased pressure to change Uikon, and

> I remember certain changes that were made at that point, we had meetings with the UI team and they said 'We can't do that. We've got no requirement to do that'. I guess there was skepticism about how we could customize this thing after it was created, and there was also I think a little pressure from the Interaction Design team, they weren't so keen on the Pearl UI.

The Symbian Interaction Design team was led by Scott Jenson who, before coming to Symbian, had been part of the original Newton team at Apple. Interaction Design had been closely involved in the design of Quartz. Pearl (or at any rate its concrete implementation) was a thoroughly Nokia-flavored UI, its design driven by Christian Lindholm and the team in Tampere (Lindholm recounts some of the history in [Lindholm *et al.* 2003]).

> **Bob Dewolf:**
>
> I remember Scott Jenson coming up with some criticisms of the UI. But you know, finding user problems with a UI is like shooting fish in a barrel, as they say, but Scott didn't like two soft keys, he thought it was too complicated. And then, out of the blue, there was a meeting and it was decided that there should be no UI in Pearl itself, Pearl would be UI-less. At that point everything changed. There was to be no UI deliverable from Pearl, and we became overnight the Nokia 7650 UI development team.

In a new shift of direction, Pearl was redefined to be a DFRD without a user interface, or what came to be called a 'headless' DFRD. In effect it was the first step towards the post-DFRD strategy of a headless operating system. The Nokia 7650 (arguably the first true phone based on Symbian OS) launched what was branded at the time as the Nokia Series 60 UI. Pearl thus became Series 60. Dewolf sums up the situation.

> **Bob Dewolf:**
>
> Crystal was a product which was elevated to a DFRD. We had a DFRD which was deflated to a product.

16.7   Nightingale

Series 60 (now rebranded as S60) was announced at Comdex, Las Vegas, in November 2001. The Nokia 7650 itself (rumors of its launch had been circulating for several months[8]) caused quite a ripple of interest when it

---

[8] For example see online articles in the Register around November 2001.

was shown a few days later at Nokia's Multimedia Developer Conference in Barcelona, before its launch in February at 3GSM 2002 in Cannes. It was the first product on Symbian OS v6.1, the first Symbian 2.5G (i.e. it supported GPRS) phone and, with its VGA camera, it was the first camera-phone anywhere outside Japan. At that time, the Nokia 9210 had just become the best-selling PDA, unseating Palm. Symbian OS rose for the first time to the top of the platform chart for PDAs and smartphones.

> **Bob Dewolf:**
>
> Series 60 made us think about all the things Symbian OS had to worry about, like distribution policies and whether components are only published internally or were third-party published. We put in a whole new layer of control to say not only that header files export but whether they get filtered out to the SDK process. Pearl became Series 60 at that point, version 0.9, and SDKs came out soon after at version 1.0.

But above all, the Nokia 7650 was the first S60 phone. The big announcement at 3GSM in Cannes was Symbian OS v7.0. At the same time, Symbian announced a new UI strategy. The DFRD idea was dropped. In fact, Symbian no longer proposed to ship a GUI implementation at all on top of its UI Framework implementation. The field was opened to UI vendors or licensees to do their own thing. Meanwhile the new UIQ UI was launched in place of Quartz, with UIQ spun out as a separate, independent company, though it remained Symbian owned. (In late 2006, UIQ was purchased by Sony Ericsson and became completely independent of Symbian.)

A few weeks later, the first Symbian OS v7.0/UIQ phone, the Sony Ericsson P800, was announced (it launched later that year). It was a pen-based phone featuring a removable flip keypad, somewhat in the style pioneered by the Ericsson R380, a jogdial thumbwheel for navigation and, of course, a camera.

The new strategy was possibly the only strategy that made sense not just of the unique nature of the phone market, but of Symbian's unique position in it. Perhaps, to some extent, Nokia identified this most quickly (and so platformized its own UI). The moral, perhaps, is that where it is impossible to have more than partial foresight of evolving market requirements and opportunities, multiple visions are better than one. Enabling a competitive ecosystem allows multiple different visions to emerge and have a chance to succeed.

The change in strategic direction predated Colly Myers's departure as CEO in February 2002, but its implementation spanned the interregnum and was picked up and seen to fruition by the incoming CEO, David Levin, from April 2002. Fittingly, Myers's – and Nokia's – Christmas gift to the company was a Nokia 7650 phone apiece.

16.8   How to Develop a World-class GUI

The most visible mobile phone trends are in conflict: devices are becoming increasingly smaller, packing in more and more hardware and demanding longer battery life, while moving increasingly into the consumer end of the market as features and functions expand. The phone operating system is the battleground on which these conflicts are played out. Shrinking size and increasing complexity are not simple trade-offs. The evidence is that the successful phone needs both.

Perhaps it is worth citing the definition from Pekka Ketola, a usability engineer on S60 [Lindholm *et al.* 2003, p. 172].

> Smartphones are an emerging product category where communication, namely voice calls and messaging, is still the main function, but where personal information management is fundamentally improved compared to conventional mobile phones. Smartphones have good calendars, versatile contact management properties, to-do lists, address books, and so on. They are solid platforms for imaging and gaming.

In one way, phones are become increasingly personal items, statement-making lifestyle accessories. In another direction, they have evolved from being 'expensive showoff tool[s] to an everyman's communication platform' (Nieminen-Sundell and Vaananen-Vainio-Mattila in [Lindholm *et al.* 2003]). In both directions, the trend is to a normalized market, that is, a consumer market in which sophisticated users give way to naïve ones and complexity is not tolerated. Phones lose novelty value and are expected to perform as easily and predictably as TV sets and stereos. In this battleground, the UI becomes the front line.

Big User Interfaces Don't Scale

> **Bob Dewolf:**
>
> There's an awful lot to a UI. It's not just controls and their customizations, there's a lot of interaction. That's a surprise we got in the process of development when we did Pearl, how much more integrated you had to be. Integration is how the applications fit together, what keys the phone is grabbing from the window server, and who's managing the power key and what happens if you press the Phone button. There were a lot of issues like that and it was important to get them right.
>
> I remember very fundamental decisions about what happens when you press the menu key. For instance, does it just put the menu on top? In the we decided on something very simple, but there were lots of different proposals,

which led us to decide that simplest is the best and we just leave everything stacked where it is and go with the window-server policy. But it's amazing how long the debates took. In that particular case, it ended up very simple, but not in all cases.

An absolute rule is that small interfaces are different from big ones. Direct manipulation (the familiar Windows or Macintosh model), for example, does not scale down. You do not expect to drag a mouse cursor across your phone screen. The parallel model of the desktop, with multiple open windows between which you task, does not scale either. On a small phone screen parallel is out, sequential is in.

Just as even the best UIs do not scale, nor do they move easily from one device class to the next. In effect, this is exactly what Symbian's UI strategy reflects and what the architecture has been evolved to support. One size does not fit all, and there is no single right model. What works on a flip phone probably does not work on a keyboard-centric Communicator; what works on a pen-driven tablet probably does not work on a phone designed for the shirt-pocket.

The first generation Quartz devices (that is, before the evolution of Quartz into UIQ) were strongly tablet-based designs. There have been tablet phones (for example, the original XDA based on Microsoft Windows CE and, perhaps, the phone-enabled iPaq counts as tablet-like) but they have not been huge successes in the mobile phone market. The evidence suggests that consumers want phones with productivity functions, rather than PDAs (which are really productivity devices) with phone functions. There is little to suggest that the UI is a critical factor in the decline of the PDA market; neither Palm PDAs nor Microsoft PDAs have fared much differently. (Palm has now gone all the way and adopted Windows Mobile, so the point becomes moot going forward).

Given the differences between Windows on a desktop and Windows Mobile on a phone, there is not much case to be made that users are seeking the desktop-style behavior on their phones; they won't get it. What they will get is familiarity at some level, even if the behavior is different. Perhaps what they are really seeking is the unstated promise of compatibility; but in that case the value of having Microsoft running on their phones is only incidentally about the UI itself and has more to do with the promise of the brand. But as the Palm and Macintosh combination demonstrated quite ably years ago, smooth interoperation has not much to do with sharing a UI on different platforms, it's just a matter of plain old-fashioned good, careful, user-centered design.

The most interesting point about phones is that while the design point has in a sense been fixed and is the same across manufacturers, UI differentiation makes different phones quite radically different from each other (to the point that picking up someone else's phone and trying to make a call can be surprisingly difficult). Kiljander and Jarnstrom (in [Lindholm *et al.* 2003]) argue that it is because phones have no standard UI and, therefore, everyone is using their own solutions that the UI has become such a significant competitive asset. This is as true at the low end of proprietary operating systems and custom UIs as it is at the high end of Symbian OS and its competitors.

If this is the case, Nokia's commanding market share (at the time of writing anyway, but with little sign of it waning) is a testament to its UI designers and to its strategy of evolving a family of UIs, each tuned to the design point of one of the categories in the famous Nokia segmentation matrix.[9]

Usability Values

As well as becoming cameras, music players and diaries too, phones have become open platforms for third-party software of all kinds, including games. It is hard to devise a UI that is as fit for playing fast-paced shoot-'em-ups as it is for managing your daily, weekly and monthly meetings and appointments, while still allowing users to answer (and initiate) calls with a single button press. While phones have become the archetype of the omnifunctional, converged device, Symbian OS is still designed to be flexible enough to support a wide range of possible device types including more narrowly specialized categories – digital cameras, in-car navigation systems and set-top boxes have all been rumored at one time or another as possible target devices.

Arguably, however, the particular, and unique, strength of Symbian OS is its power (as well as its compactness) as an open, thoroughly GUI-centric, standards-driven application platform. The fact that phones are no longer merely phones, but complete open software platforms is a critical turn on the road. It is also a central part of Symbian's play for the market. Applications are critical.

The Application Model

A key usability issue is that of the application model exposed to users: how does the user work with files, documents and applications, and what

---

[9] Nokia segments its users according to customer categories (Balancers, Controllers, Experiencers and Impressers) and its phone models according to style categories (for example, Expression, Classic, Fashion and Premium). The fully populated matrix matches phone styles to customer segments. Typically, Nokia's different UIs (Navikey, Series 30, Series 40, Series 60 and Series 80) match their complexity to the different requirements of the segmented model. See the chapter by Kiljander and Jarnstrom in [Lindholm *et al.* 2003].

concepts (and how many and how intuitive are they) must the user grasp in order to use the device successfully and naturally.

Andrew Thoelke:

Windows has never really ever escaped being file-based. You think in terms of files. Everything is a file. Macintosh at least went one step further because you generally didn't find files: you had documents and you had applications. The Series 3 had a system screen that was application-centric and a file browser that gave you a folder view of everything. So the system screen showed you what we called the washing line of applications, and under each application it said, 'Here are all the documents this application knows about'.

With Series 5, we went to a bolder view of the world. It was document-centric, not application-centric. You could basically tap on a document and the system recognized the document. It knew exactly what application was associated with it, and it would then launch that application and give you your document. We didn't believe in file extensions as a way of identifying content, so we invented our own system of internal designation, UIDs (Unique Identifiers). Macintosh, by the way, has always been the same. But our application architecture in Symbian OS really came about from saying, 'Okay, so given a document I need to basically do something with this document', and that's what the user really cares about. The fact that there was an application doing it for you was almost like, 'Well, the user shouldn't have to worry about that. The user shouldn't have to care'. Of course you knew that if you hit the Word button it would take you to that application, but if you found a document that was just sitting there you could tap on it and ta-da! It would have the right icon and it would just all work. Well that is probably right for a PDA, because a PDA is all about creating documents. But actually that isn't right for a mobile phone.

The native, document-centric application model also runs into trouble as devices start to become increasingly connected. While documents created on the device, by the native applications, can be created with appropriate UIDs which identify their document types (stores and DBMS databases, for example), recognizing externally created documents so that the appropriate application can be launched is more problematic. The early use cases around which the Series 5 was designed were manageable, mostly requiring recognizers for email and web document types. But in a modern phone the sheer diversity of the document types which may be encountered, ranging from office-type documents to a wide variety of media formats, requires not just a comprehensive recognition system (which Symbian OS provides), it also forces at least a degree of file centricity on the UI. Thus UIQ, for example, which has a non-file-based application model, must adopt a file-based approach for dealing with media files. S60, which is file-based, runs into a slightly different problem when swapping files between devices because the built-in file manager application chooses to hide significant parts of the file system for usability reasons with novice users. (A third-party file manager is required before the user can fully explore the file system.)

In Symbian OS, the application model is, in effect, created as a collaboration between the operating system and the UI implementation. In the original architecture and in the Uikon architecture which was derived from it, the application model (i.e. the document framework) is implemented by the application framework but brokered through the UI framework (Uikon) and the UI layer which is built on top of it.

Both UIQ and S60 have evolved away from the original application architecture, for example of the original Eikon GUI, towards more task-based approaches. Underlying that evolution is the difference in the way that PDAs and phones are used.

Andrew Thoelke:

If you're using a mobile phone, the way you think is very much doing one thing at a time. The fact that a Symbian phone can multitask is something that probably most people don't realize, except that as soon as you go back to an old phone you realize that when you send a message you've got to sit there waiting for it to be sent. On an S60 or UIQ phone, you send a message and you carry on doing everything else, and eventually the phone says, 'Oh, by the way, I've sent the message'. So the multitasking in a phone is a bit less obvious. Generally on a phone you are task-oriented, so you're interested in what you're doing right now, and then in the next thing you're doing, and so on. Some UIs are better than others at being able to backtrack to what you were doing just before you interrupted it, but in that respect you really want to have a system that is based around applications and tasks. So that's where UIQ has gone, in the direction of having the view server, where a view is essentially a task, but sometimes it's one way of looking at tasks. So each application has got different views and you can flip between those views, but also flipping between views will activate different underlying applications. Whereas S60 has a quite different model of either embedding views or just jumping between applications, and it produces a different system.

In its time, the Eikon approach was quite innovative, and it is part of what made the Series 5 so easy to use.

Andrew Thoelke:

I guess it wasn't innovative in the software space in '94, but certainly in the PDA space, to have something that was document-centric and so highly integrated with the actual application framework. It was certainly unusual. And it has not been a barrier for being in the mobile phone space, but it's not necessarily been totally tuned to the way mobile phones work, and the UIs have had to address that.

Java proved to be another area in which the cleverness of the application architecture and its seamless integration into the UI caused some unexpected headaches. Java proved quite hard to integrate well with the subtly different application model of Symbian OS, although for the MIDlet writer, as for the user, the differences are transparent and Java applications 'just work'.

Andrew Thoelke:

Even today the system still thinks you're a native standard Symbian C++ application, so for Java we have to provide special 'stubs' and 'dummy' resource files to make the application architecture think its got a valid application to run. It would be better to be able to register with the application architecture and tell it there is an application and how to run it. That would be the ultimate, to have an application architecture and an application lifecycle model that is truly agnostic to the actual programming language.

16.9 Symbian OS User Interface Architecture

Symbian OS supplies the UI Framework and Application Services as its topmost layers, but does not supply the concrete user interface. Instead, phone vendor licensees create (or license) a UI which is based on the framework classes supplied by Symbian OS. Symbian OS has no notion of a console application. Native applications, by definition, present a UI which is implemented using the concrete classes of a particular GUI implementation, layered over the generic GUI framework support made available by the operating system. Applications implement the abstract classes of an object-based, MVC-style, event-based application model which is defined partly by the operating system frameworks and partly by the GUI implementation. In fact, the event model goes surprisingly deeply into the system, all the way to the Window Server.

Non-native applications, written in Java, Visual Basic, OPL or, more recently, Python and even Ruby (these are currently the most mainstream high-level languages available on Symbian OS), depend completely on the graphical support of the GUI implementation on top of which they are written. Except for low-level development work, for example system programming, writing Symbian applications is inherently GUI-dependent.

Symbian ships only non-production test application UIs for system components that require a UI for testing, configuration, and so on. In a final licensee product, licensee applications manipulate the components directly or provide the application-level user interface.

16.10   Future Directions

From 2006, smartphones are shipping with the latest releases of both the UIQ and S60 user interfaces, both of which introduce new features to support UI differentiation and specialization, aimed particularly at enabling phone vendors and operators to customize, extend, and brand phone UIs. In both cases, the latest UIs depend on the Symbian OS v9 platform.

If these releases are indicative, increased flexibility and customizability certainly seem to be the dominating trends for the future evolution of phone UIs.

UIQ 3

UIQ 3 emphasizes personalization and configurability. It includes features that offer extended flexibility and new opportunities for differentiation and specialization by phone vendors and operators, as well as new opportunities for third parties to provide end-user-installable UI customizations. In particular, UIQ 'themes' bring the notion of the skinnable user interface to UIQ. Themes may be ready-made (in other words, supplied by UIQ or created or customized by the phone vendor or other UI integrator), user-customizable, operator-specific or supplied by a third party, encouraging the market for downloadable themes.

Operator customization based on Operator Configuration Package (OCP) themes and skins allows the provision of preloaded content including applications, sounds, settings, fonts, icons, animations and so on, by vendors supplying UIQ phones to operators. There is also support for over the air (OTA) features, allowing operators to configure (and customize) phones remotely.

Supporting these features is the notion of parameterizable UI properties, with support for multiple screen sizes, multiple portrait and landscape modes, touch-screen and non-touch-screen modes, and various hard and soft key combinations. The resulting UI configurations (for example, flip-open and flip-closed styles and pen and softkey styles for different product families) are intended to help developers maintain application compatibility across a range of different devices while enabling vendors to create a wider variety of phone styles.

S60 3rd Edition

Again, much of the focus of the S60 3rd Edition UI is on enabling customization, including vendor, operator, end-user and enterprise customization. Increased flexibility is emphasized too, with UI scalability supporting a wide range of screen sizes, both portrait and landscape modes, and 'double-resolution', that is, high-resolution displays.

The most significant support for differentiation is the enabling of core application extensions, allowing operators to extend the core applications with new features while retaining basic application compatibility with other S60 phones. The emphasis is on supporting integration of operator-specific features and services, as well as service settings and operator-customized UI look and feel, to enable a high degree of sophisticated operator-specific branding to be provided by phone vendors using S60 on their products.

17

System Evolution and Renewal

17.1 Introduction

This case study looks at the forces and pressures that have driven the evolution and renewal of Symbian OS, from the earliest days of the system to the latest Symbian OS v9 releases.

A critical issue for any software company is keeping a complex software system fit for purpose in a context of rapidly changing technology. To do so requires continuous evolution and continuous renewal. Evolution and renewal are, therefore, among the strongest internal drivers for engineering and architectural change.

Customers, on the other hand, are typically more interested in features. Balancing the competing pressures, for feature growth on the one hand and internal renewal on the other, is an important part of the business. Preserving the freedom to invest in renewal against relentless feature pressure and deciding where to focus that investment are critical elements of a successful system strategy.

In reality, of course, the two things are not independent. Renewal is just another way to talk about the prerequisites for future features. But, simplistically, while the costs of new features might be charged to customers, the costs of renewal are an overhead. Left at that, features tend to win the competition for development resources.

Symbian OS faces the particular challenges of the mobile telephony context, which are acute. The pace of evolution since mobile telephony began to take-off in the mid-1990s, and particularly since the second boom around 2000, has been breathtaking. That period of a little under a decade is also the period during which Symbian OS has been licensed to phone vendors. The increase in capabilities of Symbian OS v9 compared with the original release of Symbian OS is quite dramatic.

Charles Davies is forthright about the need for continuous renewal.

> **Charles Davies:**
>
> All of Symbian OS needs to be renewed over time and it's a challenge to work out how to do that. In architectural terms, renewal is one of the biggest challenges.

In theory, five years or so from now, Symbian OS will be at the outer limit of its original intended design life of approximately 15 years (depending on when you start counting). Of course, it is a fact of life that software systems commonly outlive their design lives by several multiples. Bits do not wear out. Software does not fall apart over time. But still, the changing requirements and changing context mean that, in practice, system evolution is almost continuous. The likelihood is that, in five years time, most of the headline features of the operating system, and many of its significant architectural features, will be ones that were not present in the original releases; many of them may not even be present today.

This is the magic of renewal (or the problem, depending on how close you are to it). Unlike the products which are built on top of software systems, software design life is typically managed by forestalling obsolescence through continuous re-invention and evolution, rather than by withdrawal and replacement of products (although that difference may have much to do with the less tangible nature of software compared with hardware).

17.2   Design Lifetime

All designed systems have a design lifetime and, for any manufactured system, deciding what that lifetime should be is as much about economics as it is about technologies. For example, the design lifetime is one determinant of cost because it determines quality requirements for components. It is also a key variable in calculating return on investment and other accounting metrics and is, therefore, a key variable in deciding the business case for a system and it remains a factor in deciding how much to invest in maintenance.

Software systems are no different. However, unlike other manufactured systems, they do not wear out. Maintenance is not driven by use of the system causing parts to become defective because of 'wear and tear', but by change. Maintenance in a software context mostly means defect fixing. Typically, the defects to be fixed are either those introduced by earlier (internal) software changes, or those which are revealed by an (external) change in the way the system is deployed. Thus, software maintenance is driven by change, whether adding new components to the system, removing components, or using the system in a new product, or because some other contextual change exposes previously ignored defects.

A simplistic way to distinguish maintenance from renewal is to count changes made to keep a system viable during its design lifetime as maintenance and changes carried out to extend a system's design lifetime as renewal. It is because software systems do not wear out that renewal is possible and attractive. But increasingly, what makes renewal necessary is the rapid pace of technological change. Maintenance, therefore, shades into renewal and is necessary not to extend system design life but to ensure that a system actually survives for its design life and is not made obsolete by technology changing more rapidly than predicted around it or by new, disruptive technologies appearing and changing the external context to one in which the system cannot compete. Renewal becomes as much a matter of survival as of extending the design life.

Deciding between maintenance and renewal is by no means always easy. There is a critical balance between maintenance and renewal, and getting the balance right can be a tough challenge for any organization; deciding when to maintain and when to renew requires foresight, a clear and well understood strategy, and a good understanding of the technology context. In the short term, maintenance always looks attractive; renewal frequently looks expensive.

Determining the Design Lifetime of Symbian OS

There was never any doubt about the goals of the software project for the Series 5 operating system. It remained as it began, focused very clearly on the requirements of that very particular device. However, there was also a broader context.

Geert Bollen:

There was already explicit intent to create an operating system suitable for multiple mobile devices, with a design life of 10 to 15 years. There was long-term thinking and there was also a vision of having software licensees that included mobile-phone manufacturers.

David Wood is equally clear about the long-term nature of the original vision.

David Wood:

I think we said something akin to 'The 16-bit system was designed for five years and would last ten; EPOC32 was designed for ten years and would last for 20'. It is interesting that we have already been through ten years of history, and I think it's going to be around a lot longer than even we foresaw.

Even perfect foresight would not be enough to secure a 15-year system lifetime. There is no option about accepting the challenge of renewal. The need for renewal is a fact of life. David Wood argues that it has been fundamental to the way Symbian OS has evolved right from the beginning.

> **David Wood:**
>
> The fact is that we did redesign many things in the initial evolution of Symbian OS. The UI is an important example of that, the fact that we went through three separate UI designs to get to the Eikon UI. And there were similar changes, not quite so large-scale, but other changes in the implementation of other aspects of Symbian OS, even before it got launched.

While you cannot guarantee longevity for a system, you can certainly design for it. Designing for longevity means designing for the unknowns to which your system will be subjected in the future. What that really means is designing extensibly.

The two prerequisites for extensibility are a good conceptual design and natural extension methods. The Symbian OS architecture is built on multiple levels of extensibility. System services are provided by servers that serialize access to system resources. To provide a new service, you just create a new server. Most servers are themselves extensible, because they are implemented as frameworks. To extend a server, you just create a new plug-in. Within the system there are multiple patterns of abstraction. Active objects, for example, abstract all events, so that applications only need to know how to handle the events they care about, without worrying what their source was. Sockets abstract all communications bearers, so that applications only need to know about the message types they want to send, without worrying about message transport.

These are all examples of extensible design and there are many others. David Wood tells a story to illustrate his point about the way extensibility is designed into Symbian OS at a deep level.

> **David Wood:**
>
> Bluetooth only emerged as a technology in 1998, about the same time Symbian was formed. So when we designed the system, Bluetooth didn't exist. Now I'm not trying to diminish the efforts of Symbian engineers, but when we did want to implement Bluetooth it wasn't that hard. Of course there are many devils in the details but architecturally it slotted in quite easily as just another active object, just another event source. In broad terms, it fitted easily into our architecture; we had foreseen that need by designing to be future-proof.
>
> At one of the early developer conferences there were questions from the floor during a talk on implementing Bluetooth on Symbian OS, asking, 'How

did you manage to do this?' When the answer was given, the question came back again, 'But surely it couldn't have been as easy as that!' And the answer was, 'Well, for us it's just another active object.' That was the phrase used: just another active object. Later on it turned out those questions were from Palm engineers, from the operating system part of Palm, and clearly they had been struggling to squeeze Bluetooth into their system which hadn't been sufficiently future-proof. In the end they managed, of course, and there certainly is Bluetooth on Palm OS today, but it required much more effort I think than with Symbian OS. So I do think Symbian OS is future-proof, because we design our frameworks with extensibility in mind.

'Future-proof' is a bold claim but it was certainly the goal of the early design of Symbian OS and, in key respects, has demonstrably been achieved.

There is, perhaps inevitably, danger too in abstraction. Too many levels of abstraction can lead to code bloat and poor performance and also to over-complex and unmaintainable code. The job of the component designers and architects is to understand where extensibility is required, and to enable it appropriately, and where it is superfluous and constitutes over-design and over-engineering.

High Impedance of Change

Technology moves fast and each technology generation seems to move faster. The founder of Netscape, Jim Clark formulated his 'Internet-time' law to express that fact, the so-called law of constantly increasing acceleration which compels 'technological products to be released faster and faster' [Lewis 1999].[1]

However, systems, especially large and complex systems, evolve slowly. This is not just true of software. It is as true of social organizations (from large bureaucracies to large crowds) as of whole species (evolution) as of particular technologies (fossil-fuel-powered systems). Thus the corollary of the speed of change of technology is the high impedance of systems to change.

Entrenched systems change most slowly, as if large-critical-mass systems are slowed by systemic inertia. Operating systems become entrenched when they achieve a certain level of market saturation. This kind of entrenchment and inertia may be the biggest threat to renewal.

Two ideas which have become central to what Symbian does, and how it does it, reflect the interesting dynamic between the drive for change and the drive for stability:

---

[1] The limit is based, as he says, on the speed of light: 'How fast light moves down a fiber-optic cable.'

- Platformization is what drives Symbian's business and technology goals, the goal of creating a software platform suitable for hosting end-user-centered services across a diverse range of phone hardware

- The lead product concept is a key part of the way Symbian drives projects, so that each software iteration is closely coupled to and driven by the requirements of a particular licensee product.

Platformization requires stability. Lead products, on the other hand, drive innovation and change. In part, this approach has crystallized naturally, out of practice.

Providing an open platform for third-party development and bringing it to the emerging world of mobile phones (and other pocketable devices highly centered on communications) was one of the early rationales for Symbian OS, and remains as much so now as then. To make that strategy support multiple licensee UIs (to enable licensee differentiation), an equally basic rationale, requires a high degree of commonality in the system below the licensee UI variations (80:20 was the early rule of thumb guideline: 80% of common APIs underlying 20% of UI-specific APIs). The challenge of the strategy, therefore, is to avoid the platform fragmenting beneath the centrifugal effect of UI variation, to enable as far as possible a write-once–run-on-all-variants development model. Some customization for UI variants will always be required, but the minimal goal of common data engines beneath custom UIs is achievable so long as unnecessary fragmentation is avoided.

The lead product approach has proven a successful way to capture new requirements and drive the evolution of Symbian OS forward while supporting licensees in bringing a more or less continuous stream of leading edge products successfully to market and, in particular, while maintaining the hard product deadlines required for manufacturing.

All Symbian OS projects are tied to one or more particular lead products that define the basic critical feature set for a release, drive the release schedule, and provide early validation and real products for final testing. Once the lead product has shipped, licensees can consider that OS release a proven platform for subsequent product variants. In effect, this simply rationalizes the early approach of working closely with licensees on real products, but provides a framework for supporting multiple simultaneous licensees taking a given new release.

17.3   Renewal in Symbian OS

Typically, a renewal strategy aims to anticipate future feature requirements and evolve the system to be a better platform for those features than the existing platform. To a lesser extent it also aims to improve any areas where the platform is deficient for current features.

Renewal permeates Symbian OS. Examples range from major system-wide changes, which include UI evolution, the introduction of the real-time kernel, and platform security, to numerous small upgrades and updates, for example to track standards evolution in supported technologies (including Bluetooth, SyncML, vCard and vCalendar). Between the two extremes are many examples of significant architectural enhancements, whether to evolve services such as telephony to support new network technologies, to introduce new services or to replumb essential system services.

- The UI architecture evolved through multiple iterations to the early Eikon UI and, thereafter, to Uikon plus licensee UI variants (for example, S60, MOAP and UIQ), which are themselves continuously evolving.

- The new kernel provided a complete generational change at the heart of the operating system.

- Platform Security was a major architectural evolution with system-wide impacts and high external visibility.

- Large-scale tools and system infrastructure changes include toolchain upgrades from Visual C++, to Codewarrior, to Carbide and ARM, tools updates driven by the move towards standard C++.

- New services include the power-management and system-startup frameworks.

There are numerous examples of small architectural changes related to the above major changes, including evolution of the IPC mechanism, a new client–server architecture, and the ECom plug-in framework architecture. Other examples of significant architectural evolution include two generations of multimedia re-architecture, significant telephony evolution to support new network technologies (CDMA and 3G) and a forthcoming new communications architecture.

There are also numerous examples of continuous architectural evolution driven by performance requirements and of extending existing services as new technologies emerge, for example, flash filing systems and USB support.

The largest changes have been phased in over multiple releases, partly to manage the risks to licensees of large-scale change and partly to try to maintain an incremental approach to development and to maintain the model of frequent releases. (Symbian's earliest releases were monolithic and the project management lessons were learned the hard way.)

For example, the real-time kernel was introduced as an option for Symbian OS v8 and then became the standard kernel for Symbian OS v9, enabling licensees to evaluate the new architecture and giving them flexibility in when to migrate. The introduction of system-wide platform security was similarly phased in over two v8 releases: v8.0 and v8.1. (It overlapped with the introduction of the new kernel because it depended on the new kernel to police the security policy implementation.) Both are examples of a major re-architecture of the system and both cases reflect Symbian's perception of the needs of the market. Interestingly, in both cases there was some short-term customer resistance, even if customers have later embraced the changes.

> **David Wood:**
>
> It has taken a long time for EKA2 to come to fruition, but that has been an internal renewal of Symbian OS and that is an extremely important initiative that's happened, because it has addressed some of the things that, over the course of time, it became clear to us were shortcomings in the original design. So it's very good that we've been able to do that. I don't think the full significance of the new kernel will be appreciated for some time, but we will see it employed not just in smartphones, but in a large number of other mobile devices too in my view, because of what we were able to do to refresh the design.

Platform Security is a highly significant change, which introduces a security capability model and data caging supported by a certification scheme, and requires some significant kernel changes including changes to IPC. The overall impact has been high and, in platform terms, is highly visible, because it has immediate (external) developer impact. Again, the project had to weather some significant early resistance. To some extent, the external impact may not yet have been fully felt. Charles Davies is sanguine.

> **Charles Davies:**
>
> It's a big thing, you know. There is no industry-wide accepted best practice for how to secure a system, I think nobody knows exactly what the best thing is to do. There is no 'right' answer. But PlatSec is our answer and time will tell.

17.4   Evolution in the Kernel

The move to the new EKA2 kernel is perhaps the most significant evolutionary step Symbian OS has taken. However, as Martin Tasker points out, architecturally it remains in some ways quite a local change.

> **Martin Tasker:**
>
> At the system-design level, it hasn't actually radically changed the system design. It's still either application processes or server processes and that design was pioneered way back in SIBO and hasn't changed much since the earliest releases of EPOC. One reason it hasn't changed much is that it's a proven design.

In fact, the beginnings of the kernel re-architecture go all the way back to the first collaboration with Nokia on the project to bring the 9210 Communicator to market. This was not Symbian's first licensee project and it was not the first Symbian OS phone project.[2] However, at the time it was certainly the largest and most complex Symbian project to date.

The design approach was based on a two-operating-system, two-processor solution, the so-called 'partner operating system' approach. Symbian OS, since it was not then real-time-capable, was not capable of hosting the GSM telephony stack. The phone side and the application side were therefore separated, with a dedicated RTOS running on a dedicated phone-side processor and hosting the baseband software, while Symbian OS hosted the application side, running separately on an application-side processor (see Chapter 15). While Symbian OS has evolved hugely in the intervening years (probably almost as much as the phone hardware has evolved from the first, brick-like Communicator design to the latest sleek devices), the problems of that early solution are instructive. The most immediate problem was how to force the phone-side RTOS and Symbian OS to cooperate.

> **Morgan Henry:**
>
> To integrate the two OS schedulers and interrupt handing, there was only a small amount of management that EPOC did first and then it would call one of these hooks to hand-off to the RTOS. The RTOS would only give control back once it felt it had done its job. I believe there have been similar solutions for things like Linux running with a partner RTOS. But the challenges aren't over yet and you have to decide what owns which bit of hardware. Some of the hardware was owned by the RTOS and some of it was owned by EPOC. I think the MMC card, for example, was owned by EPOC and all of the peripheral drivers, but all of the baseband hardware was on the RTOS side, so it owned the power management, and it owned the real-time clock which caused some interesting problems.
>
> Coming from the Series 5, where EPOC always had direct access to every bit of hardware, especially the real-time clock and the timers, and now finding

---

[2] The Philips Ilium/Accent and the Ericsson R380 projects both preceded it, though only the R380 came to market.

that it had to go over a communications protocol to get the time from the real-time clock was a bit of an architectural shock for it. So there were lots of work-arounds, but where we had the opportunity to do it we tried to redesign in a sane way. This is what started the move to try and push a lot of things outside the kernel and shrink its responsibilities. For example, the kernel being able to persist system settings is something that was possible on the Series 5 because it had battery-backed RAM available to it, so it was persisting data in the superpage and so on. This wasn't really possible in the Nokia 9210 model where a lot of that data was owned by the RTOS and persistent storage required writing to flash memory. So there were improvements in how we did the HAL [Hardware Abstraction Layer] to move from the Series 5 model, where a lot of the responsibility for persisting was with the kernel and the User Library. In the new world a lot of that either got pushed up higher into the operating system or got pushed over to the RTOS.

Those were just some of the challenges of the partner OS approach, because it was a fairly difficult piece of hardware sharing. For example, because we didn't have visibility of what was going on inside these RTOS hooks we'd get defect reports or see problems on the hardware which were kind of inexplicable. You quickly realized you needed someone in the kernel team with a big brain to think hard about the problem and discover that it would be that the RTOS was affecting the processor state in ways it shouldn't, or blocking interrupts when it shouldn't.

Other basic differences between the Nokia 9210 hardware and the original Series 5 hardware design also required some significant changes.

Morgan Henry:

In terms of architectural changes, new functionality was certainly added; things like generic support for the DMA controller and power management, as well as drawing the delineation between the kernel's responsibilities and higher level responsibilities, for example, in terms of persistence of data. Those kinds of architectural changes were happening then and those have been carried forward into the new kernel as well. So those decisions were the right ones at the time.

Another explicit goal of the EKA2 kernel architecture was to improve portability of the operating system by improving the modularity of the kernel design, so that hardware dependencies were isolated from common kernel code and so that different levels of hardware dependencies were isolated from each other, for example, to distinguish between more generic dependencies and the specific dependencies of particular devices.

The so-called 'partner operating system' solution is of course only one approach to solving the phone problem. Another goal of the new kernel architecture was to enable single-operating-system and, therefore, single-processor-core solutions.

Morgan Henry:

If you look at the problem we were trying to solve with partner OS, now with the new kernel we are in a situation where we've solved it in a better way. It certainly is a lot more architecturally sound. So the reference design team in Symbian recently announced that they'd got their first single-core-solution running using Symbian OS as the real-time OS with a personality layer. In terms of functionality, they're at the same place but with a better solution, a more robust solution.

In this approach, a 'personality' layer is used to interface the baseband stack directly to the EKA2 real-time nanokernel. The personality layer mimics the interface of the RTOS for which the particular baseband stack was written. Since the nanokernel has true real-time performance, this solution allows the baseband to be hosted on Symbian OS along with the application side, for which the extended EKA2 kernel (i.e. the nanokernel plus kernel) provides the interface, enabling a single-operating-system, single-core solution.

As well as the new kernel architecture, there have been significant other additions to the lowest levels of the system over multiple releases. A new framework for power-state management has been added to support the latest generation of phones which incorporate hardware previously found only on high-end laptops and dedicated devices such as digital cameras and camcorders, but still need to provide phone-style extended battery life.

Radio technologies, such as Bluetooth, Wi-Fi and 3G, are extremely power-hungry (power drain and battery problems were among the early technical hurdles that stalled the rollout of 3G networks). Coupled with the motors needed to drive optical zoom lenses, electronic camera flashes and large LCD displays, the power demands of high-end phones really do push the limits of battery-management technology. Simplistic three-state models (on, off and standby) which were adequate for an earlier generation of phones no longer meet the requirements. Symbian OS has been very successful at maintaining its significant lead over less-well-adapted operating systems running in mobile phones.[3]

Similarly, there have been substantial changes to keep pace with evolving generations of flash-memory technology, matching the demands of the latest phones for large removable (and non-removable) drives that provide multigigabyte internal and removable memories. File system extensions supporting flash-file storage media have been added. Flash systems are not byte-addressable, so conventional read–write mechanisms have to

---

[3] It has close to 80% of the market, see Chapter 2.

be completely redesigned. Flash systems also have a more limited lifetime than other memory technologies, supporting a lower, fixed number of accesses before wearing out, which means that memory accesses must be evenly distributed across all sectors of the card and not concentrated in one physical location. These all imply significant behavioral differences which must be abstracted by the file-system server to provide a common file-system interface. (The behavioral differences between NOR and NAND flash have also become significant as phone vendors have adopted the cheaper NAND technology as a way of boosting internal memory size in a cost-effective way.) Similarly, solutions such as demand paging have been explored as solutions both to RAM inflation and to mobile disk-based storage options.

Performance improvements have also been made throughout the system, at all levels from application support to the kernel and other low-level system areas. At the low level, file seek time and inter-process-communication timings have been improved, in addition to the fundamental work of managing interrupt and other low-level process latencies as part of the real-time reengineering of the kernel. In the application support layers, significant effort has been spent to maintain and improve the core PIM application engines, to keep them up to speed with current technologies as well as with evolving expectations. Thus Agenda, Contacts and Messaging have all had their share of renewal through recent releases of the operating system, particularly to improve performance.

17.5   Telephony Evolution

Telephony is an obvious area in which technology has evolved at a relentless pace since the first Symbian smartphones came to market. GSM has been enhanced with successive generations of go-faster technologies such as EDGE and with half-way house technologies such as GPRS which extend the basic voice capabilities of GSM with packet-based data services. Finally, it has evolved to full 3G, with packet-based voice and data services. 3.5G network technologies are now reaching the market and 4G will no doubt soon start to emerge (although it is not yet clear what 4G networks will mean). As well as this kind of generational network change, there is the further complication of a global market divided between competing network technologies, although GSM has, to date, dominated globally over the North American CDMA alternative. Unfortunately, 3G perpetuates the divisions. 3G evolution is based on standards but while GSM evolves to the 3GPP standard, CDMA evolves to CDMA2000 and remains as incompatible as its 2G counterpart.

Further disruptive change is promised as competing high-speed wireless technologies such as Wi-Fi and WiMAX converge with the latest telephony technologies. It is not clear what mix of standards will succeed or which technologies will dominate. But clearly, to retain its current advantage, Symbian OS needs to be flexible, adaptable and to support this technology evolution seamlessly.

To some extent, the beginnings of the telephony architecture evolution go back a long way, to Symbian OS v7 and even earlier.

Andy Cloke:

We have invested a lot of time and effort in multimoding the ETel API to support the North American market where the CDMA2000 specification, the Qualcomm-influenced specification to put it like that, was prevalent. Our original ETel API was very GSM-centric. Since 3GPP specs are based on GSM, that translates nicely across to 3G in Europe and Japan and anywhere else with a GSM footprint that is upgrading to 3G, but it doesn't translate well to the North American and Korean markets. So that was the ambition, to support both GSM and CDMA, with a view toward 3GPP and CDMA2000.

Since the lead time for getting these sorts of software components right is quite long, the work started off quite early. So we had to scour the Qualcomm CDMA specifications and work to align data formats, data structures, requests, responses and notifications, and sort out where we could align and where we had to create two separate functions, one which would be used in GSM and one which would be used in CDMA2000. And, of course, the design principle is to support keeping the phone engine as simple as possible. You don't want to say 'Dial a Qualcomm voice call' or 'Dial a GSM voice call'; you just want to say, 'Dial a voice call and here's the number I want you to dial'.

The aim was not so much to support multimode phones capable of both GSM and CDMA operation, as to support phone vendors with portfolios of phones targeting all markets. For the vendor, being able to build both GSM and CDMA phones, or their 3G equivalents, from a single source base is a significant advantage (even though the source base is more complex), allowing maximum reuse of software across a phone portfolio.

Andy Cloke:

The core ETel stuff actually is very thin. So, for example, SMS transmission and reception is an ETel extension. And it turns out that the multimode ETel extension is almost so standard that you'd always have to use it. But the point is that it did give us another opportunity to redesign. For example, when we created the original GSM API, we had a split between Basic and Advanced. The idea was that you would have the core ETel which would be suitable for standard wireline modems, and the Basic ETel extensions would be suitable for a phone connected by wire or infrared, a two-box solution, and Advanced

> would extend on top of that and would be suitable for a phone where the signaling stack is built in. That was the original design intent, but we never really ended up using just Basic ETel, so it was good to be able to deprecate that.

It may seem an obvious point, but in the early days of Symbian, phone expertise was hard to find. The company's own background had been strongly PDA-based, with a particular flair for squeezing powerful UIs and applications into diminutive machines. Conventional networking technologies were well understood and fixed-line telephony was a well-understood technology. But mobile telephony was still highly specialized and mostly disjoint from more conventional computing. Embedded systems was a similarly highly specialized field. Few individuals (and companies) had expertise across all those boundaries.

> **Andy Cloke:**
>
> By the time we got to develop Multimode ETel, we were a lot more experienced in the whole phone area. By then, we had people who had come in from companies where they had been developing GSM phones, so they thoroughly understood all this stuff.

Supporting all the available global telephony standards is a prerequisite for a genuinely capable phone operating system. Because those standards continuously evolve, the operating-system support for them must continuously evolve. Without an extensible telephony architecture that kind of evolution is impossibly hard. The goal of an extensible architecture is not simply to make evolution possible, it is also to make it as easy as possible and as safe as possible since, in any system, change is expensive; in a complex system, all change is also a risk to the stability of the system.

Another rather more minor example of the way that the telephony architecture has changed in unforeseeable ways is exemplified by the introduction of the ETel third-party API in Symbian OS v6.1. In the original release of Symbian OS v6.0, ETel was substantially open to third-party developers. The release of Symbian OS v6.0 SDKs which documented in detail the ETel APIs for the first time caused concern for some licensees,[4] reflecting extreme operator nervousness about possible implications for network security. It should be remembered that Symbian OS was not just the first, it was also the only open operating system being used as a platform for phones. (And, arguably, it still is.) Operators

---

[4] At that time I was managing the documentation team which had published the controversial APIs and I saw very clearly the upheavals it caused.

had simply never been in a position in which their networks were even potentially open to third-party developers. A solution was rapidly agreed, the API documentation was withdrawn and an open subset API was introduced (and documented) to abstract public telephony behavior to enable applications to control phone calls, while the underlying API was locked down for internal use only.[5]

Again, future change is foreseeable in principle but never in detail. Extensible design is the only assurance against future disruption.

17.6   Sound and Vision Evolution

Another area of rapid recent technology evolution is multimedia, and this is again an area in which Symbian OS has undergone significant architectural change. Symbian OS has always supported sophisticated sound- and image-based applications, with Symbian OS v6.1, for example, being the platform for the first Nokia camera-phone in Europe, the Nokia 7650. Multimedia really arrived for the first time in Symbian OS v7.0s which introduced a new multimedia server. In Symbian OS v8.x, the architecture was reinvented to support a full suite of sophisticated phone functions. Between the releases of Symbian OS v6.0 and v8.1 therefore the change has been dramatic, from still camera to movie camera; from instamatic-like snapshots to full-function, multimegapixel digital camera replacement; from ring tones to music player. With image- and sound-recording, -editing and -manipulation capabilities, Symbian OS is evolving to support capabilities that until recently were limited to multimedia workstations.

The underlying rationale for the multimedia architecture redesign was support for the increasingly complex built-in hardware of the new generations of phones, with their dedicated multimedia components such as graphics hardware accelerators, MIDI-tone generators, dedicated DSPs and voice synthesizers, as well as high-end imaging and audio hardware parts.

Evolution of the licensee UIs for Symbian OS has been another big driver for change in the graphics services which underlie the UI framework support. Numerous UI enhancements aimed at supporting specific UI effects, such as fading, transparency and multimillion-entry color palettes, have required a steady stream of changes, mostly focused at the level of either the Window Server or GDI and BitGDI. The Window Server is a central component of the operating system, in many senses underpinning the GUI application model of the operating system, responsible not just for the windowing model, managing screen real estate and serving display

---

[5] Incidentally, incompatible changes were back-ported to Symbian OS v6 to ensure that later Symbian OS v6 products could not be 'hacked' based on the previously published documentation.

regions to applications, but also responsible for the system-event model. An example of the complexities which can be involved are the challenge to the fundamentally single-threaded Window Server model by explicit multithreading in the UI. It is easy to arrive at a position in which a clash of programming models (the native asynchronous model of Symbian OS based on Active Objects versus explicit multithreading) turns into an architecture problem.

There are other challenges too. A good example of the unpredictable emergence of new technologies is the emergence of vector graphics as the future display technology of choice (or at least a plausible candidate) within the wider industry. Traditional display technologies are bitmap-based (or have been for a good many computer generations), but vector graphics for display is certainly not new. Since the days of the NeXTstep Unix machine which put Display PostScript to work to manage the phone display, it has remained an interesting possibility.

However, recently there has been a vector revolution. The big driver has been 3D and naturalistic graphics for games. 3D graphics are calculation driven and vector graphics are a natural vehicle for their expression.

> **Charles Davies:**
>
> Now we have the challenge of how to deal with moving to Open GL and SVG, and what happens to Window Server in that context. There's a huge amount of technology and value in the Window Server and in the GDI and, for example, none of that's available in Linux; people have to license Trolltech above Linux. That makes the Window Server quite a critical asset. But it's under threat, because the world might be moving on to vector graphics models of drawing. So we had better be careful about that. It's not going to happen tomorrow, but it's one of the challenges.

17.7 Defining the Skin

One of the trickier problems Symbian OS faces is the boundary problem, or as Charles Davies puts it, the problem of defining the skin of the operating system, the boundary which determines what is in the operating system and what is not, what value Symbian creates and what value Symbian retains compared with the value that licensees themselves create and retain. There is a complex relationship between the architecture of the operating system and the business model, which is designed to encourage platform unity while enabling variation. At the same time, Symbian has deliberately sought to create an open platform capable of supporting a broad ecosystem of partners, third-party developers, software and tools vendors, enterprise customizers and hardware vendors, as well as licensees. Against external competition from competing systems such as Linux and Microsoft's Windows Mobile, Symbian must maximize the value it creates for its customers, thus maximizing Symbian benefit, without doing so at the expense of the value chain, which includes the ecosystem of Symbian's own partners as well as the partners of Symbian OS licensees. The essential rule, of course, is that Symbian must maximize the value it delivers without cannibalizing the value chain.

Charles Davies:

When I was at Psion, when we were building a PDA, I understood where the PDA ended and where the things outside the PDA began. I knew the boundaries of the product. I came into Symbian OS and I thought, 'Where are the boundaries?'

One of the things I've done since being here is to try to identify the scope of Symbian OS. We have had arguments like 'Should MMS have been inside or outside?' and I would have liked, dearly loved, to say, 'We do all the APIs, but we don't do the implementation'. That would be a simple thing to say, but the moment's gone. S60, MOAP and UIQ have extensive APIs. So that's not the boundary and it's really tough for both our customers and us to know where the boundary is.

To look at it another way, we've got, say, 750 people in Software Engineering doing Symbian OS and we can't make that 1500 overnight and we're certainly not going to make that 200. So with 750 people, what boundary can we draw that matches a decent product?

There are no easy answers. Not only is the system complex, the technologies are complex and the market is complex. The Symbian OS licensee model is complex.

Charles Davies:

For example, we could do an OMA DRM plug-in. Or is that an application? Is that in the scope of the OS or is that something for the UI? And the answer is borderline, borderline. With the borderline cases we will work with our customers when our capacity doesn't match it.

Keith de Mendonca has been on the engineering side of similar decisions in the past. MMS is an example. Writing MMS messaging plug-ins is complex. Providing support for it within the operating system clearly has an immediate benefit for licensees. The problem is that if some licensees take the Symbian OS solution and some do not, then both fragmentation and duplication result.

Keith de Mendonca:

We did produce an MMS solution when MMS first appeared. But we ended up with multiple solutions, with us and licensees each investing and producing basically the same technology. It was a waste of resources. Nowadays, Symbian OS no longer supports MMS directly at all, and so our customers have to provide those MTMs themselves if they want them, and they obviously do want to support MMS. So although our MMS solution is still in some products, say the Motorola M1000 phones or the v7.0 phones that are shipping, we don't supply it any more. I think the truth is, it was one of those situations where it's hard to decide who should really innovate in areas that seem to be very hot, where the technology was moving very quickly.

Avoiding the trap of duplication of effort, however, is hard. Arguably, precisely because customization is required to adapt the operating system to a given hardware platform, it may be unavoidable toward the bottom of the operating system in the hardware adaptation interfaces. On the one hand, the risk is of under-providing, which leaves too much work for licensees in porting the operating system to their hardware, while on the other hand the risk is over-providing, duplicating solutions that at least some licensees wish to supply for themselves.

17.8 Moving Towards Standard C++

A quite different, but no less important, perspective on renewal emerges from considering recent moves in Symbian OS towards a more standard use of C++, as discussed in Chapter 15.

David Wood:

Historically, we were biased towards coping with programmers who could handle quite a lot of complexity themselves, so while the operating system hides many, many complexities, still the bits that spill out are pretty difficult, and it requires a serious and very capable C++ software engineer to deal with it.

We used to say, or some of us used to say, that if you're a Java programmer stick with Java; if you're a C++ programmer, convert to Java unless you happen to be a very good C++ programmer, in which case you can stick with C++ when you're programming Symbian OS. I'm not sure we quite said it like that but that was the implication. We didn't make concessions for the 'mid-range' C++ programmer.

This is changing now, because we are at the stage where Symbian OS is going mainstream. When I say mainstream, I don't just mean that more phones are running Symbian OS, I mean that mainstream developers want to get their applications ported onto Symbian OS and they're not prepared to put up with the level of complexity that early adopters have.

We are looking at alternative run-time environments that hide more of the complexity. And you don't get quite so much control, certainly, and you probably end up with programs that are less efficient in several ways, but the added power of the machine probably makes that a more acceptable trade off than before.

Python may be one of the examples David Wood has in mind. But providing alternative language environments for application developments does not address the fundamental problem, that for those system programmers who must use C++ (whether they are porting the operating system or extending it), Symbian OS C++ sometimes appears arcane and willfully difficult in a world where standards rule. Symbian C++ idioms include its Leave() mechanism and cleanup stack, Active Objects rather than explicit multithreading, descriptors rather than C++ string classes, ordinal-based function calling into DLLs rather than more conventional name-based calling, and so on.

David Wood:

We did our own implementation of exception handling, because we took the view that exception handling was insufficiently standard, which at the time it was, and insufficiently efficient. As it happens that efficiency condition was borne out to be correct too, because when we eventually we turned the compiler flag that said 'Right, enable native exception handling of C++' in Symbian OS v9, something like 10% or more in bulk was added to the code. Not only does the native mechanism require more code, it's also slower to unwind the stack. We did various things in Symbian C++ where we would unwind the stack in one whole lot just by calling User::Leave(). So we avoided going up the stack layer by layer by layer, unlike the corresponding function inside native C++ – I think it is throw(). When you throw(), it doesn't jump up all at once, it goes up layer by layer of the stack.

Standard C++ error handling also requires memory allocation, which, of course, fails if the error being handled is an out-of-memory error.

David Wood:

There is occasional memory allocation needed by our cleanup stack as well, but it's done in such a way that each individual call to the cleanup stack will succeed, although if it's going to run out of memory for the next allocation it then unwinds. But whatever you push onto the cleanup stack is guaranteed to still be there. So that was very carefully designed in.

It is easy enough now to look at these mechanisms and condemn them as idiosyncratic, quirky, and frustrating. However, the fact is that at the time there was no C++ standard and there were no standard mechanisms. While memory leakage on desktop systems is still frequently taken for granted as part of the programming context, that approach was simply not acceptable in the context in which Symbian OS emerged, and nor is it for the mobile-phone market as it exists today. Power, memory and CPU cycles remain scarce resources in mobile devices and desktop assumptions do not hold.

Andrew Thoelke:

Memory allocation and out of memory was very important on the Series 3 and it was going to continue to be a big deal on Series 5 and it still is a big deal on any mobile device, even though some of the assumptions have changed. So we assumed that the operating system had to run for days or months or years without memory leakage or failure resulting from managing memory because you never switched off a Series 5. C++ isn't a garbage-collected language, so a lot of the rules actually came about from asking, 'How do we help programmers write code and review code to make sure that it is memory correct?'

It also predated exception handling in C++, so the cleanup stack, which is an idea from Series 3, was brought across, as was the actual exception mechanism for error propagation. This was followed by the naming convention too, basically to say this is a class that you should only ever allocate on the heap (or sometimes as an embedded member of another object on the heap) and this is a class you can safely have as a stack item, but if you've got to close it you need to make sure that you always close it, or you have to find some way of ensuring that it can be closed as part of an exception propagation.

Martin Tasker argues that while the detailed rationale may have changed as Symbian OS has evolved for new classes of device, the design choices were sound, and remain sound. But there is no doubting the barrier they pose for entry for developers coming to Symbian OS for the first time.

Martin Tasker:

There are barriers to entry, and there are what you might call ongoing costs. What happens in a lot of programming is that people assume that clean up is either an issue that you never have to handle, or that you can terminate your program if ever a resource is unavailable. It's very easy to write that kind of code. But what gets difficult is if exceptions can occur pretty much in every function you call and then the old-style programming gets cumbersome. At this point you have got to invest some thought in asking 'What happens if any of the functions you are about to call fail? Should you set up some kind of trap?' Whatever system you have, you've got some thinking to do which doesn't come quite so naturally. I think the solution we chose was actually quite simple compared to other solutions for the same problem. However, from the perspective of a third-party application programmer or a developer within a Symbian OS licensee, that is one of the more tricky areas. But we had to do this and our design choices were actually pretty good.

On this argument, discounting the high initial cost of the Symbian OS idiom against the resulting low ongoing costs may be a better option for developers than the opposite choice, that of using a standard mechanism which is immediately familiar but which is not optimized for the device class. Like it or not, the reality is that programming for small devices, and for phones in particular, is a specialist discipline, whether at application level (working with the different UI considerations imposed by mobility, the small footprint constraints, and so on) or at system level (ROM-based driveless devices, hardware with awkward properties such as NAND flash memory and intermittent connections). Desktop assumptions and techniques do not carry across.

Bob Dewolf:

I quite like the Symbian OS programming culture. Of course it seemed odd at the beginning because you had to learn all these strange new ways of working, like two-stage construction and the cleanup stack, but I'm quite a fan of that now; I quite like it. And I don't think that's held us back, I think that's been a good story, because it gives clear rules about how to deal with memory allocations. It's also a very strong OO-design pattern, which I also think is good in general.

Nonetheless, the pressure to standardize is being responded to.

Andrew Thoelke:

In some cases, our constraints go beyond the advisory, 'This is advisable because of the way the system's written', and become, 'If you don't do it like this your code won't work'. We definitely intend to move away from this now, and try to support a much more mainstream C++ programming environment. This is partly because there is such a thing now, which there wasn't when we started, but also because it will aid more programmers to get onto the platform and it will allow them to take code that already exists outside the platform and more easily to deploy on the platform.

The investment in already-written software has often proved decisive in relation to whether new systems succeed or not. Ease of porting is an essential requirement for an operating system which aims to become the standard in its market. To enable Symbian OS to continue to grow as a software platform requires providing better options to those looking to port existing code from other operating systems, whether at system or application level. Requirements have been framed and the changes will come.

Of course, it is an open question how much existing Unix or Windows code would make suitable candidates for running on a phone. The porting market is almost certainly not for standard third-party code, except for games perhaps, so much as for proprietary solutions. Languages may turn out to be as significant as mechanisms.

The company's language strategy has not always been consistent. While there has always been a commitment to native C++ development, at times Java seems to have been promoted as the programming language 'for the rest of us' working on Symbian OS as a platform. Martin Tasker takes some pains to explain the truth.

Martin Tasker:

It was always assumed that C++ would be used externally and there were conscious design decisions around that. In fact, Colly Myers used to feel quite strongly about it. He would say, 'We can't assume that everybody understands operating systems and we cannot expose an API in such-and-such a category to people who don't understand operating systems'. That was because the C++ APIs were exposed. And, if you look at active objects, if you look at the C and T types which offer a very very simple guide to the programmer as to how to use these types operationally, as simple as Java objects and built-ins, then in some ways we are as simple as Java. We don't do garbage collection, because C++ doesn't do it, so programmers have to do that stuff manually. But otherwise we're as simple as Java.

For application developers, languages such as Java, Visual Basic and Python are obvious options for enabling cross-platform portability, rapid development and improved productivity on Symbian OS.

David Wood:

Java is only one option. There are now other run-time environments as well, including Visual Basic from AppForge and the run-time environment they call Crossfire. They also support a. NET library or run-time environment for Symbian OS. And then there's OPL, a very old and venerable programming environment and there are more recent contenders such as Python. The common theme of these interpreted languages is that they are less efficient in terms of manipulating the bare metal of the device, but they require less of a learning curve to become productive.

What I've seen is evidence that you don't need to be anything like such an experienced programmer to use them, indeed non-programmers such as journalists and students of journalism or students of media or students of arts and technology are able to learn how to create quite impressive programs in Python from a course based on just one lesson per week, and that is part of Symbian OS coming to the mainstream.

18

Creative Zoo or Software Factory?

18.1   Introduction

This case study takes a step back from Symbian OS itself to explore
some of the broader questions about how software is created. One
consequence of the success of Symbian OS has been the rapid growth
of Symbian, the company, and of its software engineering organization.
From its small-company origins, Symbian has become a middle-sized
company, established in a global market. Success, inevitably, brings its
own particular slant to the perennial problems of how best to make
software.

18.2   The Software Problem

Like all software companies, Symbian wrestles with the problems of
efficiency, effectiveness and predictability. Every software development
organization faces the same basic question: what is the right way to
organize so as to be as effective as possible at making and shipping great
software? In other words, how should development teams be organized
(or how they should organize themselves), with respect to the software
base and the need to continuously maintain, evolve and improve it? And
there are softer questions too: what should the project culture be and
how should it feel to work in a project?

These are not new problems and a lot has been written about them, but
nonetheless they are difficult problems. One reason they are difficult is
that they are ill-defined problems, not the kinds of problems with which
software companies like to deal and problems with which, typically,
software companies are bad at dealing. (It is always a mistake to seek
engineering solutions to non-engineering problems.) They are not made
easier by the fact that there is disagreement among practitioners about
even the basics.

Typically, software creation is counted as an engineering practice, which is to say one based on measurable, formal procedures and systems, although not necessarily formally mathematical. (A dissident, formalist view is that it would better be classed as mathematical practice). Development methodologies therefore are engineering methodologies (process engineering and product engineering) which should deliver validated solutions to well-specified problems (customer requirements), predictably and repeatably.

There is a minority view that the very use of the word engineering in this context is a category mistake. Software creation is not an engineering discipline but something more like a craft practice, carried out by skilled professionals making intelligent, intuitive, but necessarily 'soft' or 'fuzzy' or simply underdetermined decisions. Worse, it is not just the case that requirements are often poorly specified; they are, in many cases, unable to be specified before the development activity (i.e. outside the context of a proposed solution). Development methodologies, in this view, should be designed around these basic matters of fact.

The business demands imposed by the commercial context do not help to reconcile the differences. When software companies are product companies (it is easier when they are simply internal suppliers), they are subject to the same commercial disciplines as any other business. Less orthodox approaches to requirements capture and design, based on prototyping and experimentation, trial-and-error and iteration, because they are inherently uncertain and therefore high risk, are in immediate conflict with a command and control business culture. As much at issue as methodological questions, then, are cultural and sociological ones and the underlying questions of control and where it resides in the organization.

A final difficulty of the pure engineering approach is that whatever the other merits of the argument, there is no doubt that the underlying activity of actually writing software looks as much like an art as a science. It is full of subtleties, is strictly non-deterministic, is highly context-sensitive, lends itself to multiple possible solutions, and requires experience, expertise, imagination and inspiration. These are the facts that underlie the familiar statistics about individual programmer productivity. No matter what the business needs, it simply is not possible to mandate software productivity.

If proliferation of theory is an indicator that a research area is underdetermined by the available facts, then software practice is up there with the best of them. Development methodologies proliferate and it can be hard to sort quack remedies from principled alternative practices. Theories are rapidly inflated into fully marketed, patent medicines for all software development ills. Gurus abound and none of them agree about much at all. Metaphors abound too, from 'design factories' and 'software factories' to 'total quality' and 'Sigma 6' to 'Scrum' and 'Extreme'.

18.3   Too Many Dragons

[Aho, Sethi and Ullman 1986] has a fire-breathing monster bearing the label 'Complexity of Compiler Design' on its cover. There are many dragons to slay in software development, of which the innate complexity of the endeavor is the fiercest and fieriest. Presumably, complexity is also what Stanley Lippman, the author of more than one well-regarded C++ primer, is alluding to when he chooses Durer's engraving of the knight and the devil for a frontispiece [Lippman 1996].

But almost as fierce and fiery, and certainly as famous, is the dragon 'of poor programming productivity', the problem to which [Brooks 1976] first drew attention and which, for example, [Gabriel 1996] confronts in an influential essay, without reaching any very hopeful conclusions. Estimates of programmer productivity vary from 10 to a few hundred lines per month, or perhaps 1000 to 2000 'non-commentary source lines per programmer per year.' As [Gabriel 1996, p. 127] points out, that is about four lines a day – 'There is a software crisis.'

The phrase 'software crisis' was coined as long ago as 1968,[1] and most of the current software creation infrastructure has evolved in its shadow, including the dominant operating systems, programming and modeling languages, and analysis methodologies, not to mention the modern software–hardware infrastructure.[2] All can be seen as part of the same calculated effort to move the practice of making software from a black-art to a well-founded science. The 'crisis' was created by the impossibility of reliably planning, implementing and maintaining systems beyond a certain size and complexity threshold. Interestingly, another phrase coined at the same conference, by Doug McIlroy, one of the pioneer creators of Unix, was 'software engineering'. McIlroy also observed that without some evidence of a components industry, there was no sound basis for thinking of software production even as an industry (let alone an engineering industry) [Assmann 2003].

Of course, what McIlroy had in mind was a research and engineering program that might lead to an industry founded on the manufacturing of software out of components. Since components and composition of components explicitly underlie and underwrite today's object-oriented approaches (and, in fact, most other recent programming methodology research too), it seems fair to say that McIlroy's call has been heeded. But still, the striking fact is that, close to 30 years later and what should therefore be a good way beyond the black-art and wizardry stage, the crisis seems as strong as ever. It's not so much that no solutions were

---

[1] At an international conference of software professionals and academics called to address the question, 'How can software be produced systematically?' [Assmann 2003, p. 6]

[2] Take your pick, but Unix first appeared in 1969; the PC in 1981; the C language in the early 1970s; C++ in the early 1980s; and Java not until 1995.

found, as that the problem has simply continued to grow exponentially. For every solution, it seems, there is an immediate test case problem for which, in one dimension or another, the solution is inadequate.

There is extensive literature on the software crisis and on the proposed solutions to it, from structured techniques to object techniques and reuse, along with a host of related approaches and methodologies, from Class Responsibilities Collaborators (CRC) cards[3] and patterns, to feature teams, to Extreme and Agile programming. Abstracting the detailed differences, the common core of the radical solutions is an emphasis on incremental and iterative development and the freeing of programmers to think, approaches which are promoted by a generally broad and probably wise consensus. But, as [Gabriel 1996] somewhat regretfully concludes, there is little evidence that the industry overall is either pleased to hear the remedies or attempts to apply them in practice on any large scale. On the whole, they do not offer solutions that a traditionally managed and typically 'over-managed and under-led' industry wants to hear. It carries on as it always has, even though eventually, as [Gabriel 1996, p. 128] puts it, 'we'll find that traditional software development methodologies are among the least productive and possibly produce the lowest quality.'

What drives the traditional software development methodologies are the basic commercial imperatives of management control and process repeatability and predictability. The problem is that these imperatives seem to be at odds with the practices which actually deliver better software productivity and quality. Perhaps software is 'just different', but if Gabriel and others are right, achieving productivity and quality seem to require non-traditional approaches to control. Traditional management doesn't want to let go.

18.4   Software Development Approaches

One argument, or perhaps it is more strictly a metaphor or another analogy, which has proved very popular is that making software is a bit like making buildings. Prefabrication and componentization are only small parts of any answer as to how to do it better, more reliably or more predictably. Following the analogy of 'habitability', making 'better' software means making software which is more elegant, more long-lived and more adaptable; on the analogy of buildings which do not fall down, reliable software does not fail unexpectedly; and predictability means completing the job in something like the predicted time.

This metaphor is at the heart of the software patterns movement, which argues that not only is making software like making buildings, but that it is an irreducibly human activity and team activity. Relationships and

---

[3] A good introduction is [Beck 1999].

interactions between individuals, and individual and team behaviors and all the subtleties and difficulties associated with them, come into play and must be managed by the software creation process in order for the process to succeed [Rising 1998, p. 143]. Different organizations have tried different approaches, including Extreme and Scrum[4] programming approaches and similar team-based systems, which along with ideas such as 'creative chaos' (as a tool for fostering innovation and creativity within the industrial organization) have been the subject of industrial experimentation across many different industries in Japan.[5] Teams, of course, are just groups of collaborating individuals plus leadership (teams without leaders are groups) and the health of a team is a function of the health of its individuals plus the quality of its leadership.

The common contention behind all of these approaches is that the core processes of software development cannot be understood from a purely task-based perspective. A broader perspective which explicitly makes room for the 'people' dimension, as well as mapping task outputs (for example, an 'artifacts, roles, actors, and agents' model [Rising 1998, p. 122]) is essential to understanding what is an essentially dynamic activity. Traditional process models, especially those such as ISO9000 but including also the popular Capability Maturity Model (CMM), have their place but are not perfect at capturing software development as it is actually practiced and, indeed, as it must be practiced.

Another point that these approaches all stress is that if making software is a creative process then it depends for its success (presumably) on enabling creative individuals to do the creating, and creativity is not easily prescribed. A preponderance of strong personalities among the creative people also means that command-and-control approaches are doomed to fail, if not absolutely (some software will get made) then relatively (team potential will have been poorly exploited and software quality will be poorer than it could have been).

Old-fashioned, waterfall development is very thoroughly and almost universally deprecated. Full specification followed by full design followed by full 'coding' does not work, because 'full specification' and 'full design' have both proven unachievable in practice. Iterative development, in some form or another, seems to be the only rational alternative [Rising 1998, p. 148], allowing a full specification of the problem and a fully designed solution to emerge through successive cycles of partial design and implementation. However, in practice, software organizations like most others have an almost insatiable desire for detailed specification and up-front design, not to mention planning and budgeting, ahead of any resource commitment. The organization defeats itself, because it is risk-averse and wants certainty, or relative certainty, when in fact it should

---

[4] 'Scrum', or 'relay', is an attempt to translate the Japanese word 'sashimi'.
[5] At Nissan, Fuji Xerox and Matsushita for example. See [Nonaka and Takeuchi 1995], especially for the theory of creative chaos.

be organizing to manage uncertainty. But to many planners, uncertainty does not sound much like engineering. It is also easy to fall into the trap of seeking to control process artifacts, rather than managing processes through the people who implement them.

Another problem is that iteration depends on short cycles, but often enough the short cycle is subverted by the planning process. The implementation phase is repeatedly postponed for the sake of a little more planning certainty. The result is not a short cycle at all, but a traditional long one (six to nine months, say), with a planning front-end which lasts for six months out of nine, followed by a hasty development tail. Inevitably the tail turns out not to be short at all, development takes as long as it takes, and the overall cycle reverts to being a one-year or 18-month cycle.

Organizational fear of 'randomness' and the indeterminate or merely underdetermined fuels the urge to centralize and control, to legislate, plan and create metrics (define and measure!). Randomness, for want of a better word, is a necessary part of the process of exploration.[6] Uncertainty, whether we like or not, is a given of creativity. The creative factory is probably an impossibility.[7]

Building construction mixes art and science, but it also has another important dimension. 'Ethical' architecture is essential, because the buildings that architects create determine our physical spaces and, done badly, despoil our towns and cities. 'Ethical software' might be something we had better start to consider too. Software increasingly permeates our lives and, in some cases, is beginning to dominate and control them, not necessarily for the good. Identity cards and 'big' databases are one example; software-driven munitions are another;[8] as is the 'Google in China' question.[9] An ethical-software manifesto, if there was such a thing, would fit well with the original goals of the software pattern movement for habitability and would fit well with the original aims of Christopher Alexander in his architectural patterns work.[10]

---

[6] Error, indeed, is objectively indistinguishable from creativity, in so far as both are underdetermined and only subjective measurement against a goal can tell one from another. To stretch the point only a little, the software on the Ariane rocket made an error, not a creative leap; but in other circumstances, departure from the norm might be called inspiration.

[7] A factory is a place where mechanized production takes place, originally organized around the principle of machine-minding ('satanic mills', for example, with steam-driven looms), re-imagined by Henry Ford and others around the idea of the production line and the factoring of the production task into simple actions performed by highly specialized but unskilled labor. A factory has only lately been reinvented as a place where teams work together to meet their production targets.

[8] Mobile phones are not ethically neutral. Mobile phone access to emergency and rescue services saves lives. But equally, mobile phones are increasingly being used to trigger bombs and as homing devices for remotely delivered munitions to target individuals.

[9] How can we achieve the 'borderless Internet' when it runs up against state censorship?

[10] The classic text is [Alexander 1979].

In some senses, the open-source and free-software movement is an ethical movement, but in other senses it is more obviously mercantile and market-led and markets know no ethics. But the hacker ethic (see [Himanen *et al.* 2002]) proposes a rather different work ethic from the conventional one: work is fun, programming is play and the passion for the machine has a moral dimension.

18.5   What Making Software Is Really About

'Shipping great software on time', as [McCarthy 1995, p. 2] puts it, is what making software is all about. But as he emphasizes, software is peculiarly intangible. It is not simply 'stuff'; it is embodied thought, idea plus design plus implementation; each of which is an intellectual (not mechanical) process. What makes the process interestingly different from other intellectual processes (thinking, writing and painting) is that beyond a certain level of size and complexity, it is necessarily a team activity.

For McCarthy at least, the word 'development' (as in 'software development') is a clue to the nature of the enterprise, a dynamic process of maturation, in which what matures is precisely that embodied thought: as he writes in [McCarthy 1995, p. 85], 'It's the team ideation, gradually migrating from highly individual (even private) notions toward a group articulation in the shipping code.' This way of putting it will resonate with anyone who has been involved in the software development process (and that means, as he says, 'everybody on the team', whether planning, scheduling, creating or validating the software product). The essential development act is this development of individual ideas into embodied intellect, productized thought: intellectual property, in other words.

This is the view, of course, that sees making software not as an engineering process (although there are engineering aspects to it, as there are in any construction process) but 'as primarily a sociological or cultural phenomenon' [McCarthy 1995, p. 87]. Perhaps more even than engineering 'aspects', there are some engineering fundamentals involved. Machines are engineered constructs and all software is in some sense 'soft machine'. But, nonetheless, if the software industry is fundamentally a creative industry, then there are necessary limits to how far industrialization (and even formalization) can go. Every developer is familiar with the notion of the 'death march'[11] and it is hard to imagine that anyone would willingly adopt it as a project model. But without a development methodology that understands, and serves to support, the reality of the software creation process, it is probably the inevitable end for all software projects.

---

[11] The 'death march' is the bringing to final completion of a long and difficult project and its repeated slippage as an 'exhausted, physically and emotionally spent' team marches, stumbles, and lurches to final shipping of the product [McCarthy 1995, p. 33].

Software is not the only creative industry to have attempted a 'factory' approach. Hollywood is probably the most famous, if not the most obvious, example. Thus at Warner Brothers studios in the 1930s (see [Bordwell *et al.* 1985, p. 326]), all creative people (directors and writers included) had set working hours and a per-day expected piece-rate production (which may not in the end have been so much more than the equivalent of four lines of code per day). In one sense, it worked and the films are there to prove it, but it almost certainly did not work for the simplistic reasons that Jack Warner thought that it worked. More likely, it worked because a hot-house of talented people did good things despite the more foolish aspects of the regime.[12] In an updated context, it might also be legitimate to ask whether the independent cinema sector produces higher quality films than the modern Hollywood factory (surely it does!) and, if so, how much of the difference is due to their different production models. The analogous comparison between conventionally and open-source produced software might be similarly instructive and for similar reasons.

Software development, says [McCarthy 1995, p. 87], is more like a jam session than an orchestrated event and, in so far as that is true, it cannot be factoryized and cannot be scripted. The jam session requires motivated, independent, skilled, autonomous individuals to work together with a common aim. What the development model can enable is a team context which makes it likely that the goal will be achieved, including time goals if time-boundedness is part of the project specification.

Flavor of the Times

The stereotype of the early days of the Protea project which created the precursor to Symbian OS is, on the one hand, Charles Davies and his elite band of application and 'middleware' programmers with their object-oriented cloud and, on the other, Colly Myers and his lieutenants with their heads down in the practicalities of leaner, meaner and faster code, only vaguely aware of the cloud but exploring some interesting object-oriented ideas of their own. Somewhere between the two elites is David Wood, throwing new recruits at the interesting problems of creating a complete object-oriented GUI from the ground up and laboring mightily

---

[12] According to legend, Jack Warner was accustomed to taking a mid-afternoon stroll around the studio lot every day. If when he passed the Writers' Building there were any offices from which he did not hear the sound of a typewriter from the window, he would have his assistant enquire why the writers in question had not been writing.

(and often singlehandedly) into the night to integrate the raw code they created into a complete and elegant system. Bill Batchelor stalks the corridors with a furiously scribbled and rescribbled back-of-envelope project plan looking distracted. In the basement, meanwhile, Richard Harrison presides over a rack of home-made-looking black boxes (ROM writers) and a heap of yellow, later green, 'Banana' and 'Lime' Psion Series 5 prototype devices.

Davies, according to the stereotype, is the cerebral purist and Myers the bulldog-like pragmatist, the one who is actually building the system and needs therefore to know what to build. Bollen meanwhile, as a relative newcomer, was neither exactly in Myers's corner nor in Davies's corner.

Geert Bollen:

I was 'piggy in the middle'. I had arguments with both.

Others who were there tell a broadly similar story. Most strikingly, for all its informality, it was not a particularly relaxed environment. As Martin Tasker puts it, it was a frontier atmosphere, exciting, charged but driven.

Martin Tasker:

Colly Myers and Charles Davies led the project, and they had three trusted lieutenants who they put in key positions, namely Nick Healey, David Wood and Bill Batchelor, but Colly and Charles were formally Psion Group directors and they directed the software on the project. They had absolutely poles apart different management styles. With Colly, everything came from within and was asserted. His style was to assert everything more or less out of his own brain and his own experience, you know, an 'I've already written three OSs, this is my fourth and I know what I'm doing' kind of mentality. And people who followed that had an easy time, and people who thought otherwise had a hard time, which didn't mean to say that Colly couldn't be challenged, but it was difficult.

Whereas Charles's style was very much to read books about other projects and basically to design the interfaces, and then to train up people and get them to implement stuff to the specification, and then he would iterate. He had a group of people who he basically taught Rose clouds to and he architected iteratively with them. I mean Charles was quite assertive but he had broader input than Colly and was gentler in the way he let things out as well. And David Wood, meanwhile, although he was slightly under Charles's guidance, he gave a huge degree of autonomy to his people, basically he used to say, 'these guys are bright and they can do it, I won't give them any brief at

> all, I'll just comment on what they've done when I integrate it', whereas the people who worked directly for Charles got quite a lot of attention at a certain interface level.
>
> There were a lot of different styles, and I would say overall there was a lot of creative tension around. I think the atmosphere was really very frontier. There were probably people who didn't have the experience to deliver on their potential, for whom that environment wasn't the right environment. And there was also a class of people who kind of came in and always wore a white shirt and would have been a lot happier with Symbian as it is today than Psion as it was then. You know, the frontier mentality didn't suit.

There was an architecture committee, ArchComm, and Bollen was on it, as were Davies and Wood. There was also a UI design committee, presided over by Nick Healey and Bill Batchelor, the two UI gurus, dedicated to keeping down the number of key strokes required to drive the system in keyboard-shortcut modes and to keeping the application UIs clean and consistent.

Howard Price:

There was a lot of counting of key presses. Navigating with the keyboard, you want to avoid having more than the strictly minimum number of key-presses.

Peter Jackson, who had joined the company well before the Protea project began and was by then almost an old hand, also has another, quite specific recollection of the transition from the early design activity to the construction of a real system.

Peter Jackson:

I have a very clear recollection about what life in Psion was like at the end of 1994. There was a lot of quite deep thought going on about architecture and the way things should be modeled and so on, and then one day Colly made his world-changing submission. He heroically wrote all this code and arrived with an implicit statement of, 'Here we are, here's EPOC32'. And then all hell broke loose, because basically all this deep thought stopped and everyone started coding.

You can't argue against it, because time to market is everything. But at the same time, what Colly catalyzed also included some poorly considered design. At the same time, we were recruiting a lot of people and letting them loose on difficult problems. One example was David Wood assigning some relatively new people to write a menu system which would include things like drop-down-menu technology.

> I remember that I had spent a lot of time, because of the lack of documentation in SIBO, reverse-engineering the design of the menu system to implement on the MC400 laptop, so I knew how difficult it was to actually get it right. And all these people were kind of fresh out of university! David Wood would collate all their output, and he would make a huge effort to put it all together to produce a user-interface system. And the good thing about it was that we made progress and it cheered people up. The bad thing about it was we had to throw a lot of it away that wasn't right.

Lotus Notes had been adopted by the company quite early on and provided an important part of the mechanism which allowed strongly decentralized team-working without completely abandoning control of the high-level design. The result was a kind of hub-and-spokes model, in which strong design directions were transmitted from the hub to the individual teams at the end of the spokes but without much direct communication between the different teams. But the Lotus Notes culture enabled a lot of direct communication between individual developers, through the medium of the databases.

> **Howard Price:**
>
> Design was pretty much a central activity, though you didn't get people showing their Rose diagrams to each other especially. I remember going to Charles Davies with my design for review just once. My design was only really seen then and the rest of it was up to me. But there was a design database which was very active. Charles Davies had a central Rose design on the network where each team was supposed to design their own Rose diagrams that fitted into it and you could link them together.
>
> But there was a way you did things, definitely. There was a strong Notes culture. There would be a lot of activity on databases discussing ways to do things and it would be hard to miss the right way to do it, at least if you were coding something standard anyway. So there was some very interactive development discussion on the databases.

In fact, for a system in which in-house idioms were important (for example, the idioms of descriptors and active objects, client–server and framework–plug-ins), Notes was an essential mechanism for providing guidance on how to use the idioms effectively and, indeed, for enforcing their proper use.

While high-level design was always to a large extent autonomous, although beneath the eye of an overall Architecture Committee, coding standards were always strongly defined (they had to be, in a company which was getting to grips with a new language, for which many engineers had only had the brief exposure of a formal training course and no hands-on experience).

Howard Price:

Client–server, for example, was treated as something that everybody had to know. The way it worked was quite particular, the way you package and pass data across the server boundary. You couldn't simply invent your own way of doing that, you had to stick to the right ways of doing it and it was enforced. If you hadn't done it before, you had to find out how to do it.

The mechanism was very much developer to developer, either within teams by direct coaching and code review, or directly between individual developers using Lotus Notes as a universal read–write discussion medium. The Lotus Notes culture was always extremely strong (and remains so, although to some extent it has been diluted more recently by the scale of expansion of the company and by new ways, for example Wiki, leaking in).[13] But among the older guard, it remains strong.

Culture, Culture, Culture

The strong culture of autonomy inherited by Symbian dates back to Psion's beginnings as a small company of highly technically capable individuals, with not much hierarchy but with highly visible leaders, in which everyone contributed to make things work. In particular the company was engineer-led, with a deep culture of engineer-led design, pulled together by a centralized but loose design process.

Peter Jackson:

Culture, not process, is what is important. And it really was the case that everyone was given permission to do what it took.

In many respects the early engineering culture in Symbian seems close to what has more recently (and fashionably) become formalized as 'agile' programming: strongly team-based, strongly team-meeting- and review-based, strongly decentralized, highly localized and organized around clear goals. There was a strong concept of team ownership of code and, indeed, not just of code but more strongly of ownership of the design and implementation of a well-defined, discrete piece of the system.

---

[13] Wiki also has great potential to give the culture of collaboration a new lease of life and take it in new directions.

Howard Price:

I think we were Agile, but we didn't have the culture of daily meetings Agile recommends, though we did have weekly meetings. For instance, I think it was for the Series 3, Charles Davies would organize a weekly meeting and all the team leads would go in, and the meeting would go round the table and you'd say how you were doing and maybe Charles would decide that some team should implement a certain feature or that some other team shouldn't implement some other feature. And everyone would agree, but maybe that team would go and do it anyway, because they thought it was the right thing to do. So you had a lot of ownership of your own code, which was good, and you could decide largely what you wanted to do. However, you had to answer for what you'd done if it turned out to be a bad decision.

Charles Davies was very much the driver of the culture of design, a believer in design if only because, as he puts it, 'It helps you be dumb'.

Charles Davies:

So I'm a believer in design, which I tried to promote using design tools. It wasn't UML, it was Rose at that time, so I was maintaining Rose diagrams. I think you do have to have a design ethic. If you just end up putting code where you happen to have the editor open... well, that's a bit too harsh... but you do need to take grasp of it and keep simplifying it. I believe you're a much better programmer if you're a bit forgetful or can't remember things easily, because then you have to simplify until anyone can remember how it's done. I'm a great proponent of having design idioms so that you can recognize designs. You can see that there is a design and that it makes sense.

Martin Tasker thinks the design culture was particularly important in enabling an informal culture nevertheless to be particularly effective.

Martin Tasker:

Charles Davies led the design and trained people who had been in related disciplines – he got them to do object-oriented software. In any management view there are massive benefits out there if you take the enabling steps to achieve them and Charles did this extremely consistently and well. Charles, in particular, paid minute attention to the details of his APIs. He used their explanatory power to motivate his people and he almost didn't need to look at what they produced in terms of implementation code or test code. He basically believed that if the code met the requirements of the API and if you felt it was correct then he trusted you that it was correct. Charles used that to massively good effect. He used object orientation as a means of controlling a sea of junior programmers very successfully.

Perhaps of all teams in the company, the Base team (which developed the kernel and low-level systems) had the capacity to survive longest before compromising the quite particular culture which had evolved: informal, devolved, expert and committed, attributes which it retained certainly well into the Symbian OS v6 release projects and beyond into Symbian OS v7. Morgan Henry was working in the Base team on the original port of EPOC32 to Nokia's new 9210 Communicator hardware and what eventually became Symbian OS v6.0.

Morgan Henry:

Certainly in the Base team there was a mentality that you get the best people you can find, people who are interested and excited by the hardware and software interaction, and broadly you let them get on with it and they come up with good code. And it was led by people who were interested in the technology and enjoyed what it was doing. I think they understood that if there's enough space there, that if you allow talented people to do the things they're interested in, you end up with a good quality product. That's not to say there were no processes and no project management, but, that it's all about the balance.

Transition of the Development Model

Formalization of the software development model was probably an inevitable consequence of success and growth and, in particular, of the evolution from a software company making a software product for a single customer (and, to all intents and purposes, a customer within the same company), to one which was suddenly faced with multiple licensees, all competing in a high-growth and relatively new, global consumer market. What's more, compared with Psion these companies were industry giants, the likes of Nokia, Ericsson, Motorola and Matsushita. They had established markets, established practices, global reach, and strong (and quite different) internal cultures. The small-scale and home-grown practices which had evolved within Psion and carried over into Symbian were suddenly confronted by a very different external reality.

Certainly the reality is that the company now is very different to the one it used to be. Having grown by several multiples, the approach to software creation is inevitably very different.

> **Morgan Henry:**
>
> Even during the process of Psion Software becoming Symbian, there were many attempts to regularize development processes. The Base team was always a bit different, but now it seems very much like the technology architects have less freedom to make the architectural choices that they need to make, and they don't get the time to write any code or do any proper design. So Symbian as a company has begun to regulate more, which is no bad thing, because you have to have reliability of delivery over everything else if that's what your customer is asking for. Back in the Nokia Communicator and Psion Series 5 days, we were a technology-driven company, and this transition to a marketing-led company has probably contributed to the difference.

The most immediate loss is probably that of developer autonomy or, at least, the sense of autonomy.

> **Morgan Henry:**
>
> With the Nokia Communicator, you'd see a problem and you'd be responsible for it, up to the point where you'd be responsible for going to see Nokia. For instance, I'd be sent out to Finland and I'd be put in a room with ten Nokia engineers and they'd ask lots of questions and it would be my responsibility to come up with a solution, and possibly even go back and implement the solution.
>
> There was no project manager doing it for you, it was a case of having to manage your time on that and work out when it was going to be done, which meant you had much more freedom and you had much more contact with the hardware. So I was prepared to put in a lot of time and a lot of effort, doing long nights and weekends if I needed to, just to make sure that it worked, because you cared about it and you wanted to see it ship.

As Symbian has evolved, driven by the need to scale to many more licensees and to support a multitude of licensee projects, responsibility for porting to new hardware has moved away from the core engineering teams to a dedicated porting team and dedicated licensee support teams. Inevitable and necessary as it may be, one consequence is the distancing of the core software-engineering teams from real mobile-phone products.

> **Morgan Henry:**
>
> The Base team has been detached from the base porting activity and delivering a base port for a 'real' product. At best they're delivering a base port for a hardware reference platform. There are many reasons for this, but certainly the desire to regularize and the desire to control the time investments are part of it. The fact is that they are less in contact with the final product.

> Having said that, there are some parts of the organization which still have a close relationship with the final product and they still do the long weekends and they still go and see the customer and get bombarded with questions, so that's still there. I'd argue it's probably the best way of doing that kind of job, as opposed to the process-driven activity that is common in Symbian.

While the Lotus Notes database and discussion culture remains strong, the design culture that used to underlie it has been dissipated as the company has grown.

> **Howard Price:**
>
> The Lotus Notes culture has survived to a fair extent and there's a lot of discussion of how to do things. However, there doesn't seem to be huge amounts of design discussion going on, it's more programming discussion. Maybe that's because there's a feeling that everything is stable and there's not really that much new design and big new design now goes through the System Design Authority.

There is no doubt that the design culture is very different in Symbian now from what it was in the early days.

> **Charles Davies:**
>
> Do we do enough design now? Of course programmers like doing design, but it's whether you do design as an explicit activity. I'm a believer in modeling. I'm not up to the UML standards these days but I've found UML useful. But even if you just draw it on flipcharts, you do need a design. Design gets you to reusable patterns, so if you've got a community of people doing design it's very useful, and if you leave a programmer on their own they won't get there. You need a culture of design, of people sharing designs and talking about it. So I did used to do that. Design, ultimately, is the thing that cuts your defect cost down, because good designs have fewer defects. But then of course you need good requirements. It's all 'Page One' of the software-engineering manual. So we know that, but it's not just knowing it, it's actually doing it.

Control and flexibility are not easy to reconcile. The deadline pressures of time-to-market inevitably conflict with the need to refactor and redesign as part of the daily software development approach. David Wood still considers iterative design an essential part of any successful development style.

> **David Wood:**
>
> I do think it's worthwhile having enough project time set aside to allow for some degree of refactoring of designs, because you can never be sure of getting it right first time. Even with the cleverest people, the most experienced people, there are things that only become clear as the design evolves, so you should, if possible, allow yourselves the freedom to put that into future products. It gets tough when you've got to maintain compatibility as well, so it may be that for a while you've got to run the two systems in parallel, you've got to run the old system for older products on new machines, but then also you can have the new APIs which gradually people can switch to.

Size tends to work against flexibility, but in fact flexibility is not the only thing that gets lost with increased size. Increased size also works against the consistency and design elegance of the system, tending to cause dilution. Peter Jackson argues that design dilution is an almost inevitable consequence of a traditionally devolved approach to design confronted by rapid expansion.

> **Peter Jackson:**
>
> A general problem that has been faced as Symbian OS development teams have grown is the varied experience of the people you employ to work on the operating system. If you try to do something new, if you try and invent a sophisticated locale system, the problem is that somebody from outside who may have, say, a Unix background, doesn't understand the design principles of what you're doing and may get it wrong or may take longer to learn what you're up to.

Jackson's own area of specialization in the system in the early days, internationalization and localization, is a case in point.

> **Peter Jackson:**
>
> The problem with internationalization is that typical programmers like to believe that everybody is American! They can ignore basic issues, like for example sometimes you can't assume that pluralizing a word means putting an 's' on the end of it. So we were always compromising saying, 'this would be the elegant thing to do, but everybody expects it to happen a different way', and we have to do it the way everyone expects. It is a fact of life working in this industry, that software tends to devolve towards the lowest common denominator rather than towards the most elegant thing.

'Worse is better' is the label Richard Gabriel coined to describe the problem that inferior systems or designs tend to beat superior ones in the market by getting to the market quicker and occupying the market niche with a system that is just good enough to make the switching cost to the better system unattractive.

> **Peter Jackson:**
>
> It's probably a particular case of the 'Worse Is Better' thesis, which basically says, 'You can put a lot of effort into doing the right thing at all times, but meanwhile your competitor will have ignored all the difficult cases and got to the market first, and the difficult cases happen so rarely that people can be conned into accepting an inferior product.' That's a paraphrase of Worse Is Better, but that's how I think of it.

It is a truism that success can be a dangerous thing in its own ways. Success for Symbian has meant near continuous growth and expansion.

> **Peter Jackson:**
>
> The software development process determines your success in producing software. I'm thinking of things like the role of testing in software development, the question of where defect-fixing comes at the end. Issues like these are cultural issues. You want to institutionalize something that's good, the right thing, and I don't think it's easy at all, because as soon as you have started doing the right thing, you lose focus and you change the focus to something else. Meanwhile, the company grows a bit more and suddenly the thing you thought was taken care of isn't taken care of at all. So it's quite hard even to work out what you have to do to institutionalize the things you care about.

But the biggest impact of growth is in a way the most obvious one, if also the most puzzling one. With responsibility for managing the code repository within the integration and build organization within the company, Jackson is well placed to observe it.

> **Peter Jackson:**
>
> The number of submissions into the repository is enormous. You watch it, and it just scrolls up the whole time, you just see it, submission after submission after submission. And yet we produced the first version of EPOC with an engineering community of closer to 100, it was of that order of size. So sometimes you think, What is all this work going on? What is it? Why do we need so much of it? So looking back, the scale really is quite different.

The Value of 'Whole Product' Development

Charles Davies identifies a particular problem which has become highly relevant for the company, and which has deep roots not so much in the details of the development model, but in the wider business model and the place in which Symbian finds itself in the market. The evolution from being a product company to being a pure software company (and indeed, a pure operating-system company) tends to work against the holistic, whole-product understanding of what the company is producing, which makes validation much more complex. In the worst case, focus on validation can be lost altogether.

> **Charles Davies:**
>
> In our context, validation is hard. It used to be much easier in Psion, because we would provide an API with our OS hat on, and then applications would use it in the same organization, and then the device would use the application. If it wasn't any good, people would say so! And the application is working this way because the API works this way, so we'd change the API to work better. That cycle doesn't exist now. We deliver APIs to our customers, and there isn't that natural process of validation.
>
> Pick up a book on software engineering. All of them say that verification means, in software terms, that you've met the specification, and validation means that the specification that you agreed with the customer was actually useful in the event. And we used to do that in Psion without realizing it, because it was part of being a product company. Now we have become too remote from the final product for that.

The problem is that Symbian, by the nature of its business model, is at least one step removed from the true product cycle of making and shipping phones.

> **Charles Davies:**
>
> So, we've produced an API! It's in the release and the job is finished. But the job is really only finished when the value has been delivered, not when the API has been delivered, and that's what validation means. It means a customer conversation to say, 'Was that valuable? Did it do it for you?' If the API wasn't used, the answer is, 'No!'

In other words, creating the OS is not just a matter of delivering software that works and delivering well-designed, well-abstracted and well-implemented APIs. It is also a matter of delivering the right APIs to customers, at the right time, and validating them with the customer. In a sense the problem is the familiar one of the need for iteration, to enable validation and where necessary refactoring and redesign. And the difficulty is successfully creating a project model which delivers that, while meeting the other needs of supplying into the global product manufacturing businesses of phone-vendor licensees.

> **Charles Davies:**
>
> For any new API, it would be remarkable if it was right first time. It is guaranteed to be wrong for any non-trivial API. So you should expect in the normal course of events that for a new API, you'll have defects at the requirements level and you have to rely on that iteration. For a new API, you need to factor in the extra work you get from a process that does validation.

It's all too easy, instead, simply to finish the API and ship it.

> **Charles Davies:**
>
> Imagine you ship an API, and that's it, you then move on to work on something else. What if the customer says it doesn't meet their needs? What do you do? Ignore it? In Psion they would have said, 'This API doesn't work and I'm not using this unless you fix it!' And so it would be fixed and the APIs got good because of that. And that's validation!

This is an inherent problem of the business model and one that exercises many people in the company and one on which the company is continually striving to do better.

Putting the Magic Back

The goal of a successful development methodology must be to reconcile the inevitable conflicts between the different demands of the business. Performing that balancing act is an essentially dynamic activity and it is hard to build dynamic behavior into organizations. Effective software development, as if we didn't know it already, turns out to be a hard problem to solve. It may be that there is no such thing as a software company above a critical size threshold that can get it right. Keith de Mendonca certainly believes that it is hard, but not impossible, to scale up without losing agility. Successfully scaling up probably requires a broader perspective than just that of the development methodology and raises the larger question of the way the development organization is structured.

Keith de Mendonca:

Agility is much bandied around as a hot topic, but certainly the attributes it should represent are that you are close to your customer, you understand better what you are doing, you are continuously improving, you are basically making the best decisions that you can at each stage, and being able to take each step and improve until you deliver the best version of the code to your customer.

Organizationally, we are committed to this concept and that motivates the way the software engineering organization is aligned and organized into technology streams today. We consciously said we wanted to create the organization around the technology architecture rather than just managerial units. And although we maybe haven't achieved it yet, there was meant to be a much greater degree of autonomy inside the technology streams, where they own their own roadmaps. So effectively if they were given clear instructions about what they needed to do by when, they would have much more autonomy to actually do those things in the best way that they see fit.

The key ideas are to align around a technology vision of where the product should be going and to devolve responsibility for delivering the vision to technology-focused engineering streams.

Keith de Mendonca:

A company which is still growing and still maturing must try to control itself, while market forces constantly challenge that autonomy. The struggle is to provide that flexibility and allow that flexibility to the technology streams when external issues often arise that cause changes in direction. Naturally, this makes an organization wish to have very tight control of what each component is actually doing.

The forces of control and centralization are almost inevitable in the face of commercial pressures – without which of course there would be no system at all.

Keith de Mendonca:

Of course, success in a new market needs many changes in the way that a software organization is controlled. Symbian has control of its own destiny in engineering terms, but it will take time to balance that autonomy and responsibility on the smaller units with the execution and the very predictable delivery that we demand as a company.

> But if you look at the high level, we release the OS to our customers every two weeks now. We also have regular official operating system releases (about three a year compared to the original model of one huge release every 18 months). The shorter release cycle delivers regular improvements to the product and gets faster feedback from the customer.

Possibly there is no way back to the kind of flexible and autonomous development model of the early days of the company. At the end of the EPOC Release 5 project, 48 or so senior developers including architects and team leads spent two days at an off-site debriefing with the company management. Management had mapped out a two-day program, but on the first morning it was overturned by the engineering participants, who had a different view of where the obstacles to progress were and what should be done about them. It is unlikely that such a thing could happen now – if only because the likelihood of engineers and managers having sufficient time to spend two days out at the same debriefing session is remote.

> **Keith de Mendonca:**
>
> I think that physically the tools needed for controlling a large organization tend to automatically restrain that kind of flexibility and fast movement. But to some extent you can build agility into even medium- or large-sized companies, and I suppose that's what we are currently planning to do, to put that agility back.

It may also be that perhaps the company was never really a 'creative zoo' at all.

> **Bob Dewolf:**
>
> Development methodology is a really tricky issue we have. But creative zoo, is that what we used to have? I don't know. I don't think so.

Symbian is practiced at innovation. The nature of the development model and software engineering organization remain open questions. Watch this space.

Appendix A: Symbian OS Component Reference

Introduction

This appendix attempts to provide a definitive component reference for Symbian OS, based on Symbian OS v9.3, the most current available release at the time of publication. Although associated with a specific release, for the most part the component information will be useful for developers working on any release.

The goal is to make available to external developers (including those working with licensee or partner companies, as well as independent third-party developers), a minimum level of information about the system at component level.

Note that the component set should be interpreted as a superset of all possible components, and not as a definitive guide to the components present on a given Symbian OS device.

A.1 Explanation of Fields

Each component is documented in a simple format. The meaning of the fields is explained below.

Development Name

The development name is the short name by which the component is commonly known. For example, the Plug-In Framework component is commonly known as ECOM; the Data Comms Server is commonly known as C32, and so on. Typically, the short name is either identical to or a derivation from the name of the binary that the component builds (where it builds to a single binary), the principal significant binary (where it builds several binaries), or the name of the directory in which the component source is located in the source tree.

Build File Location

The build file location is the path at which the build file for the component can be found in the OS source tree. Note that external developers do not have access to the full source tree.

System Model Location

Where the component appears in the System Model, its location is given as *Layer* (in all cases), *Block* (where applicable), *Sub-block* (where applicable), and *Component collection* (in all cases).

Licence Categorization

In the license that defines the terms of use of Symbian OS by licensees, all components intended for production deployment are categorized as either *Common* or *Optional*, and as either *Symbian* or *Replaceable* components, thus creating four possible categories. Components which are not intended for production deployment are categorized as *Reference* or *Test* components.

From an external developer perspective, the most meaningful interpretation of these categories is as follows:

	Symbian	Replaceable
Common	Always present in a vendor platform. Symbian supplied, defines Symbian OS APIs.	Always present in a vendor platform. Symbian or licensee supplied, preserves (but may extend) the Symbian OS APIs.
Optional	Optionally present in a vendor platform. Symbian supplied, if present, defines Symbian OS APIs.	Optionally present in a vendor platform. Symbian or licensee supplied, if present. Not guaranteed to preserve the Symbian OS APIs.

Reference and Test components are not intended for production deployment, although licensees may choose to deploy reference components. Test components must not be deployed on licensee devices.

Exposes Third-Party APIs

If this is YES, the component exposes public APIs which any developer can use. They are classified as `publishedAll` in the source code. If it is NO, the APIs are restricted.

Restricted APIs are classified as `publishedPartner`, `internal-Technology` or `internalAll`. Although partner APIs are not supported in public SDKs, they are supported in the kits shipped to partner developers and can be used by them. Internal APIs are reserved for use by Symbian development teams.

It is important to note that header files available either in kits or SDKs may contain mixed categories of APIs. The classification rules are intended to help guide developers towards the APIs they can safely use.

Present in OS Releases

Data is provided for all releases from Symbian OS v7.0 to v9.3 as an aid to developers migrating code from earlier releases.

Description

This is a brief plain text description of the functionality provided by the component and its role in the system.

A.2   Agenda Model

Development Name: AGNMODEL

Build File Location: `/common/generic/app-engines/agnmodel/group/`

System Model Location:

Layer: Application Services

Component collection: PIM Application Services

License Classification: Optional Symbian

Exposes Third-Party APIs: YES

Present in OS Releases: 7.0, 7.0s, 8.0, 8.1, 9.1, 9.2, 9.3

Description: Public APIs are limited to enumerations in `calnotification.h`. Legacy API, replaced in Symbian OS v9 with the new Calendar API which is more suitable for a phone. Agenda Model is maintained for compatibility.

A.3 Alarm Server

Development Name: ALARMSERVER

Build File Location: `/common/generic/app-services/alarmserver/group/`

System Model Location:

    Layer: Application Services

    Component collection: PIM Application Support

License Classification: Optional Replaceable

Exposes Third-Party APIs: YES

Present in OS Releases: 7.0, 7.0s, 8.0, 8.1, 9.1, 9.2, 9.3

Description: Server that manages a queue of system-wide, time-based alarms that provide APIs for client applications to set, modify, query and notify alarms.

A.4 Animation

Development Name: ANIMATION

Build File Location: `/common/generic/app-framework/animation/group/`

System Model Location:

    Layer: UI Framework

    Component collection: UI Support

License Classification: Optional Replaceable

Exposes Third-Party APIs: YES

Present in OS Releases: 9.1, 9.2, 9.3

Description: Framework that supports window-, sprite-, and bitmap-based animation, enabling animation plug-ins to be created and loaded.

A.5 Application Architecture

Development Name: APPARC

Build File Location: `/common/generic/app-framework/apparc/group/`

System Model Location:

    Layer: Application Services

    Component collection: Application Framework

License Classification: Common Symbian

Exposes Third-Party APIs: YES

Present in OS Releases: 7.0, 7.0s, 8.0, 8.1, 9.1, 9.2, 9.3

Description: Framework that defines basic application responsibilities and the interactions between core application classes, broadly following the MVC pattern. Abstracted via Uikon, and ultimately by a vendor-specific variant UI.

A.6 Application Utilities

Development Name: BAFL

Build File Location: `/common/generic/syslibs/bafl/group/`

System Model Location:

    Layer: Base Services

    Component collection: Low-Level Libraries and Frameworks

License Classification: Common Symbian

Exposes Third-Party APIs: YES

Present in OS Releases: 7.0, 7.0s, 8.0, 8.1, 9.1, 9.2, 9.3

Description: Essential application utilities organized as a single library DLL. Includes system sounds, the clipboard, utility classes for resource-file handling and file finding, and implementations of string pools and descriptor arrays.

A.7 Audio Driver

Development Name: SOUNDDEV

Build File Location: `/common/generic/Multimedia/MMF/sounddev/group/`

System Model Location:

    Layer: Kernel Services and Hardware Interface

    Block: Kernel Architecture

    Component collection: Logical Device Drivers

License Classification: Common Symbian

Exposes Third-Party APIs: YES

Present in OS Releases: 8.0, 8.1, 9.1, 9.2, 9.3

Description: Hardware-abstraction layer for digital audio acceleration.

A.8 Backup and Restore Notification

Development Name: BACKUPRESTORENOTIFICATION

Build File Location: `/common/generic/app-services/BackupRestoreNotification/group/`

System Model Location:

    Layer: Application Services

Component collection: PIM Application Support

License Classification: Optional Replaceable

Exposes Third-Party APIs: NO

Present in OS Releases: 8.0, 8.1, 9.1, 9.2, 9.3

Description: Intended for internal use only. An alert service to notify backup–restore progress to PIM applications. All other applications should use the Publish and Subscribe service to achieve similar functionality.

A.9 Baseband Channel Adaptor

Development Name: BCA

Build File Location: `/common/generic/networking/BasebandAdaptation/bca/group/`

System Model Location:

Layer: OS Services

Block: Comms Services

Sub-block: Networking Services

Component collection: Link Layer Control

License Classification: Optional Replaceable

Exposes Third-Party APIs: NO

Present in OS Releases: 8.0, 8.1, 9.1, 9.2, 9.3

Description: Abstracts the channel used to communicate with the baseband processor and defines a plug-in interface for a hardware-specific interface-implementation module.

A.10 Baseband Channel Adaptor for C32

Development Name: C32BCA

Build File Location: `/common/generic/networking/BasebandAdaptation/c32bca/group/`

System Model Location:

Layer: OS Services

Block: Comms Services

Sub-block: Networking Services

Component collection: Link Layer Control

License Classification: Optional Replaceable

Exposes Third-Party APIs: NO

Present in OS Releases: 8.0, 8.1, 9.1, 9.2, 9.3

Description: Intended for partner use only. Plug-in providing a serial comms implementation of the Baseband Channel Adapter interface.

A.11 Bearer Abstraction Layer

Development Name: BALSERVER MROUTER-PLUGIN

Build File Location: `/common/generic/connectivity/BAL/group/`

System Model Location:

Layer: OS Services

Block: Connectivity Services

Component collection: Device Connection

License Classification: Optional Replaceable

Exposes Third-Party APIs: NO

Present in OS Releases: 8.0, 8.1, 9.1, 9.2, 9.3

Description: Intended for internal use only. Abstraction framework for connectivity plug-ins that encapsulate actual bearers (e.g., m-Router), providing a connection management API for use by PC link-type applications.

A.12 BIO Messaging Framework

Development Name: MSG_BIOMSG

Build File Location: `/common/generic/messaging/biomsg/group/`

System Model Location:

Layer: Application Services

Component collection: Messaging Application Support

License Classification: Optional Symbian

Exposes Third-Party APIs: YES

Present in OS Releases: 7.0, 7.0s, 8.0, 8.1, 9.1, 9.2, 9.3

Description: Messaging extension supporting 'smart' message types (Bearer Independent Objects), for example vCard or vCalendar messages, network setup messages and so on, which are not intended for end-user action but for system components or applications. Provides a mechanism for creating application-specific, bespoke message types.

A.13 BIO Messaging Parsers

Development Name: CBCP, ENP, GFP, IACP, WAPP

Build File Location: `/common/generic/messaging/biomsg/group/`

System Model Location:

Layer: Application Services

Component collection: Content Handling

License Classification: Optional Symbian

Exposes Third-Party APIs: YES

Present in OS Releases: 8.0, 8.1, 9.1, 9.2, 9.3

Description: Plug-ins to the BIO Messaging Framework for parsing specific message types including compact business card, email notification, Nokia Smart Message Internet Access, Nokia and Ericsson OTA, as well as a general file parser.

A.14 BIO Watchers

Development Name: BIOMSG, NBSWATCHER, WAPWATCHER, BIOWATCHERSCDMA

Build File Location: `/common/generic/messaging/biomsg/ BioWatchers/bld.inf, /common/generic/messaging/ biomsg/BioWatchersCdma/group/`

System Model Location:

    Layer: Application Services

    Component collection: Messaging Application Support

License Classification: Optional Symbian

Exposes Third-Party APIs: NO

Present in OS Releases: 8.0, 8.1, 9.1, 9.2, 9.3

Description: Watcher framework and service for notifying applications of message arrival.

A.15 Bit GDI

Development Name: BITGDI

Build File Location: `/common/generic/graphics/bitgdi/group/`

System Model Location:

    Layer: OS Services

    Block: Multimedia and Graphics Services

    Component collection: Graphics and Printing Services

License Classification: Common Symbian

Exposes Third-Party APIs: YES

Present in OS Releases: 7.0, 7.0s, 8.0, 8.1, 9.1, 9.2, 9.3

Description: Device- and display-mode-independent implementation of the concrete graphics context for bitmaps.

A.16 Bluetooth 1.0

Development Name: BLUETOOTH
Build File Location: `/common/generic/ME/group/`
System Model Location:
    Layer: Java ME
    Component collection: MIDP 2.0 Packages
License Classification: Optional Replaceable
Exposes Third-Party APIs: YES
Present in OS Releases: 8.0, 8.1, 9.1, 9.2, 9.3
Description: MIDP 2.0 Bluetooth 1.0 (JSR-082) APIs that support Bluetooth messaging including Push support.

A.17 Bluetooth 1.0 Push Plug-in

Development Name: BLUETOOTH
Build File Location: `/common/generic/ME/group/`
System Model Location:
    Layer: Java ME
    Component collection: Bluetooth and SMS Push
License Classification: Optional Replaceable
Exposes Third-Party APIs: YES
Present in OS Releases: 8.0, 8.1, 9.1, 9.2, 9.3
Description: Forms part of JTWI. Plug-in that binds the Bluetooth 1.0 package to the underlying system.

A.18 Bluetooth CSY

Development Name: BTCOMM
Build File Location: `/common/generic/bluetooth/`
System Model Location:
    Layer: OS Services
    Block: Comms Services
    Sub-block: Short Link Services
    Component collection: Serial Comms Server Plug-ins
License Classification: Common Replaceable
Exposes Third-Party APIs: YES
Present in OS Releases: 7.0, 7.0s, 8.0, 8.1, 9.1, 9.2, 9.3

Description: CSY plug-in to C32 Serial Server providing serial port emulation over Bluetooth.

A.19 Bluetooth HCI

Development Name: HCI
Build File Location: /common/generic/bluetooth/
System Model Location:
  Layer: OS Services
  Block: Comms Services
  Sub-block: Short Link Services
  Component collection: Short Link Protocol Plug-ins
License Classification: Reference/Test
Exposes Third-Party APIs: YES
Present in OS Releases: 8.0, 8.1, 9.1, 9.2, 9.3
Description: Host Controller Interface (HCI) firmware driver implementation for reference only, so must be replaced by the licensee. The HCI framework, for example, includes public APIs.

A.20 Bluetooth Manager

Development Name: BLUETOOTH_MANAGER
Build File Location: /common/generic/bluetooth/
System Model Location:
  Layer: OS Services
  Block: Comms Services
  Sub-block: Short Link Services
  Component collection: Short Link
License Classification: Optional Symbian
Exposes Third-Party APIs: YES
Present in OS Releases: 7.0, 7.0s, 8.0, 8.1, 9.1, 9.2, 9.3
Description: Information store for managing details of the local and remote Blueooth devices, implemented over DBMS.

A.21 Bluetooth PAN Profile

Development Name: BLUETOOTH_PAN
Build File Location: /common/generic/bluetooth/
System Model Location:

Layer: OS Services

Block: Comms Services

Sub-block: Networking Services

Component collection: Networking Plug-ins

License Classification: Optional Symbian

Exposes Third-Party APIs: YES

Present in OS Releases: 8.0, 8.1, 9.1, 9.2, 9.3

Description: Support for Bluetooth Personal Area Networking (PAN) Profile, which is analogous to a network interface agent for Bluetooth. Implements the Bluetooth Network Encapsulation Protocol (BNEP) as an Ethernet Packet Driver module, enabling PAN to behave like a regular Internet access provider.

A.22 Bluetooth Profiles

Development Name: BLUETOOTH_AVRCP, BLUETOOTH_GAVDP

Build File Location: /common/generic/bluetooth/

System Model Location:

Layer: OS Services

Block: Comms Services

Sub-block: Short Link Services

Component collection: Short Link

License Classification: Optional Symbian

Exposes Third-Party APIs: YES

Present in OS Releases: 9.1, 9.2, 9.3

Description: Support for Bluetooth Audio/Video Remote Control Profile (AVRCP) and Generic Audio/Visual Distribution Profile (GAVDP). AVRCP is implemented as a bearer plug-in to the remote-control framework. GAVDP is implemented as a thin layer over the Socket Server client APIs and allows clients to configure, send and receive data over the AVDTP protocol running inside the Bluetooth protocol plug-in.

A.23 Bluetooth Protocol Client APIs

Development Name: USER

Build File Location: /common/generic/bluetooth/

System Model Location:

Layer: OS Services

Block: Comms Services

Sub-block: Short Link Services

Component collection: Short Link

License Classification: Optional Symbian

Exposes Third-Party APIs: YES

Present in OS Releases: 7.0, 7.0s, 8.0, 8.1, 9.1, 9.2, 9.3

Description: Bluetooth-specific APIs for use by Bluetooth socket clients, providing support for low-level control of protocol parameters (e.g., packet sizes) and hardware (e.g., power modes).

A.24 Bluetooth SDP

Development Name: BLUETOOTH_SDP

Build File Location: /common/generic/bluetooth/

System Model Location:

Layer: OS Services

Block: Comms Services

Sub-block: Short Link Services

Component collection: Short Link

License Classification: Optional Symbian

Exposes Third-Party APIs: YES

Present in OS Releases: 8.0, 8.1, 9.1, 9.2, 9.3

Description: Service Discovery agent and database, used by connected Bluetooth devices to query each other and exchange and store information about the Bluetooth services they support. Note that the SDP database is not persistent.

A.25 Bluetooth Stack PRT

Development Name: BLUETOOTH_STACK

Build File Location: /common/generic/bluetooth/

System Model Location:

Layer: OS Services

Block: Comms Services

Sub-block: Short Link Services

Component collection: Short Link Protocol Plug-ins

License Classification: Optional Symbian

Exposes Third-Party APIs: NO

Present in OS Releases: 8.0, 8.1, 9.1, 9.2, 9.3

Description: Bluetooth stack implementation in the form of a PRT protocol plug-in to Socket Server.

A.26 BMP Animation

Development Name: BMPANIM

Build File Location: `/common/generic/app-framework/bmpanim/group/`

System Model Location:

Layer: UI Framework

Component collection: UI Support

License Classification: Optional Replaceable

Exposes Third-Party APIs: YES

Present in OS Releases: 7.0, 7.0s, 8.0, 8.1, 9.1, 9.2, 9.3

Description: Plug-in utility to Window Server implementing support for bitmap-based frame sequence animation. Forms part of the Animation framework.

A.27 Bookmark Support

Development Name: BOOKMARKS

Build File Location: `/common/generic/application-protocols/bookmarks/group/`

System Model Location:

Layer: Application Services

Component collection: Internet and Web Application Support

License Classification: Optional Replaceable

Exposes Third-Party APIs: NO

Present in OS Releases: 9.1, 9.2, 9.3

Description: Bookmark database support for Web browsers.

A.28 Bootstrap

Development Name: BOOTSTRAP

Build File Location: `/cedar/generic/base/e32/`

System Model Location:

Layer: Kernel Services and Hardware Interface

Block: Kernel Architecture

Component collection: Variant

License Classification: Optional Replaceable

Exposes Third-Party APIs: NO

Present in OS Releases: 7.0, 7.0s, 8.0, 8.1, 9.1, 9.2, 9.3

Description: Bootstraps the system by preparing the hardware, including memory and peripherals, mapping the virtual address space if an MMU is present and starting the kernel.

A.29 Broadcast Tuner

Development Name: TUNER
Build File Location: `/common/generic/Multimedia/Tuner/group/`
System Model Location:
    Layer: OS Services
    Block: Multimedia and Graphics Services
    Component collection: Multimedia
License Classification: Optional Replaceable
Exposes Third-Party APIs: YES
Present in OS Releases: 9.1, 9.2, 9.3
Description: Integrated broadcast tuner API for analog or digital broadcast channel reception.

A.30 C Standard Library

Development Name: STDLIB
Build File Location: `/common/generic/syslibs/stdlib/group/`
System Model Location:
    Layer: OS Services
    Block: Generic OS Services
    Component collection: Generic Libraries
License Classification: Optional Replaceable
Exposes Third-Party APIs: YES
Present in OS Releases: 7.0, 7.0s, 8.0, 8.1, 9.1, 9.2, 9.3

Description: Port of a subset of the POSIX/C programming language Standard Library. Maps C function calls to native Symbian OS C++ APIs, allowing ported C applications to interface to native services. Orginally written to support porting of binary-only Java VM.

A.31 C32 Serial Server

Development Name: C32
Build File Location: `/common/generic/ser-comms/c32/group/`
System Model Location:

Layer: OS Services
Block: Comms Services
Sub-block: Comms Framework
Component collection: Data Comms Server

License Classification: Common Symbian

Exposes Third-Party APIs: YES

Present in OS Releases: 7.0, 7.1, 8.0, 8.1, 9.1, 9.2, 9.3

Description: Provides the client-session APIs for serial communications and the framework for creating and loading CSY plug-in modules which implement serial port abstractions, enabling clients to access virtual serial ports independently of the underlying hardware.

A.32 Calendar

Development Name: CALINTERIMAPI

Build File Location: `/common/generic/app-engines/calinterimapi/group/`

System Model Location:

Layer: Application Services
Component collection: PIM Application Services

License Classification: Optional Symbian

Exposes Third-Party APIs: YES

Present in OS Releases: 9.2, 9.3

Description: Calendar API replacement for Agenda Model, partially supports the iCalendar standard.

A.33 Camera

Development Name: ECAM

Build File Location: `/common/generic/Multimedia/Ecam/group/`

System Model Location:

Layer: Multimedia and Graphics Services
Component collection: Multimedia

License Classification: Optional Replaceable

Exposes Third-Party APIs: YES

Present in OS Releases: 7.0s, 8.0, 8.1, 9.1, 9.2, 9.3

Description: Onboard camera API to provide compatibility for camera client applications between devices.

A.34 CDMA MTM

Development Name: CDMASMSMTM

Build File Location: `/common/generic/messaging/sms/`
`multimode/group/`

System Model Location:

    Layer: Application Services

    Component collection: Messaging Application Support

License Classification: Optional Symbian

Exposes Third-Party APIs: YES

Present in OS Releases: 8.0, 8.1, 9.1, 9.2, 9.3

Description: MTM plug-in to Messaging Store framework that implements messaging support for CDMA.

A.35 CDMA SMS Plug-ins

Development Name: CDMASMSSTACK

Build File Location: `/common/generic/nbprotocols/`
`cdmasmsstack/group/`

System Model Location:

    Layer: OS Services

    Block: Comms Services

    Sub-block: Telephony Services

    Component collection: SMS Protocol Plug-ins

License Classification: Optional Symbian

Exposes Third-Party APIs: YES

Present in OS Releases: 8.0, 8.1, 9.1, 9.2, 9.3

Description: CDMA SMS protocol implementation supporting CDMA teleservices.

A.36 CDMA TSY

Development Name: CDMATSY

Build File Location: `/common/generic/telephony/cdmatsy/`
`group/`

System Model Location:

    Layer: OS Services

    Block: Comms Services

    Sub-block: Telephony Services

Component collection: Telephony Server Plug-ins

License Classification: Reference/Test

Exposes Third-Party APIs: NO

Present in OS Releases: 7.0, 7.0s, 8.0, 8.1, 9.1, 9.2, 9.3

Description: Reference-only TSY implementation of CDMA telephony extensions, replaced by licensees with a hardware specific implementation. Plug-in to ETel Telephony Server framework.

A.37 Central Repository

Development Name: CENTRALREPOSITORY

Build File Location: `/common/generic/syslibs/ centralrepository/group/`

System Model Location:

Layer: Base Services

Component collection: Low-Level Libraries and Frameworks

License Classification: Common Symbian

Exposes Third-Party APIs: YES

Present in OS Releases: 8.0, 8.1, 9.1, 9.2, 9.3

Description: Persistent store for global settings that provides a notification mechanism allowing clients to register interest when settings change.

A.38 Certificate and Key Management

Development Name: CERTMAN

Build File Location: `/common/generic/security/certman/ group/`

System Model Location:

Layer: OS Services

Block: Generic OS Services

Component collection: Generic Libraries

License Classification: Optional Replaceable

Exposes Third-Party APIs: YES

Present in OS Releases: 7.0, 7.0s, 8.0, 8.1, 9.1, 9.2, 9.3

Description: Framework for managing and storing security certificates and keys supporting storage and retrieval, assignment of trust status, certificate chain construction, and certificate validation and revocation.

A.39 Certificate Store

Development Name: CERTSTORE
Build File Location: /common/generic/security/certman/certstore/
System Model Location:
  Layer: OS Services
  Block: Generic OS Services
  Component collection: Generic Libraries
License Classification: Optional Replaceable
Exposes Third-Party APIs: YES
Present in OS Releases: 7.0, 7.0s, 8.0, 8.1, 9.1, 9.2, 9.3
Description: Unified certificate store that provides clients with a single point of access to certificates stored on the device.

A.40 Character Encoding and Conversion Framework

Development Name: CHARCONV
Build File Location: /common/generic/syslibs/charconv/
System Model Location:
  Layer: Base Services
  Component collection: Low-Level Libraries and Frameworks
License Classification: Optional Replaceable
Exposes Third-Party APIs: YES
Present in OS Releases: 7.0, 7.0s, 8.0, 8.1, 9.1, 9.2, 9.3
Description: Extensible framework supporting text conversion between Unicode and non-Unicode character sets (Symbian OS native text formats are Unicode).

A.41 Character Encoding and Conversion Plug-ins

Development Name: CHARCONV
Build File Location: /common/generic/syslibs/charconv/
System Model Location:
  Layer: Base Services
  Component collection: Low-Level Libraries and Frameworks
License Classification: Optional Replaceable
Exposes Third-Party APIs: YES
Present in OS Releases: 7.0, 7.0s, 8.0, 8.1, 9.1, 9.2, 9.3

Description: Converter plug-ins to the Character Encoding and Conversion Framework that support conversion to and from a variety of ASCII, UTF-7 and UTF-8 text formats, including JIS/ShiftJis.

A.42 Chinese Calendar Converter

Development Name: CALCON

Build File Location: `/common/generic/app-services/calcon/`

System Model Location:

   Layer: Application Services

   Component collection: PIM Application Support

License Classification: Optional Replaceable

Exposes Third-Party APIs: YES

Present in OS Releases: 7.0, 7.0s, 8.0, 8.1, 9.1, 9.2, 9.3

Description: API for converting between Gregorian and Chinese lunar calendar dates.

A.43 CLDC HI 1.1

Development Name: CLDCHI

Build File Location: `/common/generic/ME/group/`

System Model Location:

   Layer: Java ME

   Component collection: Virtual Machine

License Classification: Common Replaceable

Exposes Third-Party APIs: YES

Present in OS Releases: 8.0, 8.1, 9.1, 9.2, 9.3

Description: Symbian OS port of the Sun CLDC HotSpot Implementation VM (CLDC HI) that forms part of the CLDC 1.1 specification (JSR-139).

A.44 Client Provisioning Adaptors

Development Name: DEVPROV_CLIENTPROV_ADAPTERS

Build File Location: `/common/generic/DevProv/Adapters/ClientProv/group/`

System Model Location:

   Layer: Application Services

   Component collection: Client Provisioning

License Classification: Optional Replaceable
Exposes Third-Party APIs: NO
Present in OS Releases: 9.1, 9.2, 9.3
Description: Adaptor plug-ins to the Client Provisioning Framework.

A.45 Client Provisioning Framework

Development Name: DEVPROV_CLIENTPROV_FRAMEWORK
Build File Location: `/common/generic/DevProv/ClientProv/group/`
System Model Location:
   Layer: Application Services
   Component collection: Client Provisioning
License Classification: Common Replaceable
Exposes Third-Party APIs: NO
Present in OS Releases: 9.1, 9.2, 9.3
Description: Framework implementing the OMA Client Provisioning standards and supporting provisioning of devices by network operators.

A.46 Clock

Development Name: CLOCK
Build File Location: `/common/generic/app-framework/clock/group/`
System Model Location:
   Layer: UI Framework
   Component collection: UI Support
License Classification: Optional Replaceable
Exposes Third-Party APIs: YES
Present in OS Releases: 7.0, 7.0s, 8.0, 8.1, 9.1, 9.2, 9.3
Description: Shared library plug-in to the Window Server that supports creation of animation-based digital and analog clocks, used by UIs and applications.

A.47 Color Palette

Development Name: PALETTE
Build File Location: `/common/generic/graphics/palette/group/`

System Model Location:

> Layer: OS Services
>
> Block: Multimedia and Graphics Services
>
> Component collection: Graphics Device Interface

License Classification: Common Replaceable

Exposes Third-Party APIs: YES

Present in OS Releases: 8.0, 8.1, 9.1, 9.2, 9.3

Description: Low-level graphics support for licensee implementations of color-scheme switching.

A.48 Comms Database

Development Name: COMMDB

Build File Location: `/common/generic/comms-infras/commdb/group/`

System Model Location:

> Layer: OS Services
>
> Block: Comms Services
>
> Sub-block: Comms Framework
>
> Component collection: Comms Configuration Utilities

License Classification: Common Symbian

Exposes Third-Party APIs: YES

Present in OS Releases: 7.0, 7.0s, 8.0, 8.1, 9.1, 9.2, 9.3

Description: Legacy repository for storing communications settings. It is replaced in Symbian OS v9 by the CommsDat interface to the Central Repository, although the original CommDB API is preserved for compatibility.

A.49 Comms Debug Utility

Development Name: COMMSDEBUGUTILITY

Build File Location: `/common/generic/comms-infras/commsdebugutility/group/`

System Model Location: N/A

License Classification: Test/Reference

Exposes Third-Party APIs: YES

Present in OS Releases: N/A

Description: Reimplementation of File Logger intended only for use by communications programs.

A.50 Comms Framework

Development Name: COMMSFW

Build File Location: `/common/generic/comms-infras/commsfw/group/`

System Model Location:

   Layer: OS Services

   Block: Comms Services

   Sub-block: Comms Framework

   Component collection: Comms Framework Utilities

License Classification: Common Symbian

Exposes Third-Party APIs: NO

Present in OS Releases: 8.0, 8.1, 9.1, 9.2, 9.3

Description: Framework providing the base classes used to create communications servers, and the communication mechanisms used to communicate between communications server threads.

A.51 Comms Root Server

Development Name: ROOTSERVER

Build File Location: `/common/generic/comms-infras/rootserver/group/`

System Model Location:

   Layer: OS Services

   Block: Comms Services

   Sub-block: Comms Framework

   Component collection: Comms Process and Settings

License Classification: Common Symbian

Exposes Third-Party APIs: NO

Present in OS Releases: 8.0, 8.1, 9.1, 9.2, 9.3

Description: Provides the main thread in the communications process that is responsible for starting and managing all other communications server threads (i.e. for starting communications servers as threads within the root server process). From Symbian OS v8, communications servers are started when the device boots up, rather than on demand.

A.52 Connection Provider Plug-in

Development Name: IPCPR

Build File Location: `/common/generic/networking/iprpr/group/`

System Model Location:

Layer: OS Services

Block: Comms Services

Sub-block: Networking Services

Component collection: Networking Plug-ins

License Classification: Optional Replaceable

Exposes Third-Party APIs: NO

Present in OS Releases: 8.0, 8.1, 9.1, 9.2, 9.3

Description: Plug-in providing IP connections to clients, supporting bearer mobility.

A.53 Contacts Model

Development Name: CNTMODEL

Build File Location: /common/generic/app-engines/cntmodel/group/

System Model Location:

Layer: Application Services

Component collection: PIM Application Services

License Classification: Optional Symbian

Exposes Third-Party APIs: YES

Present in OS Releases: 7.0, 7.0s, 8.0, 8.1, 9.1, 9.2, 9.3

Description: Application model (i.e. data store plus APIs) for a common contact or address-book application.

A.54 Content-Access Framework for DRM

Development Name: CAF2

Build File Location: /common/generic/syslibs/caf2/group/

System Model Location:

Layer: Application Services

Component collection: Content Handling

License Classification: Optional Symbian

Exposes Third-Party APIs: NO

Present in OS Releases: 8.0, 8.1, 9.1, 9.2, 9.3

Description: Framework for brokering DRM-protected content between agents (DRM applications) and consumers (e.g. media players). Includes a Reference DRM Agent implementation.

A.55 Content-Handling Framework

Development Name: CONTENT_HANDLING

Build File Location: `/common/generic/content-handling/framework/group/`

System Model Location:

    Layer: Application Services

    Component collection: Application Framework

License Classification: Optional Replaceable

Exposes Third-Party APIs: YES

Present in OS Releases: 8.0, 8.1, 9.1, 9.2, 9.3

Description: Provides a framework for content handlers that implements finding, loading, processing, and displaying of typed content on behalf of applications.

A.56 Control Environment (CONE)

Development Name: CONE

Build File Location: `/common/generic/app-framework/cone/group/`

System Model Location:

    Layer: UI Framework

    Component collection: UI Application Framework

License Classification: Common Symbian

Exposes Third-Party APIs: YES

Present in OS Releases: 7.0, 7.0s, 8.0, 8.1, 9.1, 9.2, 9.3

Description: Control hierarchy and environment, providing UI-policy-free abstract controls, i.e. interactive screen elements and control context. Derived concrete controls are provided by the variant UI.

A.57 Core IPSec PRT

Development Name: IPSEC6

Build File Location: `/common/generic/networking/ipsec/ipsec6/group/`

System Model Location:

    Layer: OS Services

    Block: Comms Services

    Sub-block: Networking Services

Component collection: Network Protocol Plug-ins

License Classification: Optional Replaceable

Exposes Third-Party APIs: YES

Present in OS Releases: 7.0, 7.0s, 8.0, 8.1, 9.1, 9.2, 9.3

Description: IPSec for IP v4 and v6 including Authentication Header (AH) and Encapsulating Security Payload (ESP) cryptographic protocols. Implemented as a sockets server PRT plug-in module.

A.58 Cryptographic Token Framework

Development Name: CRYPTOTOKENS, FILETOKENS

Build File Location: `/common/generic/security/cryptotokens/group/`

System Model Location:

Layer: OS Services

Block: Generic OS Services

Component collection: Generic Libraries

License Classification: Optional Symbian

Exposes Third-Party APIs: YES

Present in OS Releases: 8.0, 8.1, 9.1, 9.2, 9.3

Description: Framework supporting use of secure hardware tokens (or their equivalent software emulations). Defines certificate and key storage and authentication APIs for secure hardware tokens, for example SD memory cards.

A.59 Cryptography Library

Development Name: CRYPTOGRAPHY

Build File Location: `/common/generic/security/crypto/group/`

System Model Location:

Layer: Base Services

Component collection: Low-Level Libraries and Frameworks

License Classification: Optional Symbian

Exposes Third-Party APIs: YES

Present in OS Releases: 7.0, 7.0s, 8.0, 8.1, 9.1, 9.2, 9.3

Description: Non-RSA-based cryptographic algorithms including symmetric and asymmetric ciphers, hash functions and a cryptographic random-number generator. Supersedes RSA-based implementations. Implemented in 'strong' and 'weak' versions, of which the strong version is export-restricted.

A.60 CSD AGT

Development Name: CSDAGT

Build File Location: `/common/generic/networking/csdagt/group/`

System Model Location:

Layer: Comms Services

Sub-block: Networking Services

Component collection: Networking Plug-ins

License Classification: Optional Replaceable

Exposes Third-Party APIs: YES

Present in OS Releases: 7.0s, 8.0, 8.1, 9.1, 9.2, 9.3

Description: AGT agent plug-in to Connection Agent framework that negotiates circuit-switched connections e.g. to GSM and CDMA networks, supporting dial-up networking services.

A.61 Data Engine

Development Name: DAMODEL

Build File Location: `/common/generic/app-engines/damodel/group/`

System Model Location:

Layer: Application Services

Component collection: Office Application Engines

License Classification: Optional Replaceable

Exposes Third-Party APIs: YES

Present in OS Releases: 7.0, 7.0s, 8.0, 8.1, 9.1, 9.2, 9.3

Description: Legacy application model for a free-form database application, originating from the early EPOC built-in application set.

A.62 DBMS

Development Name: DBMS

Build File Location: `/common/generic/syslibs/dbms/group/`

System Model Location:

Layer: Base Services

Component collection: Low-Level Libraries and Frameworks

License Classification: Optional Symbian

Exposes Third-Party APIs: YES

Present in OS Releases: 7.0, 7.0s, 8.0, 8.1, 9.1, 9.2, 9.3

Description: Relational database manager and server that defines database-access APIs and implements a proprietary database format. Supports both exclusive-access and shared-access databases.

A.63 Device Management Adaptors

Development Name: DEVPROV_DEVMAN_ADAPTERS
Build File Location: `/common/generic/SyncML/DevMan/group/`
System Model Location:
  Layer: Application Services
  Component collection: Device Management
License Classification: Optional Replaceable
Exposes Third-Party APIs: NO
Present in OS Releases: 8.0, 8.1, 9.1, 9.2, 9.3
Description: Adaptor plug-ins to Device Management Framework, supporting application management, browser bookmarks, data synchronization and Accounts, Device Information, Device Management Accounts, Email, MMS, Network Access Points, SMS, WAP Proxies.

A.64 Device Management Framework

Development Name: DEVPROV_DEVMAN_FRAMEWORK
Build File Location: `/common/generic/SyncML/DevMan/group/`
System Model Location:
  Layer: Application Services
  Component collection: Device Management
License Classification: Optional Replaceable
Exposes Third-Party APIs: NO
Present in OS Releases: 8.0, 8.1, 9.1, 9.2, 9.3
Description: Framework that implements OMA Device Management based on SyncML and supports remote provisioning of devices by network operators.

A.65 DHCP

Development Name: DHCP
Build File Location: `/common/generic/networking/dhcp/group/`
System Model Location:
  Layer: OS Services
  Block: Comms Services

Sub-block: Networking Services

Component collection: TCP/IP Utilities

License Classification: Optional Replaceable

Exposes Third-Party APIs: NO

Present in OS Releases: 8.0, 8.1, 9.1, 9.2, 9.3

Description: For internal use only, this Dynamic Host Configuration Protocol (DHCP) implementation is used by networking components.

A.66 Dial

Development Name: Dial

Build File Location: `/generic/telephony/dial/group/`

System Model Location:

Layer: OS Services

Block: Comms Services

Sub-block: Telephony Services

Component collection: Telephony Utilities

License Classification: Optional Replaceable

Exposes Third-Party APIs: N/A

Present in OS Releases: 7.0, 7.0s, 8.0, 8.1, 9.1, 9.2, 9.3

Description: Deprecated utility APIs for phone-number manipulation.

A.67 Dialog

Development Name: DIALOG

Build File Location: `/common/generic/networking/DIALOG/group/`

System Model Location:

License Classification: Optional Replaceable

Exposes Third-Party APIs: YES

Present in OS Releases:

Description: Generic connection-agent dialog server that allows user interaction where appropriate when setting up a connection to the Internet.

A.68 DND

Development Name: DND

Build File Location: `/common/generic/networking/dnd/group/`

System Model Location:

Layer: OS Services

Block: Comms Services

Sub-block: Networking Services

Component collection: TCP/IP Utilities

License Classification: Common Symbian

Exposes Third-Party APIs: NO

Present in OS Releases: 7.0, 7.0s, 9.1, 9.2, 9.3

Description: For internal use only, this Domain Name Service (DNS) implementation is used by networking components.

A.69 Emulator

Development Name: WINS_VARIANT_EKA2

Build File Location: `/cedar/generic/base/wins/`

System Model Location:

Layer: Kernel Services and Hardware Interface

Block: Kernel Architecture

Component collection: Variant

License Classification: Optional Replaceable

Exposes Third-Party APIs: NO

Present in OS Releases: 7.0, 7.0s, 8.0, 8.1, 9.1, 9.2, 9.3

Description: Symbian OS emulator for Microsoft Windows platforms. In EKA2, this is implemented as a hardware target variant.

A.70 ESock Socket Server

Development Name: ESOCK

Build File Location: `/common/generic/comms-infras/esock/group/`

System Model Location:

Layer: OS Services

Block: Comms Services

Sub-block: Networking Services

Component collection: Sockets Server

License Classification: Common Symbian

Exposes Third-Party APIs: YES

Present in OS Releases: 7.0, 7.0s, 8.0, 8.1, 9.1, 9.2, 9.3

Description: Server and framework for sockets-based communications. Loads socket implementations from PRT protocol plug-in modules.

A.71 ETel Third-Party API

Development Name: ETEL3RDPARTY

Build File Location: `/common/generic/telephony/etel3rdparty/group/`

System Model Location:

    Layer: OS Services

    Block: Comms Services

    Sub-block: Telephony Services

    Component collection: Telephony Server

License Classification: Common Symbian

Exposes Third-Party APIs: YES

Present in OS Releases: 7.0, 7.0s, 8.0, 8.1, 9.1, 9.2, 9.3

Description: Telephony API subset intended for third-party developer use.

A.72 ETel CDMA

Development Name: ETELCDMA

Build File Location: `/common/generic/telephony/etelcdma/group/`

System Model Location:

    Layer: OS Services

    Block: Comms Services

    Sub-block: Telephony Services

    Component collection: Telephony Server

License Classification: Optional Symbian

Exposes Third-Party APIs: NO

Present in OS Releases: 8.0, 8.1, 9.1, 9.2, 9.3

Description: CDMA extensions to the ETel Telephony Server.

A.73 ETel Multimode

Development Name: ETELMM

Build File Location: `/common/generic/telephony/etelmm/group/`

System Model Location:

    Layer: OS Services

    Block: Comms Services

    Sub-block: Telephony Services

Component collection: Telephony Server

License Classification: Common Symbian

Exposes Third-Party APIs: YES

Present in OS Releases: 7.0, 7.0s, 8.0, 8.1, 9.1, 9.2, 9.3

Description: ETel Telephony Server API extensions that provide network-agnostic APIs for voice, fax, data and multimedia calls and that, therefore, abstract the differences between GSM and CDMA networks.

A.74 ETel Packet Data

Development Name: ETELPCKT

Build File Location: /common/generic/telephony/etelpckt/group/

System Model Location:

  Layer: OS Services

  Block: Comms Services

  Sub-block: Telephony Services

  Component collection: Telephony Server

License Classification: Common Symbian

Exposes Third-Party APIs: NO

Present in OS Releases: 7.0, 7.0s, 8.0, 8.1, 9.1, 9.2, 9.3

Description: ETel Telephony Server API extensions that provide access to packet services on GPRS, UMTS and CDMA/CDMA2000 networks.

A.75 ETel Server and Core

Development Name: ETEL

Build File Location: /common/generic/telephony/etel/group/

System Model Location:

  Layer: OS Services

  Block: Comms Services

  Sub-block: Telephony Services

  Component collection: Telephony Server

License Classification: Common Symbian

Exposes Third-Party APIs: YES

Present in OS Releases: 8.0, 8.1, 9.1, 9.2, 9.3

Description: ETel Telephony Server and core APIs.

A.76 ETel SIM Toolkit

Development Name: ETELSAT
Build File Location: /common/generic/telephony/etelsat/group/
System Model Location:
  Layer: OS Services
  Block: Comms Services
  Sub-block: Telephony Services
  Component collection: Telephony Server
License Classification: Common Symbian
Exposes Third-Party APIs: NO
Present in OS Releases: 7.0, 7.0s, 8.0, 8.1, 9.1, 9.2, 9.3
Description: ETel Telephony Server API extension that provides access to the GSM/WCDMA (U)SIM Application Toolkit.

A.77 Ethernet Driver

Development Name: ETHERDRV
Build File Location: /common/generic/networking/etherdrv/group/
System Model Location:
  Layer: Kernel Services and Hardware Interface
  Block: Kernel Architecture
  Component collection: Logical Device Drivers
License Classification: Common Replaceable
Exposes Third-Party APIs: YES
Present in OS Releases: 7.0, 7.0s, 8.0, 8.1, 9.1, 9.2, 9.3
Description: LDD and PDD implementations for Ethernet cards and emulators.

A.78 Ethernet NIF

Development Name: ETHINT
Build File Location: /common/generic/networking/ether802/group/
System Model Location:
  Layer: OS Services
  Block: Comms Services

Sub-block: Networking Services

Component collection: Link Layer Control

License Classification: Optional Replaceable

Exposes Third-Party APIs: NO

Present in OS Releases: 7.0, 7.0s, 8.0, 8.1, 9.1, 9.2, 9.3

Description: Ethernet protocol NIF plug-in to Network Interface Manager supporting wired Ethernet.

A.79 Ethernet Over IR Packet Driver

Development Name: IRLANPACKETDRIVERS

Build File Location: `/common/generic/networking/ether802/group/`

System Model Location:

Layer: OS Services

Block: Comms Services

Sub-block: Networking Services

Component collection: Link Layer Control

License Classification: Optional Replaceable

Exposes Third-Party APIs: NO

Present in OS Releases: 7.0, 7.0s, 8.0, 8.1, 9.1, 9.2, 9.3

Description: Logical and physical device drivers providing Ethernet-framing services for networking over infrared.

A.80 Ethernet Packet Driver

Development Name: ETHER802

Build File Location: `/common/generic/networking/ether802/group/`

System Model Location:

Layer: Kernel Services and Hardware Interface

Block: Kernel Architecture

Component collection: Logical Device Drivers

License Classification: Optional Replaceable

Exposes Third-Party APIs: YES

Present in OS Releases: 7.0, 7.0s, 8.0, 8.1, 9.1, 9.2, 9.3

Description: Logical and physical LAN packet drivers providing Ethernet framing to generic-networking services.

A.81 Event Logger

Development Name: LOGENG

Build File Location: /common/generic/syslibs/logeng/group/

System Model Location:

    Layer: OS Services

    Block: Generic OS Services

    Component collection: Generic Services

License Classification: Optional Replaceable

Exposes Third-Party APIs: YES

Present in OS Releases: 7.0, 7.0s, 8.0, 8.1, 9.1, 9.2, 9.3

Description: Interface that supports logging events to a logging engine and retrieval, filtering and viewing of logged events by clients.

A.82 FAT Filename Conversion Plug-ins

Development Name: FATCHARSETCONV

Build File Location: /common/generic/syslibs/FATCharsetConv/group/

System Model Location:

    Layer: Base Services

    Component collection: User Library and File Server

License Classification: Common Replaceable

Exposes Third-Party APIs: NO

Present in OS Releases: 9.1, 9.2, 9.3

Description: File Server plug-ins that support converting FAT file names from and to Unicode.

A.83 Fax Client and Server

Development Name: FAX

Build File Location: /common/generic/telephony/FAX/group/

System Model Location:

    Layer: OS Services

    Block: Comms Services

    Sub-block: Telephony Services

    Component collection: Telephony Server

License Classification: Optional Replaceable

Exposes Third-Party APIs: NO

Present in OS Releases: 7.0, 7.0s, 8.0, 8.1, 9.1, 9.2, 9.3

Description: Fax server and protocol stack together with client-side APIs. Extension to ETel Telephony Server that manages fax transmission and reception requests from application clients.

A.84 Feature Registry

Development Name: FEATREG

Build File Location: /common/generic/syslibs/featreg/group/

System Model Location:

Layer: Base Services

Component collection: Low-Level Libraries and Frameworks

License Classification: Common Symbian

Exposes Third-Party APIs: YES

Present in OS Releases: 9.2, 9.3

Description: APIs for run-time discovery of supported 'features' on a given platform.

A.85 FEP Base

Development Name: FEPBASE

Build File Location: /common/generic/app-framework/fepbase/group/

System Model Location:

Layer: UI Framework

Component collection: UI Application Framework

License Classification: Common Symbian

Exposes Third-Party APIs: YES

Present in OS Releases: 7.0, 7.0s, 8.0, 8.1, 9.1, 9.2, 9.3

Description: Framework for front-end processors (FEPs) that enable application-independent input preprocessing to support keyboard mapping, multitap-keyboard input, handwriting recognition, voice recognition, and so on.

A.86 File Converter Framework

Development Name: CONARC

Build File Location: /common/generic/app-framework/conarc/group/

System Model Location:

Layer: Application Services

Component collection: Application Framework

License Classification: Optional Replaceable

Exposes Third-Party APIs: YES

Present in OS Releases: 7.0, 7.0s, 8.0, 8.1, 9.1, 9.2, 9.3

Description: Framework for converter plug-ins that enable file conversions based on MIME type.

A.87 File Converter Plug-ins

Development Name: CHTMLTOCRTCONVERTER, CONVERT, RICH-TEXTTOHTMLCONV

Build File Location: `/common/generic/app-services/chtmltocrtconv/group/,/common/generic/app-engines/convert/group/,/common/generic/app-services/richtexttohtmlconv/group/`

System Model Location:

Layer: Application Services

Component collection: PIM Application Support

License Classification: Optional Replaceable

Exposes Third-Party APIs: NO

Present in OS Releases: 7.0, 7.0s, 8.0, 8.1, 9.1, 9.2, 9.3

Description: Plug-ins to File Converter Framework that support conversions between Symbian OS rich-text objects, HTML files and Microsoft Office file formats.

A.88 File Logger

Development Name: FLOGGER

Build File Location: `/common/generic/comms-infras/flogger/group/`

System Model Location:

Layer: OS Services

Block: Generic OS Services

Component collection: Generic Services

License Classification: Optional Replaceable

Exposes Third-Party APIs: YES

Present in OS Releases: 7.0, 7.0s, 8.0, 8.1, 9.1, 9.2, 9.3

Description: Legacy utility that enables logging of events to file.

A.89 File Server

Development Name: F32_EKA2

Build File Location: `/cedar/generic/base/f32/group/`

System Model Location:

Layer: Base Services

Component collection: User Library and File Server

License Classification: Common Symbian

Exposes Third-Party APIs: YES

Present in OS Releases: 7.0, 7.0s, 8.0, 8.1, 9.1, 9.2, 9.3

Description: Generic file-system server and framework providing an extension interface that supports the implementation of custom file systems. File systems are implemented as loadable FSY plug-ins. All file-system access is managed through client sessions with the file server.

A.90 File Systems

Development Name: FILESYS

Build File Location: `/cedar/generic/base/f32/group/`

System Model Location:

Layer: Base Services

Component collection: User Library and File Server

License Classification: Common Replaceable

Exposes Third-Party APIs: NO

Present in OS Releases: 7.0, 7.0s, 8.0, 8.1, 9.1, 9.2, 9.3

Description: File system plug-in extensions implementing LFFS and FAT file systems.

A.91 Flash Translation Layer

Development Name: UNISTORE2_DRIVERS

Build File Location: `/cedar/generic/base/omap/`

System Model Location:

Layer: Kernel Services and Hardware Interface

Block: Kernel Architecture

Component collection: Variant

License Classification: Optional Replaceable

Exposes Third-Party APIs: NO

Present in OS Releases: 9.1, 9.2, 9.3

Description: File-system plug-in implementation of flash driver support.

A.92 Font and Bitmap Server

Development Name: FBSERV

Build File Location: /common/generic/graphics/fbserv/group/

System Model Location:

Layer: OS Services

Block: Multimedia and Graphics Services

Component collection: Graphics and Printing Services

License Classification: Common Symbian

Exposes Third-Party APIs: YES

Present in OS Releases: 7.0, 7.0s, 8.0, 8.1, 9.1, 9.2, 9.3

Description: Server that manages system-wide shared access to single-instance fonts and bitmaps, providing bitmap and font services for native bitmap fonts and vector fonts through its client-side APIs.

A.93 Font Store

Development Name: FNTSTORE

Build File Location: /common/generic/graphics/fntstore/group/

System Model Location:

Layer: OS Services

Block: Multimedia and Graphics Services

Component collection: Graphics and Printing Services

License Classification: Common Symbian

Exposes Third-Party APIs: YES

Present in OS Releases: 7.0, 7.0s, 8.0, 8.1, 9.1, 9.2, 9.3

Description: Provides font storage and font-file loading, including closest-fit matching of font requests. Supports the Open Font specification for vector fonts as well as proprietary Symbian OS bitmap fonts.

A.94 FreeType Font Rasterizer

Development Name: FREETYPE

Build File Location: /common/generic/graphics/freetype/group/

System Model Location:

Layer: OS Services

Block: Multimedia and Graphics Services

Component collection: Graphics and Printing Services

License Classification: Optional Replaceable

Exposes Third-Party APIs: YES

Present in OS Releases: 7.0, 7.0s, 8.0, 8.1, 9.1, 9.2, 9.3

Description: Reference implementation port of the FreeType TrueType font rasterizer, supporting FreeType 2 TrueType font descriptions and the Open Font interface.

A.95 FTP Engine

Development Name: FTP_E

Build File Location: `/common/generic/networking/ftp_e/group/`

System Model Location:

Layer: Application Services

Component collection: Internet and Web Application Support

License Classification: Optional Replaceable

Exposes Third-Party APIs: NO

Present in OS Releases: 7.0, 7.0s, 8.0, 8.1, 9.1, 9.2, 9.3

Description: For internal use only. Symbian OS File Transfer Protocol (FTP) daemon implementation used by networking components.

A.96 GDI

Development Name: GDI

Build File Location: `/common/generic/graphics/gdi/group/`

System Model Location:

Layer: OS Services

Block: Multimedia and Graphics Services

Component collection: Graphics Device Interface

License Classification: Common Symbian

Exposes Third-Party APIs: YES

Present in OS Releases: 7.0, 7.0s, 8.0, 8.1, 9.1, 9.2, 9.3

Description: Device-independent graphics context abstraction that supports drawing to various devices including screens and printers, which are treated as specialized graphics contexts.

A.97 GPRS/UMTS QoS PRT

Development Name: GUQOS

Build File Location: `/common/generic/networking/guqos/group/`

System Model Location:

Layer: OS Services

Block: Comms Services

Sub-block: Networking Services

Component collection: Networking Plug-ins

License Classification: Optional Replaceable

Exposes Third-Party APIs: NO

Present in OS Releases: 7.0, 7.0s, 8.0, 8.1, 9.1, 9.2, 9.3

Description: PRT socket protocol module used by the QoS Framework and network-interface components to implement 3GPP parameters.

A.98 Graphics Effects

Development Name: GFXTRANSEFFECT

Build File Location: `/common/app-framework/gfxtranseffect/group/`

System Model Location:

Layer: UI Framework

Component collection: UI Support

License Classification: Optional Replaceable

Exposes Third-Party APIs: NO

Present in OS Releases: 9.1, 9.2, 9.3

Description: Support for flicker-free window and window-contents animation and graphics-composition effects, for example to support animated-menu 'transition effects'.

A.99 Grid

Development Name: GRID

Build File Location: `/common/generic/app-framework/grid/group/`

System Model Location:

Layer: UI Framework

Component collection: UI Support

License Classification: Optional Replaceable

Exposes Third-Party APIs: YES

Present in OS Releases: 7.0, 7.0s, 8.0, 8.1, 9.1, 9.2, 9.3

Description: Layout engine for spreadsheet-style grid layout, presentation, print preview and printing. Now considered a legacy component.

A.100 GSM Utilities

Development Name: GSMU

Build File Location: `/common/generic/nbprotocols/smsstackv2/gsmu/`

System Model Location:

    Layer: OS Services

    Block: Comms Services

    Sub-block: Telephony Services

    Component collection: SMS Utilities

License Classification: Common Symbian

Exposes Third-Party APIs: YES

Present in OS Releases: 7.0, 7.0s, 8.0, 8.1, 9.1, 9.2, 9.3

Description: Utilities for processing GSM SMS messages, including encoding and decoding routines, used by SMS PRT and its clients.

A.101 HCI Framework

Development Name: HCI_V2_FRAMEWORK

Build File Location: `/common/generic/bluetooth/`

System Model Location:

    Layer: OS Services

    Block: Comms Services

    Sub-block: Short Link Services

    Component collection: Short Link

License Classification: Optional Symbian

Exposes Third-Party APIs: NO

Present in OS Releases: 8.0, 8.1, 9.1, 9.2, 9.3

Description: Bluetooth Host Controller Interface (HCI) implementation.

A.102 Help

Development Name: HLPMODEL

Build File Location: `/common/generic/app-services/hlpmodel/group/`

System Model Location:

    Layer: Application Services

    Component collection: Other Application Services

License Classification: Optional Symbian

Exposes Third-Party APIs: YES

Present in OS Releases: 7.0, 7.0s, 8.0, 8.1, 9.1, 9.2, 9.3

Description: Context-sensitive help engine providing a read-only interface to all help files on a Symbian OS device.

A.103 HTTP Filter Plug-ins

Development Name: HTTP

Build File Location: `/common/generic/application-protocols/http/group/`

System Model Location:

    Layer: Application Services

    Component collection: Other Application Services

License Classification: Optional Replaceable

Exposes Third-Party APIs: YES

Present in OS Releases: 7.0, 7.0s, 8.0, 8.1, 9.1, 9.2, 9.3

Description: Plug-ins to HTTP Transport Framework that are dynamically loaded to configure a transport session before use. Filters encapsulate responses to session events, for example, client authentication, message validation, and message redirection.

A.104 HTTP Protocol Plug-ins

Development Name: HTTP

Build File Location: `/common/generic/application-protocols/http/group/`

System Model Location:

    Layer: Application Services

    Component collection: Internet and Web Application Support

License Classification: Optional Replaceable

Exposes Third-Party APIs: YES

Present in OS Releases: 7.0, 7.0s, 8.0, 8.1, 9.1, 9.2, 9.3

Description: Plug-ins to HTTP Transport Framework that are dynamically loaded. Application and network-protocol handlers including TCP/IP, HTTP, and WSP.

A.105 HTTP Transport Framework

Development Name: HTTP

Build File Location: `/common/generic/application-`

`protocols/http/group/`

System Model Location:

Layer: Application Services

Component collection: Internet and Web Application Support

License Classification: Optional Replaceable

Exposes Third-Party APIs: YES

Present in OS Releases: 7.0, 7.0s, 8.0, 8.1, 9.1, 9.2, 9.3

Description: Framework that enables clients to establish a transport session for HTTP-like protocols providing core APIs for Transport Sessions, Transactions and Messages.

A.106 HTTP Utilities Library

Development Name: INETPROTUTIL

Build File Location: `/common/generic/application-protocols/inetprotutil/group/`

System Model Location:

Layer: Application Services

Component collection: Internet and Web Application Support

License Classification: Optional Replaceable

Exposes Third-Party APIs: YES

Present in OS Releases: 8.0, 8.1, 9.1, 9.2, 9.3

Description: Convenience component for storing utility classes commonly used by Internet-protocol parsing components.

A.107 Image Conversion Library

Development Name: ICL, ICL_IMAGEDISPLAY, IMAGETRANSFORM

Build File Location: `/common/generic/Multimedia/ICL/group/`

System Model Location:

Layer: OS Services

Block: Multimedia and Graphics Services

Component collection: Multimedia

License Classification: Common Symbian

Exposes Third-Party APIs: YES

Present in OS Releases: 7.0s, 8.0, 8.1, 9.1, 9.2, 9.3

Description: Extensible framework that integrates image conversion into the Multimedia Framework.

A.108 Image Conversion Library Plug-ins

Development Name: IMAGETRANSFORM, GIFSCALER

Build File Location: `/common/generic/Multimedia/ICL/ImageTransform/group/,/common/generic/Multimedia/ICL/GIFSCALER/group/bld.inf`

System Model Location:

    Layer: OS Services

    Block: Multimedia and Graphics Services

    Component collection: Multimedia

License Classification: Common Symbian

Exposes Third-Party APIs: YES

Present in OS Releases: 7.0s, 8.0, 8.1, 9.1, 9.2, 9.3

Description: Default reference codecs for common still image formats including Gif, Jpeg, Png, Bmp, Mbm, and others.

A.109 IMAP4 MTM

Development Name: IMAPSERVERMTM

Build File Location: `/common/generic/messaging/email/group/`

System Model Location:

    Layer: Application Services

    Component collection: Messaging Application Support

License Classification: Optional Symbian

Exposes Third-Party APIs: NO

Present in OS Releases: 7.0s, 8.0, 8.1, 9.1, 9.2, 9.3

Description: MTM plug-in to Messaging Store framework supporting sending, receiving and editing of IMAP4 (HTML mail) email messages.

A.110 Internet Sockets

Development Name: INSOCK

Build File Location: `/common/generic/networking/insock/group/`

System Model Location:

    Layer: OS Services

    Block: Comms Services

    Sub-block: Networking Services

    Component collection: ESock API Extensions

License Classification: Common Symbian

Exposes Third-Party APIs: YES

Present in OS Releases: 7.0s, 8.0, 8.1, 9.1, 9.2, 9.3

Description: Plug-in library to Socket Server that specializes generic socket-server address classes for TCP/IP v4 or v6 protocols to implement sockets over TCP/IP.

A.111 IP Event Notifier

Development Name: IPEVENTNOTIFIER

Build File Location: `/common/generic/networking/IPEventNotifier/group/`

System Model Location:

    Layer: OS Services

    Block: Comms Services

    Sub-block: Networking Services

    Component collection: Network Protocol Plug-ins

License Classification: Common Symbian

Exposes Third-Party APIs: NO

Present in OS Releases: 9.1, 9.2, 9.3

Description: Catches events occuring within the IP stack and publishes them to registered subscribers. Implemented as an IP Hook.

A.112 IP Hook

Development Name: INHOOK6

Build File Location: `/common/generic/networking/inhook6/group/`

System Model Location:

    Layer: OS Services

    Block: Comms Services

    Sub-block: Networking Services

    Component collection: Network Protocol Plug-ins

License Classification: Common Symbian

Exposes Third-Party APIs: YES

Present in OS Releases: 7.0, 7.0s, 8.0, 8.1, 9.1, 9.2, 9.3

Description: Implements a TCP/IP Hook interface to which modules bind to perform transformations on inbound and outbound IP packets.

A.113 IPSec

Development Name: IPSEC
Build File Location: `/common/generic/networking/ipsec/group/`
System Model Location:
  Layer: OS Services
  Block: Comms Services
  Sub-block: Networking Services
  Component collection: TCP/IP Security
License Classification: Optional Replaceable
Exposes Third-Party APIs: NO
Present in OS Releases: 8.0, 8.1, 9.1, 9.2, 9.3
Description: IPSec key negotiation for IPv4/v6 that enables policy-based secure networks, for example virtual private networks (VPNs).

A.114 IrDA CSY

Development Name: IRCOMM
Build File Location: `/common/generic/infra-red/irda/group/`
System Model Location:
  Layer: OS Services
  Block: Comms Services
  Sub-block: Short Link Services
  Component collection: Serial Comms Server Plug-ins
License Classification: Common Replaceable
Exposes Third-Party APIs: YES
Present in OS Releases: 7.0, 7.0s, 8.0, 8.1, 9.1, 9.2, 9.3
Description: CSY plug-in to C32 Serial Server providing serial port emulation over an IrDA link.

A.115 IrDA PRT

Development Name: IRDA
Build File Location: `/common/generic/infra-red/irda/`
System Model Location:
  Layer: OS Services
  Block: Comms Services
  Sub-block: Short Link Services

Component collection: Short Link Protocol Plug-ins

License Classification: Optional Symbian

Exposes Third-Party APIs: YES

Present in OS Releases: 7.0, 7.0s, 8.0, 8.1, 9.1, 9.2, 9.3

Description: IrDA protocol stack implemented as a PRT Socket Server protocol plug-in.

A.116 Java IO

Development Name: JAVA.IO

Build File Location: `/common/generic/ME/group/`

System Model Location:

    Layer: Java ME

    Component collection: CLDC 1.1

License Classification: Common Replaceable

Exposes Third-Party APIs: YES

Present in OS Releases: 8.0, 8.1, 9.1, 9.2, 9.3

Description: CLDC 1.1 Java I/O libraries that define the data-stream-based input and output APIs and APIs for reading and writing bytes and basic Java types.

A.117 Java Lang

Development Name: JAVA.LANG

Build File Location: `/common/generic/ME/group/`

System Model Location:

    Layer: Java ME

    Component collection: CLDC 1.1

License Classification: Common Replaceable

Exposes Third-Party APIs: YES

Present in OS Releases: 8.0, 8.1, 9.1, 9.2, 9.3

Description: CLDC 1.1 language libraries that define basic Java types and objects, including Byte, Integer, Object and Thread.

A.118 Java MIDlet Installer

Development Name: JAVAMIDLETINSTALLER

Build File Location: `/common/generic/security/`
`JavaMIDletInstaller/group/`

System Model Location:
   Layer: Application Services
   Component collection: Application Framework
License Classification: Optional Symbian
Exposes Third-Party APIs: YES
Present in OS Releases: 7.0s, 8.0, 8.1, 9.1, 9.2, 9.3
Description: Responsible for installation, removal and management of MIDP JAR files and MIDlets, including OTA support.

A.119 Java Utilities

Development Name: JAVA.UTIL
Build File Location: `/common/generic/ME/group/`
System Model Location:
   Layer: Java ME
   Component collection: CLDC 1.1
License Classification: Common Replaceable
Exposes Third-Party APIs: YES
Present in OS Releases: 8.0, 8.1, 9.1, 9.2, 9.3
Description: CLDC 1.1 utilities library that supplies basic utility classes, including Date and Time, and collection classes, including Hashtable, Stack and Vector.

A.120 JTWI 1.0

Development Name: J2ME9.2
Build File Location: `/common/generic/ME/group/`
System Model Location:
   Layer: Java ME
   Component collection: MIDP 2.0 Packages
License Classification: Optional Replaceable
Exposes Third-Party APIs: YES
Present in OS Releases: 8.0, 8.1, 9.1, 9.2, 9.3
Description: Wireless Messaging API for JTW 1.0 (JSR185).

A.121 Kernel Architecture 2

Development Name: E32_EKA2
Build File Location: `/cedar/generic/base/e32/`

System Model Location:

    Layer: Kernel Services and Hardware Interface

    Block: Kernel Architecture

    Component collection: Kernel Services

License Classification: Common Symbian

Exposes Third-Party APIs: YES

Present in OS Releases: 8.1b, 9.1, 9.2, 9.3

Description: Symbian OS EKA2 (real-time) kernel, delivered in Symbian OS v8.1b and in all releases from Symbian OS v9.

A.122 Key Store

Development Name: KEYSTORE

Build File Location: `/common/generic/security/certman/certstore/`

System Model Location:

    Layer: OS Services

    Block: Generic OS Services

    Component collection: Generic Libraries

License Classification: Optional Replaceable

YES

Present in OS Releases: 7.0, 7.0s, 8.0, 8.1, 9.1, 9.2, 9.3

Description: Provides a repository of private PKI keys and APIs for storing and retrieving keys and for managing the store itself.

A.123 LCDUI Plug-in

Development Name: LCDUIB

Build File Location: `/common/generic/ME/group/`

System Model Location:

    Layer: Java ME

    Component collection: Low-Level Plug-ins

License Classification: Common Replaceable

Exposes Third-Party APIs: YES

Present in OS Releases: 8.0, 8.1, 9.1, 9.2, 9.3

Description: Low-level graphics APIs with direct screen access, implemented as a plug-in that may be replaced with an alternative implementation.

A.124 Locale Support

Development Name: LOCE32, ELOCL, EKTRAN
Build File Location: `/common/generic/base/loce32/`
System Model Location:
    Layer: Kernel Services and Hardware Interface
    Component collection: Localization
License Classification: Optional Replaceable
Exposes Third-Party APIs: NO
Present in OS Releases: 7.0, 7.0s, 8.0, 8.1, 9.1, 9.2, 9.3
Description: Customizable plug-in that implements a library of locale-specific settings and standard strings including currency symbol and date format, used by both the Kernel and the User Library.

A.125 Lubbock Variant

Development Name: LUBBOCK_EKA2
Build File Location: `/cedar/generic/base/lubbock/`
System Model Location:
    Layer: Kernel Services and Hardware Interface
    Block: Kernel Architecture
    Component collection: Variant
License Classification: Optional Replaceable
Exposes Third-Party APIs: NO
Present in OS Releases: 8.0, 8.1, 9.1, 9.2, 9.3
Description: Variant code for the Intel Lubbock development board.

A.126 MBuf Manager

Development Name: MBUFMAN
Build File Location: `/common/generic/comms-infras/mbufmgr/group/`
System Model Location:
    Layer: OS Services
    Block: Comms Services
    Sub-block: Comms Framework
    Component collection: Comms Framework Utilities
License Classification: Common Symbian
Exposes Third-Party APIs: NO
Present in OS Releases: 7.0, 7.0s, 8.0, 8.1, 9.1, 9.2, 9.3

Description: MBuf implementation enabling efficient inter-thread communications within the communications process.

A.127 Media Drivers

Development Name: MEDUSII, MEDUSII_CRASHLOG, MEDUSIIS

Build File Location: `/cedar/generic/base/e32/drivers/unistore2/,/cedar/generic/base/integrator/logic/lmnand2/`

System Model Location:

Layer: Kernel Services and Hardware Interface

Block: Kernel Architecture

Component collection: Logical Device Drivers

License Classification: Common Symbian

Exposes Third-Party APIs: NO

Present in OS Releases: 9.1, 9.2, 9.3

Description: NAND-flash-media driver libraries, replacing the Flash Translation Layer implementation.

A.128 Message Store

Development Name: MSG, MSG_FRAMEWORK

Build File Location: `/common/generic/messaging/group/,/common/generic/messaging/framework/group/`

System Model Location:

Layer: Application Services

Component collection: Messaging Application Support

License Classification: Optional Symbian

Exposes Third-Party APIs: YES

Present in OS Releases: 7.0, 7.0s, 8.0, 8.1, 9.1, 9.2, 9.3

Description: The message server, store and framework. Supports standard message types (e.g. email, SMS) and defines the plug-in interface for Message Type Modules (MTMs) that implement message handling.

A.129 MIDI Driver

Development Name: MMF_DEVMIDI

Build File Location: `/common/generic/Multimedia/MMF/MIDI/group/`

System Model Location:

Layer: Kernel Services and Hardware Interface

Block: Kernel Architecture

Component collection: Logical Device Drivers

License Classification: Common Symbian

Exposes Third-Party APIs: YES

Present in OS Releases: 8.0, 8.1, 9.1, 9.2, 9.3

Description: API to support hardware-accelerated MIDI engines.

A.130 MIDP Device Control

Development Name: MIDP

Build File Location: `/common/generic/ME/group/`

System Model Location:

Layer: Java ME

Component collection: MIDP 2.0 Profile

License Classification: Common Replaceable

Exposes Third-Party APIs: YES

Present in OS Releases: 8.0, 8.1, 9.1, 9.2, 9.3

Description: Enabler for MIDP device-control APIs (JSR-118), for example, to control device vibration and backlight, and to enable platform requests (e.g. opening a browser page), triggered by device events.

A.131 MIDP File GCF

Development Name: GCF

Build File Location: `/common/generic/ME/group/`

System Model Location:

Layer: Java ME

Component collection: MIDP 2.0 Packages

License Classification: Optional Replaceable

Exposes Third-Party APIs: YES

Present in OS Releases: 8.0, 8.1, 9.1, 9.2, 9.3

Description: MIDP File Connection APIs (JSR-075) implemented through the GCF-communications framework.

A.132 MIDP GSM Security Recommended Policy

Development Name: MIDP2 SECURITY RP

Build File Location: `/common/generic/ME/midp2security/`

System Model Location:

Layer: Java ME

Component collection: MIDP 2.0 Profile

License Classification: Common Replaceable

Exposes Third-Party APIs: YES

Present in OS Releases: 8.0, 8.1, 9.1, 9.2, 9.3

Description: Enabler for MIDP 2.0 Security Recommended Policy enabling domain-based protection.

A.133 MIDP IO

Development Name: JAVAX.MICROEDITION.IO

Build File Location: /common/generic/ME/group/

System Model Location:

Layer: Java ME

Component collection: MIDP 2.0 Profile

License Classification: Common Replaceable

Exposes Third-Party APIs: YES

Present in OS Releases: 8.0, 8.1, 9.1, 9.2, 9.3

Description: MIDP high-level input/output APIs, including networking support and HTTP connections.

A.134 MIDP LCDUI

Development Name: JAVAX.MICROEDITION.LCDUI

Build File Location: /common/generic/ME/group/

System Model Location:

Layer: Java ME

Component collection: MIDP 2.0 Profile

License Classification: Common Replaceable

Exposes Third-Party APIs: YES

Present in OS Releases: 8.0, 8.1, 9.1, 9.2, 9.3

Description: MIDP-graphics APIs that use UIKON native controls to acquire platform-specific look and feel through the UI Application Framework LAF implementation, which is customized by the UI variant.

A.135 MIDP MIDlet

Development Name: MIDP2

Build File Location: /common/generic/ME/group/

System Model Location:

Layer: Java ME
Block: Java
Sub-block: ME
Component collection: MIDP 2.0 Profile
License Classification: Common Replaceable
Exposes Third-Party APIs: YES
Present in OS Releases: 7.0, 7.0s, 8.0, 8.1, 9.1, 9.2, 9.3
Description: MIDlet lifecycle implementation.

A.136 MIDP PIM

Development Name: MIDP
Build File Location: `/common/generic/ME/group/`
System Model Location:
Layer: Java ME
Component collection: MIDP 2.0 Packages
License Classification: Optional Replaceable
Exposes Third-Party APIs: YES
Present in OS Releases: 8.0, 8.1, 9.1, 9.2, 9.3
Description: MIDP Personal Information Management (PIM) APIs (JSR-075).

A.137 MIDP RMS

Development Name: JAVAX.MICROEDITION.RMS
Build File Location: `/common/generic/ME/group/`
System Model Location:
Layer: Java ME
Component collection: MIDP 2.0 Profile
License Classification: Common Replaceable
Exposes Third-Party APIs: YES
Present in OS Releases: 8.0, 8.1, 9.1, 9.2, 9.3
Description: MIDP persistence APIs implemented over native DBMS.

A.138 MIME Recognizer Framework

Development Name: EMIME
Build File Location: `/common/generic/app-framework/emime/group/`
System Model Location:

Layer: Application Services

Component collection: Content Handling

License Classification: Common Symbian

Exposes Third-Party APIs: YES

Present in OS Releases: 7.0, 7.0s, 8.0, 8.1, 9.1, 9.2, 9.3

Description: Recognizer framework for MIME data types.

A.139 MMF Recognizers

Development Name: RECMMF

Build File Location: `/common/generic/Multimedia/MMF/group/`

System Model Location:

Layer: Application Services

Component collection: Content Handling

License Classification: Common Symbian

Exposes Third-Party APIs: YES

Present in OS Releases: 7.0, 7.0s, 8.0, 8.1, 9.1, 9.2, 9.3

Description: Recognizer framework plug-ins to recognize specific multimedia data and document types.

A.140 MMS MTM

Development Name: MMS

Build File Location: `/common/generic/messaging/group/`

System Model Location:

Layer: Application Services

Component collection: Messaging Application Support

License Classification: Optional Symbian

Exposes Third-Party APIs: YES

Present in OS Releases: 7.0s, 8.0, 8.1, 9.1, 9.2, 9.3

Description: MTM plug-in to Messaging Store framework supporting sending, receiving and editing of MMS messages. From Symbian OS v9, legacy component and licensees provide their own implementations, if any.

A.141 MMS Settings

Development Name: MMSSETTINGS

Build File Location: `/common/generic/messaging/mmsettings/group/`

System Model Location:

    Layer: Application Services

    Component collection: Messaging Application Support

License Classification: Optional Symbian

Exposes Third-Party APIs: YES

Present in OS Releases: 8.0, 8.1, 9.1, 9.2, 9.3

Description: Encapsulation of MMS settings that are stored in the message store.

A.142 Mobile 3D 1.0

Development Name: M3GIO

Build File Location: `/common/generic/ME/group/`

System Model Location:

    Layer: Java ME

    Component collection: MIDP 2.0 Packages

License Classification: Optional Replaceable

Exposes Third-Party APIs: YES

Present in OS Releases: 8.0, 8.1, 9.1, 9.2, 9.3

Description: 3D-graphics APIs for scalable, small-footprint, devices (JSR-184).

A.143 Mobile Media API 1.1

Development Name: MMAPI11

Build File Location: `/common/generic/ME/group/`

System Model Location:

    Layer: Java ME

    Component collection: MIDP 2.0 Packages

License Classification: Optional Replaceable

Exposes Third-Party APIs: YES

Present in OS Releases: 8.0, 8.1, 9.1, 9.2, 9.3

Description: Mobile Media APIs (JSR-135).

A.144 m-Router

Development Name: MROUTERSECURE

Build File Location: `/common/generic/connectivity/BAL/`
`Plugins/mRouter3/`
`group/`

System Model Location:

Layer: OS Services

Block: Connectivity Services

Component collection: Device Connection

License Classification: Optional Replaceable

Exposes Third-Party APIs: NO

Present in OS Releases: 7.0, 7.0s, 8.0, 8.1, 9.1, 9.2, 9.3

Description: Plug-in to Bearer Abstraction Layer component providing m-Router connectivity link layer.

A.145 Multimedia Framework

Development Name: MMF, COMMON

Build File Location: `/common/generic/Multimedia/MMF/group/`

System Model Location:

Layer: Multimedia and Graphics Services

Component collection: Multimedia

License Classification: Common Symbian

Exposes Third-Party APIs: YES

Present in OS Releases: 8.0, 8.1, 9.1, 9.2, 9.3

Description: Lightweight multi-threaded plug-in framework for handling multimedia data that provides client APIs for audio playback, recording and conversion, tone playback, video playback and recording, MIDI playback, and speech recognition. Supports hardware acceleration via Media Device Framework.

A.146 Multimedia Framework Plug-ins

Development Name: MMFAUDIOCONTROLLER, MMFSTDSOURCE-ANDSINKPLUGIN, MMFLINEARAUDIOCODECS, GSM610, MMFAU-DIOOUTPUT, MMFAUDIOINPUT, MMFFORMATBASECLASSES, MMFWAVFORMAT, MMFRAWFORMAT, MMFAUFORMAT

Build File Location: `/common/generic/Multimedia/MMF/MMPfiles/plugin/`

System Model Location:

Layer: Multimedia and Graphics Services

Component collection: Multimedia

License Classification: Common Symbian

Exposes Third-Party APIs: YES

Present in OS Releases: 8.0, 8.1, 9.1, 9.2, 9.3
Description: Reference audio controller plug-ins.

A.147 MultiMode TSY

Development Name: MMTSY
Build File Location: /common/generic/telephony/mmtsy/group/
System Model Location:
    Layer: OS Services
    Block: Comms Services
    Sub-block: Telephony Services
    Component collection: Telephony Server Plug-ins
License Classification: Reference/Test
Exposes Third-Party APIs: NO
Present in OS Releases: 7.0, 7.0s, 8.0, 8.1, 9.1, 9.2, 9.3
Description: Reference ETel Telephony Server TSY plug-in implementing GSM and GPRS specific extensions. Replaced on an actual device by a hardware-specific licensee TSY.

A.148 Network Controller

Development Name: NETCON
Build File Location: /common/generic/networking/netcon/group/
System Model Location:
    Layer: OS Services
    Block: Comms Services
    Sub-block: Comms Framework
    Component collection: Data Comms Server
License Classification: Optional Replaceable
Exposes Third-Party APIs: NO
Present in OS Releases: 7.0, 7.0s, 8.0, 8.1, 9.1, 9.2, 9.3
Description: Used by Network Interface Manager to select an appropriate network interface agent to create an outgoing network interface.

A.149 Network Interface Manager

Development Name: NIFMAN
Build File Location: /common/generic/comms-infras/nifman/group/

System Model Location:

    Layer: OS Services

    Block: Comms Services

    Sub-block: Comms Framework

    Component collection: Data Comms Server

License Classification: Common Symbian

Exposes Third-Party APIs: YES

Present in OS Releases: 7.0, 7.0s, 8.0, 8.1, 9.1, 9.2, 9.3

Description: Framework for creating, loading and managing interface agent (AGT) and interface (NIF) plug-ins to create bearer-level network connections.

A.150 Null AGT

Development Name: NULLAGT

Build File Location: `/common/generic/networking/NULLAGT/group/`

System Model Location:

    Layer: OS Services

    Block: Comms Services

    Sub-block: Networking Services

    Component collection: Networking Plug-ins

License Classification: Optional Replaceable

Exposes Third-Party APIs: NO

Present in OS Releases: 7.0, 7.0s, 8.0, 8.1, 9.1, 9.2, 9.3

Description: Minimal connection AGT agent plug-in, enables connection to an Ethernet LAN.

A.151 OBEX Extension API

Development Name: OBEX_EXTENSIONAPIS

Build File Location: `/common/generic/obex/obexextensionapis/group/`

System Model Location:

    Layer: OS Services

    Block: Comms Services

    Sub-block: Short Link Services

    Component collection: OBEX

License Classification: Optional Symbian

Exposes Third-Party APIs: NO

Present in OS Releases: 9.1, 9.2, 9.3
Description: Packet extensions for the OBEX implementation.

A.152 OBEX MTMs

Development Name: MSG_OBEXMTM
Build File Location: `/common/generic/messaging/obex/group/`
System Model Location:
  Layer: Application Services
  Component collection: Messaging Application Support
License Classification: Optional Symbian
Exposes Third-Party APIs: YES
Present in OS Releases: 7.0s, 8.0, 8.1, 9.1, 9.2, 9.3
Description: MTM plug-ins to Messaging Store framework supporting OBEX messages over Bluetooth and infrared.

A.153 OBEX Protocol

Development Name: OBEX
Build File Location: `/common/generic/obex/group/`
System Model Location:
  Layer: OS Services
  Block: Comms Services
  Sub-block: Short Link Services
  Component collection: OBEX
License Classification: Optional Symbian
Exposes Third-Party APIs: YES
Present in OS Releases: 7.0, 7.0s, 8.0, 8.1, 9.1, 9.2, 9.3
Description: OBEX-session protocol implementation for IrDA, Bluetooth and USB transports, supporting connections from simple beaming all the way to fully fledged synchronization technologies such as SyncML.

A.154 OMA Data Sync

Development Name: SYNCMLDSCLIENT
Build File Location: `/common/generic/SyncML/group/`
System Model Location:
  Layer: Application Services
  Component collection: Data Sync Services
License Classification: Optional Replaceable

Exposes Third-Party APIs: NO

Present in OS Releases: 9.1, 9.2, 9.3

Description: SyncML client that manages the device side of a SyncML data-exchange session.

A.155 OMA SyncML DM Interface

Development Name: SYNCMLDMCLIENT

Build File Location: `/common/generic/SyncML/group/`

System Model Location:

Layer: Application Services

Component collection: Data Sync Services

License Classification: Optional Replaceable

Exposes Third-Party APIs: NO

Present in OS Releases: 9.1, 9.2, 9.3

Description: Supplies an interface to the SyncML Framework for use during a SyncML device-management session.

A.156 OMA SyncML Framework

Development Name: SYNCMLCLIENT

Build File Location: `common/generic/SyncML/group/`

System Model Location:

Layer: Application Services

Component collection: Data Sync Services

License Classification: Optional Replaceable

Exposes Third-Party APIs: YES

Present in OS Releases: 8.0, 8.1, 9.1, 9.2, 9.3

Description: Client–server-based framework that supports SyncML data synchronization and device management over HTTP, WSP and OBEX. Follows the SyncML v1.1.2 specification including large object support, Server Alerted Notification and transactional behavior. Clients may provide plug-ins to manage device-management settings.

A.157 OMAP 2420

Development Name: OMAP2420

Build File Location: `/cedar/generic/base/omap_hrp/assp/`

System Model Location:

Layer: Kernel Services and Hardware Interface

Block: Kernel Architecture

Component collection: ASSP

License Classification: Optional Replaceable

Exposes Third-Party APIs: NO

Present in OS Releases: 9.2, 9.3

Description: ASSP support for the Texas Instruments H4 development board with OMAP 2420 (ARMv6-based core). Hardware reference platform for Symbian OS releases from Symbian OS v9.2.

A.158 OMAP H2

Development Name: OMAP_H2

Build File Location: `/cedar/generic/base/omap/h2/`

System Model Location:

Layer: Kernel Services and Hardware Interface

Block: Kernel Architecture

Component collection: Variant

License Classification: Optional Replaceable

Exposes Third-Party APIs: NO

Present in OS Releases: 9.1, 9.2, 9.3

Description: Variant code for the Texas Instruments H2 development board.

A.159 OMAP H4

Development Name: OMAPH4HRP

Build File Location: `/cedar/generic/base/omap_hrp/h4/`

System Model Location:

Layer: Kernel Services and Hardware Interface

Block: Kernel Architecture

Component collection: Variant

License Classification: Optional Replaceable

Exposes Third-Party APIs: NO

Present in OS Releases: 9.1, 9.2, 9.3

Description: Variant code for the Texas Instruments H4 development board.

A.160 OpenGL ES

Development Name: OPENGLES9.X

Build File Location: `/common/generic/graphics/OpenGLES/group/`

System Model Location:

Layer: OS Services

Block: Multimedia and Graphics Services

Component collection: OpenGL ES

License Classification: Reference/Test

Exposes Third-Party APIs: NO

Present in OS Releases: 8.0, 8.1, 9.1, 9.2, 9.3

Description: Reference implementation of OpenGL ES, replaced by licensees. Provides multi-client access to screen, keyboard, and pointer or digitizer for GUI applications.

A.161 OpenGL ES Display Properties

Development Name: OPENGLESDISPLAYPROPERTY

Build File Location: `/common/generic/graphics/ OpenGLESDisplayProperty/group/`

System Model Location:

Layer: OS Services

Block: Multimedia and Graphics Services

Component collection: OpenGL ES

License Classification: Optional Symbian

Exposes Third-Party APIs: NO

Present in OS Releases: 8.0, 8.1, 9.1, 9.2, 9.3

Description: Encapsulates display-drawing properties (e.g. display rectangles and clipping regions), enabling window-surface access, i.e. drawing, to clients from non-window-owning threads.

A.162 OpenGL ES Headers

Development Name: OPENGLSHEADERS

Build File Location: `/common/generic/graphics/ OpenGLESHeaders/group/`

System Model Location:

Layer: OS Services

Block: Multimedia and Graphics Services

Component collection: OpenGL ES

License Classification: Optional Symbian

Exposes Third-Party APIs: YES

Present in OS Releases: 8.0, 8.1, 9.1, 9.2, 9.3

Description: Standard OpenGL ES headers and binary definition files to encourage compatibility between OpenGL ES implementations for Symbian OS.

A.163 Other LDDs

System Model Location:
   Layer: Kernel Services and Hardware Interface
   Block: Kernel Architecture
   Component collection: Logical Device Drivers

License Classification: Common Replaceable

Exposes Third-Party APIs: NO

Present in OS Releases: 7.0, 7.0s, 8.0, 8.1, 9.1, 9.2, 9.3

Description: Drivers supporting hardware devices, implemented as Symbian OS logical device drivers (LDDs).

A.164 Peripheral Bus Controllers

Development Name: EPBUS

Build File Location: `/cedar/generic/base/e32/`

System Model Location:
   Layer: Kernel Services and Hardware Interface
   Block: Kernel Architecture
   Component collection: Variant

License Classification: Common Symbian

Exposes Third-Party APIs: NO

Present in OS Releases: 7.0, 7.0s, 8.0, 8.1, 9.1, 9.2, 9.3

Description: Peripheral-bus controllers for supported variants implemented as a kernel-side DLL interfacing media and I/O device drivers to PC-card or MMC-card-socket hardware.

A.165 Phonebook Sync

Development Name: PHBKSYNC

Build File Location: `/generic/telephony/phbksync/group/`

System Model Location:
   Layer: OS Services
   Block: Comms Services

Sub-block: Telephony Services

Component collection: Telephony Utilities

License Classification: Optional Replaceable

Exposes Third-Party APIs: YES

Present in OS Releases: 7.0, 7.0s, 8.0, 8.1, 9.1, 9.2, 9.3

Description: Server enabling synchronization of contacts between a phonebook application and entries stored in the Integrated Circuit Card (ICC), or SIM card, of a device.

A.166 PLP Variant

Development Name: PLPVARIANT, PLP, BRDCST

Build File Location: `/common/generic/connectivity/legacy/plp/PLPVARIANT/`

System Model Location:

Layer: OS Services

Block: Connectivity Services

Component collection: Service Providers

License Classification: Optional Replaceable

Exposes Third-Party APIs: N/A

Present in OS Releases: 7.0, 7.0s, 8.0, 8.1, 9.1, 9.2, 9.3

Description: Deprecated legacy component previously used as the bearer for connectivity services. Retained only for compatibility with third-party components that use some of its APIs.

A.167 Plug-in Framework (ECOM)

Development Name: ECOM

Build File Location: `/common/generic/syslibs/ecom/`

System Model Location:

Layer: Base Services

Component collection: Low-Level Libraries and Frameworks

License Classification: Optional Symbian

Exposes Third-Party APIs: YES

Present in OS Releases: 7.0s, 8.0, 8.1, 9.1, 9.2, 9.3

Description: Framework and server for plug-in interface implementations. Defines the base classes used by conforming plug-ins and a client-side API used by framework clients to locate and load plug-ins on demand, in conformance with security-policy mechanisms.

A.168 POP3 MTM

Development Name: MSG_EMAIL

Build File Location: `/common/generic/messaging/email/popservermtm/group/`

System Model Location:

   Layer: Application Services

   Component collection: Messaging Application Support

License Classification: Optional Symbian

Exposes Third-Party APIs: YES

Present in OS Releases: 8.0, 8.1, 9.1, 9.2, 9.3

Description: MTM plug-in to Messaging Store framework supporting send/receive/edit of POP3 (dial-up) email messages.

A.169 Power and Shutdown Management

Development Name: PWRCLI

Build File Location: `/common/generic/syslibs/pwrcli/group/`

System Model Location:

   Layer: Base Services

   Component collection: Low-Level Libraries and Frameworks

License Classification: Common Symbian

Exposes Third-Party APIs: NO

Present in OS Releases: 8.0, 8.1, 9.1, 9.2, 9.3

Description: Customizable user-side power manager supporting policy-driven power management via power-domain 'profiles' at device switch-on and switch-off.

A.170 PPP Compression Plug-ins

Development Name: PREDCOMP, STACCOMP, MSCOMP

Build File Location: `/common/generic/networking/predcomp/group/`, `/common/generic/networking/staccomp/group/`, `/common/generic/networking/mscomp/group/`

System Model Location:

   Layer: OS Services

   Block: Comms Services

   Sub-block: Networking Services

   Component collection: Link Layer Control

License Classification: Optional Replaceable

Exposes Third-Party APIs: NO

Present in OS Releases: 8.0, 8.1, 9.1, 9.2, 9.3

Description: PPP NIF (Point to Point Protocol) plug-ins that implement Predictor, Stac and Microsoft compression algorithms.

A.171 PPP NIF

Development Name: PPP

Build File Location: /common/generic/networking/ppp/group/

System Model Location:

    Layer: OS Services

    Block: Comms Services

    Sub-block: Networking Services

    Component collection: Link Layer Control

License Classification: Optional Replaceable

Exposes Third-Party APIs: NO

Present in OS Releases: 7.0, 7.0s, 8.0, 8.1, 9.1, 9.2, 9.3

Description: Point-to-Point Protocol plug-in to Network Interface Manager. Supports TCP/IP over serial communications.

A.172 Printer Drivers

Development Name: PRINTDRV

Build File Location: /common/generic/graphics/printdrv/group/

System Model Location:

    Layer: OS Services

    Block: Multimedia and Graphics Services

    Component collection: Graphics and Printing Services

License Classification: Optional Replaceable

Exposes Third-Party APIs: NO

Present in OS Releases: 7.0, 7.0s, 8.0, 8.1, 9.1, 9.2, 9.3

Description: Reference implementation for concrete printer driver plug-ins. Considered legacy for most mobile phones.

A.173 Printing Services

Development Name: PRINT

Build File Location: /common/generic/app-framework/print/group/

System Model Location:

Layer: Application Services

Component collection: Printing Support

License Classification: Optional Replaceable

Exposes Third-Party APIs: YES

Present in OS Releases: 7.0, 7.0s, 8.0, 8.1, 9.1, 9.2, 9.3

Description: Framework providing standard print dialogs for print-job setup and control to application clients. Considered legacy for most mobile phones.

A.174 Printing Support

Development Name: PDRSTORE

Build File Location: `/common/generic/graphics/pdrstore/group/`

System Model Location:

Layer: OS Services

Block: Multimedia and Graphics Services

Component collection: Graphics and Printing Services

License Classification: Optional Replaceable

Exposes Third-Party APIs: YES

Present in OS Releases: 7.0, 7.0s, 8.0, 8.1, 9.1, 9.2, 9.3

Description: Framework that manages and loads printer drivers as bitmapped-device-context implementations and manages access to printer ports. Considered legacy for most mobile phones.

A.175 PSD AGT

Development Name: PSDAGT

Build File Location: `/common/generic/networking/psdagt/group/`

System Model Location:

Layer: OS Services

Block: Comms Services

Sub-block: Networking Services

Component collection: Networking Plug-ins

License Classification: Optional Replaceable

Exposes Third-Party APIs: N/A

Present in OS Releases: 8.0, 8.1, 9.1, 9.2, 9.3

Description: Deprecated functionality replaced by other components. AGT agent plug-in to Connection Agent framework that negotiates packet-switched connection, for example to GPRS networks.

A.176 QoS Framework PRT

Development Name: QOS

Build File Location: `/common/generic/networking/common/generic/networking/qoslib/group/`

System Model Location:

    Layer: OS Services

    Block: Comms Services

    Sub-block: Networking Services

    Component collection: Network Protocol Plug-ins

License Classification: Optional Replaceable

Exposes Third-Party APIs: NO

Present in OS Releases: 7.0s, 8.0, 8.1, 9.1, 9.2, 9.3

Description: QoS Framework and library modules, implemented as a PRT protocol plug-in to Socket Server.

A.177 Raw IP NIF

Development Name: RAWIPNIF

Build File Location: `/common/generic/networking/rawipnif/group/`

System Model Location:

    Layer: OS Services

    Block: Comms Services

    Sub-block: Networking Services

    Component collection: Link Layer Control

License Classification: Optional Replaceable

Exposes Third-Party APIs: NO

Present in OS Releases: 8.0, 8.1, 9.1, 9.2, 9.3

Description: NIF plug-in to Network Interface Manager that supports multiple primary-PDP contexts, i.e. multi-homing over GPRS, on the telephony-reference platform.

A.178 Reference DRM Agent

Development Name: DRMAGENT

Build File Location: `/common/generic/syslibs/caf2/group/`

System Model Location:

   Layer: Application Services

   Component collection: Content Handling

License Classification: Reference/Test

Exposes Third-Party APIs: NO

Present in OS Releases: 9.1, 9.2, 9.3

Description: Reference implementation of DRM-agent plug-in to Content Access Framework for DRM.

A.179 Reference Fonts

Development Name: FONTS

Build File Location: `/common/generic/graphics/fonts/group/`

System Model Location:

   Layer: OS Services

   Block: Multimedia and Graphics Services

Exposes Third-Party APIs: NO

Present in OS Releases: 8.0, 8.1, 9.1, 9.2, 9.3

Description: Reference native fonts for Symbian OS. Replaced by licensees.

A.180 Remote Control Framework

Development Name: BLUETOOTH_REMOTECONTROL

Build File Location: `/common/generic/bluetooth/`

System Model Location:

   Layer: OS Services

   Block: Comms Services

   Sub-block: Short Link Services

   Component collection: Short Link

License Classification: Optional Symbian

Exposes Third-Party APIs: YES

Present in OS Releases: 9.1, 9.2, 9.3

Description: Bearer-agnostic remote-control framework that enables sending and receiving of remote-control commands to and from remote Bluetooth devices using bearers provided as plug-ins to the framework.

A.181 Remote File Server

Development Name: SCREMOTEFILESERVER

Build File Location: `/common/generic/connectivity/`
`SCRemoteFileServer/group/`

System Model Location:

Layer: OS Services

Block: Connectivity Services

Component collection: Service Providers

License Classification: Optional Replaceable

Exposes Third-Party APIs: YES

Present in OS Releases: 8.0, 8.1, 9.1, 9.2, 9.3

Description: Named service providing on-device file-management functions to a remote (off-device) client over TCP/IP including access to backup and restore functions provided by other system components.

A.182 RTP

Development Name: RTP

Build File Location: `/common/generic/mm-protocols/rtp/`
`group/`

System Model Location:

Layer: Application Services

Component collection: Multimedia Protocols

License Classification: Optional Replaceable

Exposes Third-Party APIs: YES

Present in OS Releases: 9.1, 9.2, 9.3

Description: Server and user-side API providing socket-based access to Real-Time Transport Protocol (RTP) services, providing an IP-based real-time-network transport service.

A.183 Runtime Plug-in

Development Name: MIDP2RUNTIME

Build File Location: `/common/generic/ME/group/`

System Model Location:

Layer: Java ME

Component collection: Low-Level Plug-ins

License Classification: Common Replaceable

Exposes Third-Party APIs: YES
Present in OS Releases: 8.0, 8.1, 9.1, 9.2, 9.3
Description: Licensee-customizable MIDP 2.0 runtime plug-in module.

A.184 Scheduled Send MTM

Development Name: MSG_SCHEDULEDSEND
Build File Location: `/common/generic/messaging/schedulesend/group/`
System Model Location:
    Layer: Application Services
    Component collection: Messaging Application Support
License Classification: Optional Symbian
Exposes Third-Party APIs: YES
Present in OS Releases: 7.0, 7.0s, 8.0, 8.1, 9.1, 9.2, 9.3
Description: MTM plug-in to Messaging Store framework that supports scheduled sending of any available message type including SMS and fax and defines the scheduling parameters.

A.185 Screen Driver

Development Name: SCREENDRIVER
Build File Location: `/common/generic/graphics/screendriver/group/`
System Model Location:
    Layer: Kernel Services and Hardware Interface
    Component collection: Screen Driver
License Classification: Common Replaceable
Exposes Third-Party APIs: YES
Present in OS Releases: 7.0, 7.0s, 8.0, 8.1, 9.1, 9.2, 9.3
Description: Device-dependent component that implements the generic operations defined by the Bit GDI to manipulate the physical memory map of the device display or bitmap memory map. Note that this is not implemented as a standard Symbian OS device driver.

A.186 SD Card Driver

Development Name: SDCARD4C
Build File Location: `/cedar/generic/base/e32/drivers/`

System Model Location:

    Layer: Kernel Services and Hardware Interface

    Block: Kernel Architecture

    Component collection: Logical Device Drivers

    Component collection: Logical Device Drivers

License Classification: Common Replaceable

Exposes Third-Party APIs: NO

Present in OS Releases: 9.1, 9.2, 9.3

Description: Logical and physical device drivers supporting Secure Digital flash-memory cards.

A.187 Secondary PDP context UMTS Driver

Development Name: SPUD

Build File Location: /common/generic/networking/Spud/group/

System Model Location:

    Layer: OS Services

    Block: Comms Services

    Sub-block: Networking Services

    Component collection: Networking Plug-ins

License Classification: Optional Replaceable

Exposes Third-Party APIs: NO

Present in OS Releases: 9.1, 9.2, 9.3

Description: Network-interface-adapter component that supports primary and secondary PDP contexts (multi-homing over GPRS) on the telephony-reference platform only.

A.188 Secure Backup Engine

Development Name: SECUREBACKUPENGINE

Build File Location: /common/generic/connectivity/ SecureBackupEngine/group/

System Model Location:

    Layer: OS Services

    Block: Connectivity Services

    Component collection: Service Providers

License Classification: Optional Replaceable

Exposes Third-Party APIs: YES

Present in OS Releases: 8.0, 8.1, 9.1, 9.2, 9.3

Description: Manages backup and restore of device-side data, as controlled by the Secure Backup Socket Server.

A.189 Secure Backup Socket Server

Development Name: SBSSERVER

Build File Location: `/common/generic/connectivity/SBSocketServer/group/`

System Model Location:

  Layer: OS Services

  Block: Connectivity Services

  Component collection: Service Providers

License Classification: Optional Replaceable

Exposes Third-Party APIs: NO

Present in OS Releases: 8.0, 8.1, 9.1, 9.2, 9.3

Description: Named service providing backup–restore functions to a remote (off-device) client over TCP/IP. Communicates with a connected PC and with the Secure Backup Engine to carry out backup and restore operations to a PC.

A.190 Secure Software Install

Development Name: SECURESOFTWAREINSTALL

Build File Location: `/common/generic/security/swi/group/`

System Model Location:

  Layer: Application Services

  Component collection: Application Framework

License Classification: Optional Symbian

Exposes Third-Party APIs: NO

Present in OS Releases: 9.1, 9.2, 9.3

Description: Installers for native and Java apps.

A.191 Security Policy Reference Plug-in

Development Name: MIDP2SECURITY

Build File Location: `/common/generic/ME/group/`

System Model Location:

  Layer: Java ME

  Component collection: MIDP 2.0 Profile

License Classification: Optional Symbian

Exposes Third-Party APIs: YES

Present in OS Releases: 8.0, 8.1, 9.1, 9.2, 9.3

Description: Reference implementation of Java security policy, implemented as a replaceable plug-in.

A.192 Send As

Development Name: SENDASV2

Build File Location: `/common/generic/messaging/sendas2/group/`

System Model Location:

License Classification: Optional Symbian

Exposes Third-Party APIs: YES

Present in OS Releases: 9.1, 9.2, 9.3

Description: Not shown explicitly in the model and forming part of the messaging framework, from Symbian OS v9, a client–server architecture enabling message sending from within applications.

A.193 Serial Port CSY

Development Name: ECUART

Build File Location: `/common/generic/ser-comms/c32/group/`

System Model Location:

    Layer: OS Services

    Block: Comms Services

    Sub-block: Short Link Services

License Classification: Common Replaceable

Exposes Third-Party APIs: YES

Present in OS Releases: 7.0, 7.0s, 8.0, 8.1, 9.1, 9.2, 9.3

Description: CSY plug-in to C32 Serial Server providing virtual serial port.

A.194 Server Socket

Development Name: SERVERSOCKET

Build File Location: `/common/generic/connectivity/ServerSocket/group/`

System Model Location:

    Layer: OS Services

    Block: Connectivity Services

    Component collection: Service Providers

License Classification: Optional Replaceable

Exposes Third-Party APIs: YES

Present in OS Releases: 8.0, 8.1, 9.1, 9.2, 9.3

Description: Helper library that communicates service-port numbers and manages messages and commands, for use by the Service Broker.

A.195 Service Broker

Development Name: SERVICEBROKER

Build File Location: `/common/generic/connectivity/ServiceBroker/group/`

System Model Location:

   Layer: OS Services

   Block: Connectivity Services

   Component collection: Service Framework

License Classification: Optional Symbian

Exposes Third-Party APIs: YES

Present in OS Releases: 8.0, 8.1, 9.1, 9.2, 9.3

Description: Configuration-file-based service and port registration that enables device-side services to register a port number for use by PC-side clients.

A.196 Sheet Engine

Development Name: SHENG

Build File Location: `/common/generic/app-engines/sheng/group/`

System Model Location:

   Layer: Application Services

   Component collection: Office Application Engines

License Classification: Optional Symbian

Exposes Third-Party APIs: YES

Present in OS Releases: 7.0, 7.0s, 8.0, 8.1, 9.1, 9.2, 9.3

Description: Legacy application engine supporting a spreadsheet application, originating from the early EPOC built-in applications set.

A.197 SIM TSY

Development Name: SIMTSY

Build File Location: `/common/generic/telephony/simtsy/group/`

System Model Location:

   Layer: OS Services

   Block: Comms Services

   Sub-block: Telephony Services

   Component collection: Telephony Server Plug-ins

License Classification: Reference/Test

Exposes Third-Party APIs: NO

Present in OS Releases: 8.0, 8.1, 9.1, 9.2, 9.3

Description:TSY simulator module that uses static configuration data and dynamic system-agent notifications to simulate the presence of phone hardware. Implemented as a TSY plug-in to the ETel Telephony Server.

A.198 SIP Connection Provider Plug-ins

Development Name: SIPCPRT, SIPDUMMYPRT, SIPSTATEMAC, SIPPA-RAMS, SIPSCPRT

Build File Location: `/common/generic/mm-protocols/connprov/sipcpr/group/`

System Model Location:

   Layer: Application Services

   Component collection: Multimedia Protocols

License Classification: Optional Replaceable

Exposes Third-Party APIs: YES

Present in OS Releases: 9.2, 9.3

Description: Network-layer connection provision for Session Initiation Protocol (SIP).

A.199 SIP Framework

Development Name: SIP_COM

Build File Location: `/common/generic/mm-protocols/sip/group/`

System Model Location:

   Layer: Application Services

   Component collection: Multimedia Protocols

License Classification: Optional Replaceable

Exposes Third-Party APIs: YES

Present in OS Releases: 9.1, 9.2, 9.3

Description: Framework providing Session Initiation Protocol (SIP) support and integration into the networking infrastructure, but not the protocol implementation (which is provided as a plug-in by licensees).

A.200 SLIP NIF

Development Name: SLIP

Build File Location: `/common/generic/networking/slip/group/`

System Model Location:

   Layer: OS Services

Block: Comms Services
Sub-block: Networking Services
Component collection: Link Layer Control

License Classification: Reference/Test

Exposes Third-Party APIs: NO

Present in OS Releases: 7.0, 7.0s, 8.0, 8.1, 9.1, 9.2, 9.3

Description: Reference implementation of Serial Line Internet Protocol (SLIP) NIF plug-in to Network Interface Manager providing TCP/IP over serial communications via modem dialup.

A.201 SMIL Parser

Development Name: GMXML

Build File Location: `/common/generic/messaging/gmxml/group/`

System Model Location:

Layer: Application Services
Component collection: Content Handling

License Classification: Optional Replaceable

Exposes Third-Party APIs: YES

Present in OS Releases: 8.0, 8.1, 9.1, 9.2, 9.3

Description: Parser for SMIL (and other simple XML) content based on a generic XML Parser/Composer with 'mini-DOM' API. Replaces the SMIL Translator implementation of Symbian OS v7.0s.

A.202 SMS MTM

Development Name: MSG_SMS8.1

Build File Location: `/common/generic/messaging/sms/`

System Model Location:

Layer: Application Services
Component collection: Messaging Application Support

License Classification: Optional Symbian

Exposes Third-Party APIs: YES

Present in OS Releases: 7.0, 7.0s, 8.0, 8.1, 9.1, 9.2, 9.3

Description: MTM plug-in to Messaging Store framework that implements SMS-messaging support.

A.203 SMS PRT

Development Name: SMSSTACK

Build File Location: `/common/generic/nbprotocols/smsstackv2/smsprot/group/`

System Model Location:

Layer: OS Services
Block: Comms Services
Sub-block: Telephony Services
Component collection: SMS Protocol Plug-ins

License Classification: Common Symbian

Exposes Third-Party APIs: YES

Present in OS Releases: 7.0, 7.0s, 8.0, 8.1, 9.1, 9.2, 9.3

Description: Socket Server PRT plug-in that implements SMS protocol.

A.204 SMS Utilities

Development Name: SMSU

Build File Location: `/common/generic/nbprotocols/smsstack/smsu/group/`

System Model Location:

Layer: OS Services
Block: Comms Services
Sub-block: Telephony Services
Component collection: SMS Utilities

License Classification: Common Symbian

Exposes Third-Party APIs: YES

Present in OS Releases: 7.0, 7.0s, 8.0, 8.1, 9.1, 9.2, 9.3

Description: Utilities for processing SMS messages, includes streaming classes, logging support and interface to the backup server, used by SMS PRT and its clients.

A.205 SMTP MTM

Development Name: SMTPSERVERMTP

Build File Location: `/common/generic/messaging/email/SMTPSERVERMTm/group/`

System Model Location:

Layer: Application Services
Component collection: Messaging Application Support

License Classification: Optional Symbian

Exposes Third-Party APIs: YES

Present in OS Releases: 7.0s, 8.0, 8.1, 9.1, 9.2, 9.3

Description: MTM plug-in to Messaging Store framework supporting sending, receiving and editing of SMTP (Internet) mail.

A.206 Software Install Server

Development Name: SWINSTALLSERVER

Build File Location: `/common/generic/connectivity/SWInstallServer/group/`

System Model Location:

Layer: OS Services

Block: Connectivity Services

Component collection: Service Providers

License Classification: Optional Replaceable

Exposes Third-Party APIs: YES

Present in OS Releases: 8.0, 8.1, 9.1, 9.2, 9.3

Description: Named service that interacts with the software-installation components on the device to enable remote installation of SIS, JAR and JAD files over TCP/IP or OBEX.

A.207 Speech Driver

Development Name: DEVASR

Build File Location: `/common/generic/Multimedia/MMF/DEVASR/`

`group/`

System Model Location:

Layer: Kernel Services and Hardware Interface

Block: Kernel Architecture

Component collection: Logical Device Drivers

License Classification: Common Symbian

Exposes Third-Party APIs: YES

Present in OS Releases: 8.0, 8.1, 9.1, 9.2, 9.3

Description: Hardware-acceleration API for Automatic Speech Recognition that allows the computationally-intensive speech recognition algorithms to be performed in hardware, where present.

A.208 Subconnection Parameters

Development Name: UMTSIF

Build File Location: `/common/generic/networking/`

System Model Location:

Layer: OS Services

Block: Comms Services

Sub-block: Networking Services

Component collection: Subconnection Interface

License Classification: Common Replaceable

Exposes Third-Party APIs: NO

Present in OS Releases: 7.0, 7.0s, 8.0, 8.1, 9.1, 9.2, 9.3

Description: QoS parameters and Traffic Flow Templates for GPRS.

A.209 Store

Development Name: STORE

Build File Location: `/common/generic/syslibs/store/group/`

System Model Location:

Layer: Base Services

Component collection: Low-Level Libraries and Frameworks

License Classification: Common Replaceable

Exposes Third-Party APIs: YES

Present in OS Releases: 7.0, 7.0s, 8.0, 8.1, 9.1, 9.2, 9.3

Description: Defines the Symbian OS persistence model based on the streams and stores abstractions, providing an application data-storage model that shields applications from the underlying file-server implementation.

A.210 Sync Initiation

Development Name: SYNCMLINITSERVER

Build File Location: `/common/generic/connectivity/SyncMLInitServer/group/`

System Model Location:

Layer: Application Services

Component collection: Data Sync Services

License Classification: Optional Replaceable

Exposes Third-Party APIs: YES

Present in OS Releases: 8.0, 8.1, 9.1, 9.2, 9.3

Description: Allows a synchronization SyncML operation to be initiated from the PC. Note that, although Symbian does not supply a PC-side SyncML server, this service allows the creation of such a service by partners.

A.211 System Agent

Development Name: SYSAGENT2

Build File Location: `/common/generic/syslibs/sysagent2/group/`

System Model Location:

    Layer: OS Services

    Block: Generic OS Services

    Component collection: Generic Services

License Classification: Optional Replaceable

Exposes Third-Party APIs: NO

Present in OS Releases: 7.0, 7.0s, 8.0, 8.1, 9.1, 9.2, 9.3

Description: Legacy component whose functionality is largely replaced by Publish and Subscribe.

A.212 System Starter

Development Name: SYSSTART

Build File Location: `/common/generic/app-framework/SysStart/group/`

System Model Location:

    Layer: Application Services

    Component collection: Application Launch Services

License Classification: Optional Replaceable

Exposes Third-Party APIs: NO

Present in OS Releases: 9.1, 9.2, 9.3

Description: Server and framework enabling policy-based startup of system servers at boot time.

A.213 Task Scheduler

Development Name: SCHSVR_ONGOING

Build File Location: `/common/generic/syslibs/schsvr/`

System Model Location:

    Layer: OS Services

    Block: Generic OS Services

    Component collection: Generic Services

License Classification: Optional Replaceable

Exposes Third-Party APIs: YES

Present in OS Releases: 7.0, 7.0s, 8.0, 8.1, 9.1, 9.2, 9.3

Description: Task or executable launching service for time-based and condition-based task triggers.

A.214 TCP/IPv4/v6 PRT

Development Name: TCPIP6

Build File Location: `/common/generic/networking/tcpip6/group/`

System Model Location:

   Layer: OS Services

   Block: Comms Services

   Sub-block: Networking Services

   Component collection: Network Protocol Plug-ins

License Classification: Common Symbian

Exposes Third-Party APIs: NO

Present in OS Releases: 7.0, 7.0s, 8.0, 8.1, 9.1, 9.2, 9.3

Description: IPv4/v6 protocol implementation and TCP/IP stack and plug-in extension architecture, implemented as a PRT protocol plug-in to ESock Socket Server.

A.215 Telephony Watchers

Development Name: TELEPHONY_WATCHERS

Build File Location: `/common/generic/telephony/Watchers/group/`

System Model Location:

   Layer: OS Services

   Block: Comms Services

   Sub-block: Telephony Services

   Component collection: Telephony Utilities

License Classification: Optional Replaceable

Exposes Third-Party APIs: NO

Present in OS Releases: 8.0, 8.1, 9.1, 9.2, 9.3

Description: Watcher Framework plug-ins that monitor telephony conditions and report them as Publish and Subscribe properties.

A.216 Telnet Engine

Development Name: TELNET_E

Build File Location: `/common/generic/networking/telnet_e/group/`

System Model Location:

    Layer: Application Services

    Component collection: Internet and Web Application Support

License Classification: Optional Replaceable

Exposes Third-Party APIs: YES

Present in OS Releases: 7.0, 7.0s, 8.0, 8.1, 9.1, 9.2, 9.3

Description: Symbian OS Telnet daemon implementation that supports client sessions for communicating with a specified host.

A.217 Text Formatting (FORM)

Development Name: FORM

Build File Location: `/common/generic/app-framework/form/group/`

System Model Location:

    Layer: Application Services

    Component collection: Text Rendering

License Classification: Optional Symbian

Exposes Third-Party APIs: YES

Present in OS Releases: 7.0, 7.0s, 8.0, 8.1, 9.1, 9.2, 9.3

Description: Text view and layout classes that support the separation of display attributes (layout and drawing) from logical text attributes (styles).

A.218 Text Handling (ETEXT)

Development Name: ETEXT

Build File Location: `/common/generic/app-framework/etext/group/`

System Model Location:

    Layer: Application Services

    Component collection: Text Rendering

License Classification: Optional Symbian

Exposes Third-Party APIs: YES

Present in OS Releases: 7.0, 7.0s, 8.0, 8.1, 9.1, 9.2, 9.3

Description: Text-content framework that supports storing of editable text and its logical attributes, for example paragraph alignment and character fonts.

A.219 Text Shaper Plug-in

Development Name: ICULAYOUTENGINE
Build File Location: `/common/generic/graphics/iculayoutengine/group/`
System Model Location:
  Layer: OS Services
  Block: Multimedia and Graphics Services
  Component collection: Graphics and Printing Services
License Classification: Optional Replaceable
Exposes Third-Party APIs: NO
Present in OS Releases: 9.1, 9.2, 9.3
Description: Plug-in to Font and Bitmap Server that supports rendering text in Devanagari script.

A.220 Text Shell

Development Name: ESHELL
Build File Location: `/cedar/generic/base/f32/etshell/group/`
System Model Location:
  Layer: Base Services
  Component collection: Text Mode Shell
License Classification: Test/Reference
Exposes Third-Party APIs: YES
Present in OS Releases: 7.0, 7.0s, 8.0, 8.1, 9.1, 9.2, 9.3
Description: Text shell providing a command-line interface to the base system.

A.221 Text Window Server

Development Name: EWSRV
Build File Location: `/cedar/generic/base/e32/ewsrv/`
System Model Location:
  Layer: Base Services
  Component collection: Text Mode Shell
License Classification: Optional Replaceable
Exposes Third-Party APIs: YES
Present in OS Releases: 7.0, 7.0s, 8.0, 8.1, 9.1, 9.2, 9.3

Description: Text-mode window server supporting the Text Shell. Uses a text-mode display driver providing VT100 terminal emulation over a serial line and VGA/LCD implementations for reference hardware.

A.222 Timezone

Development Name: TZ, TZLOCALIZATIONRSCFACTORY
Build File Location: `/common/generic/app-services/tz/group/`
System Model Location:

 Layer: Application Services

 Component collection: Other Application Services

License Classification: Optional Replaceable
Exposes Third-Party APIs: YES
Present in OS Releases: 9.1, 9.2, 9.3
Description: Localization support including time-zone database for Standard, Daylight, Short Standard and Short Daylight names for time zones. Associated components not explicitly shown in the model include a time-zone compiler, database and localization tools.

A.223 TLS

Development Name: TLS
Build File Location: `/common/generic/networking/tls/group/`
System Model Location:

 Layer: OS Services

 Block: Comms Services

 Sub-block: Networking Services

 Component collection: TCP/IP Security

License Classification: Common Symbian
Exposes Third-Party APIs: YES
Present in OS Releases: 7.0, 7.0s, 8.0, 8.1, 9.1, 9.2, 9.3
Description: Implements Secure Sockets Layer (SSL 3.0) and Transport Level Security (TLS 1.0) protocols, enabling secure network connections.

A.224 TRP CSY

Development Name: TRP
Build File Location: `/common/generic/telephony/trp/csy/csy27010/group/`
System Model Location:

 Layer: OS Services

Block: Comms Services
Sub-block: Telephony Services
Component collection: Telephony Reference Platform
License Classification: Common Replaceable
Exposes Third-Party APIs: YES
Present in OS Releases: 7.0, 7.0s, 8.0, 8.1, 9.1, 9.2, 9.3
Description: CSY implementation for the Telephony Reference Platform that manages the internal channel between the telephony and application hardware as a standard serial port.

A.225 TRP TSY

Development Name: TRP
Build File Location: `/common/generic/telephony/trp/tsy/group/`
System Model Location:
   Layer: OS Services
   Block: Comms Services
   Sub-block: Telephony Services
   Component collection: Telephony Reference Platform
License Classification: Common Replaceable
Exposes Third-Party APIs: NO
Present in OS Releases: 7.0, 7.0s, 8.0, 8.1, 9.1, 9.2, 9.3
Description: TSY implementation for the Telephony Reference Platform.

A.226 Tunnel NIF

Development Name: TUNNELNIF
Build File Location: `/common/generic/networking/tunnelnif/group/`
System Model Location:
   Layer: OS Services
   Block: Comms Services
   Sub-block: Networking Services
   Component collection: Link Layer Control
License Classification: Optional Replaceable
Exposes Third-Party APIs: NO
Present in OS Releases: 8.0, 8.1, 9.1, 9.2, 9.3
Description: Ethernet protocol NIF plug-in to Network Interface Manager supporting IPSec tunneling capability. Forms part of the VPN client and is used when running IPSec in tunnel mode.

A.227 UI Graphics Utilities

Development Name: EGUL, NUMBERCONVERSION

Build File Location: `/common/generic/app-framework/egul/group/`

System Model Location:

Layer: UI Framework

Component collection: UI Support

License Classification: Common Symbian

Exposes Third-Party APIs: YES

Present in OS Releases: 7.0, 7.0s, 8.0, 8.1, 9.1, 9.2, 9.3

Description: Libraries used by UI-framework components as well as the variant UI and applications, providing color, font, icon, text, drawing and number-conversion utilities.

A.228 UI Look and Feel

Development Name: UIKLAFGT

Build File Location: `/common/generic/app-framework/uiklafGT/group/`

System Model Location:

Layer: UI Framework

Component collection: UI Application Framework

License Classification: Common Symbian

Exposes Third-Party APIs: NO

Present in OS Releases: 8.0, 8.1, 9.1, 9.2, 9.3

Description: Reference-implementation plug-in to Uikon framework that determines the look and feel of Control Environment controls by defining standard methods for which UI customizers provide a custom implementation.

A.229 Uikon

Development Name: UIKON

Build File Location: `/common/generic/app-framework/uikon/group/`

System Model Location:

Layer: UI Framework

Component collection: UI Application Framework

License Classification: Common Symbian

Exposes Third-Party APIs: YES

Present in OS Releases: 7.0, 7.0s, 8.0, 8.1, 9.1, 9.2, 9.3

Description: Concrete framework for UI and application creation, providing the foundation for licensee UI customization.

A.230 Uikon Error Resolver Plug-in

Development Name: ERRORRESGT

Build File Location: `/common/generic/app-framework/errorresgt/group/`

System Model Location:

Layer: UI Framework

Component collection: UI Application Framework

License Classification: Common Symbian

Exposes Third-Party APIs: NO

Present in OS Releases: 7.0, 7.0s, 8.0, 8.1, 9.1, 9.2, 9.3

Description: Resource file that maps system error numbers to helpful error-text strings, extended and customized by variant UIs.

A.231 USB CSY

Development Name: ECACM

Build File Location: `/generic/ser-comms/usb/CSY/group/`

System Model Location:

Layer: OS Services

Block: Comms Services

Sub-block: Short Link Services

Component collection: Serial Comms Server Plug-ins

License Classification: Common Replaceable

Exposes Third-Party APIs: NO

Present in OS Releases: 8.0, 8.1, 9.1, 9.2, 9.3

Description: CSY plug-in to C32 Serial Server providing serial-port emulation over USB.

A.232 USB Driver

Development Name: USBC

Build File Location: `/cedar/generic/base/e32/`

System Model Location:

Layer: Kernel Services and Hardware Interface

Block: Kernel Architecture

Component collection: Logical Device Drivers

License Classification: Common Replaceable

Exposes Third-Party APIs: YES

Present in OS Releases: 7.0, 7.0s, 8.0, 8.1, 9.1, 9.2, 9.3

Description: Logical Device Driver for USB supporting dynamically configurable USB 2.0 Full Speed device functionality.

A.233 USB Manager

Development Name: USB

Build File Location: /common/generic/ser-comms/usb/usbman/group/

System Model Location:

Layer: OS Services

Block: Comms Services

Sub-block: Comms Framework

Component collection: Data Comms Server

License Classification: Optional Symbian

Exposes Third-Party APIs: NO

Present in OS Releases: 7.0, 7.0s, 8.0, 8.1, 9.1, 9.2, 9.3

Description: USB class support that enables a Symbian OS device to both use and serve as a USB host.

A.234 User HAL

Development Name: HAL_EKA2

Build File Location: /cedar/generic/base/hal/

System Model Location:

Layer: Base Services

Block: User-Side Hardware Abstraction

Sub-block: User-Side Hardware Abstraction

Component collection: User-Side Hardware Abstraction

License Classification: Common Symbian

Exposes Third-Party APIs: NO

Present in OS Releases: 7.0, 7.0s, 8.0, 8.1, 9.1, 9.2, 9.3

Description: User-side access to hardware via hardware abstraction, deprecated for application use.

A.235 User Library

Development Name: EUSER

Build File Location: `/cedar/generic/base/e32/group/`

System Model Location:

Layer: Base Services

Component collection: User Library and File Server

License Classification: Common Symbian

Exposes Third-Party APIs: YES

Present in OS Releases: 7.0, 7.0s, 8.0, 8.1, 9.1, 9.2, 9.3

Description: Symbian OS user library, used by all user-side code to provide fundamental basic services.

A.236 vCal Plug-in

Development Name: AGNVERSIT

Build File Location: `/common/generic/app-engines/agnversit/group/`

System Model Location:

Layer: Application Services

Component collection: PIM Application Services

License Classification: Optional Replaceable

Exposes Third-Party APIs: Yes

Present in OS Releases: 8.0, 8.1, 9.1, 9.2, 9.3

Description: Limited public APIs, for example in `vcal.h`. Plug-in library used by Agenda Model to interact with the vCard and vCal component.

A.237 vCard and vCal

Development Name: VERSIT

Build File Location: `/common/generic/app-services/versit/group/`

System Model Location:

Layer: Application Services

Component collection: PIM Application Support

License Classification: Optional Replaceable

Exposes Third-Party APIs: YES

Present in OS Releases: 8.0, 8.1, 9.1, 9.2, 9.3

Description: Parsers that convert between vCard and vCalendar entries and Symbian OS native formats.

A.238 Video Driver

Development Name: DEVVIDEO

Build File Location: `/common/generic/Multimedia/MMF/DevVideo/group/`

System Model Location:

Layer: Kernel Services and Hardware Interface

Block: Kernel Architecture

Component collection: Logical Device Drivers

License Classification: Common Symbian

Exposes Third-Party APIs: YES

Present in OS Releases: 8.0, 8.1, 9.1, 9.2, 9.3

Description: Hardware-abstraction layer for video decoding and encoding acceleration.

A.239 View Server

Development Name: VIEWSRV

Build File Location: `/common/generic/app-framework/viewsrv/group/`

System Model Location:

Layer: Application Services

Component collection: Application Framework

License Classification: Common Symbian

Exposes Third-Party APIs: YES

Present in OS Releases: 7.0, 7.0s, 8.0, 8.1, 9.1, 9.2, 9.3

Description: Provides a mechanism for sharing and switching views between applications. A running application can switch into and use a view belonging to another application.

A.240 VPN

Development Name: VPNAPI, VPNCONNAGT, VPNMANAGER

Build File Location: `/common/generic/networking/ipsec/vpnapi/group/`, `/common/generic/ipsec/vpnconnagt/group/`, `/common/generic/ipsec/vpnmanager/group/`

System Model Location:

Layer: OS Services

Block: Comms Services

Sub-block: Networking Services

Component collection: TCP/IP Security

License Classification: Optional Replaceable

Exposes Third-Party APIs: YES

Present in OS Releases: 8.0, 8.1, 9.1, 9.2, 9.3

Description: Key negotiation and tunneling using IPSec for VPN connections, enabling users to connect to VPNs.

A.241 WAP Message API

Development Name: WAPMESSAGE

Build File Location: `/common/generic/wap-stack/wapmessage/group/`

System Model Location:

Layer: OS Services

Block: Comms Services

Sub-block: Networking Services

Component collection: WAP Stack

License Classification: Optional Replaceable

Exposes Third-Party APIs: YES

Present in OS Releases: 7.0, 7.0s, 8.0, 8.1, 9.1, 9.2, 9.3

Description: APIs for WAP Push, connectionless WSP and WDP datagrams.

A.242 WAP Push Framework

Development Name: WAPPUSH

Build File Location: `/common/generic/wap-browser/wappush/group/`

System Model Location:

Layer: Application Services

Component collection: Internet and Web Application Support

License Classification: Optional Replaceable

Exposes Third-Party APIs: NO

Present in OS Releases: 7.0, 7.0s, 8.0, 8.1, 9.1, 9.2, 9.3

Description: Provides an interface between the WAP stack and the messaging infrastructure to support WAP as a messaging transport. WAP Push embeds links to WAP addresses within SMS messages.

A.243 WAP Push Handlers

Development Name: WAPPUSHSUPPORT

Build File Location: `/common/generic/wap-browser/WapPushSupport/group/`

System Model Location:

   Layer: Application Services

   Component collection: Content Handling

License Classification: Optional Replaceable

Exposes Third-Party APIs: NO

Present in OS Releases: 8.0, 8.1, 9.1, 9.2, 9.3

Description: Plug-in handlers, including Several Interfaces Single Logic (SISL), that use the WAP message API.

A.244 WAP Push MTM

Development Name: WAP-BROWSER

Build File Location: `/common/generic/wap-browser/wappush/pushmtm/group/`

System Model Location:

   Layer: Application Services

   Component collection: Messaging Application Support

License Classification: Optional Replaceable

Exposes Third-Party APIs: YES

Present in OS Releases: 7.0s, 8.0, 8.1, 9.1, 9.2, 9.3

Description: MTM plug-in to Messaging Store framework supporting WAP messaging.

A.245 WAP Short Stack

Development Name: WAPSTACK

Build File Location: `/common/generic/wap-stack/wapmessage/group/`

System Model Location:

   Layer: OS Services

   Block: Comms Services

   Sub-block: Networking Services

   Component collection: WAP Stack

License Classification: Optional Replaceable

Exposes Third-Party APIs: NO

Present in OS Releases: 8.0, 8.1, 9.1, 9.2, 9.3

Description: Shortened WAP stack supporting WDP, WAP Push and Connectionless WSP over IP or SMS, implemented as an ESock Socket Server plug-in. May be replaced by licensee implementation.

A.246 WBXML Parser

Development Name: WBXMLPARSER

Build File Location: `/common/generic/syslibs/xml/group/`

System Model Location:

Layer: Base Services

Component collection: Low-Level Libraries and Frameworks

License Classification: Optional Replaceable

Exposes Third-Party APIs: YES

Present in OS Releases: 8.0, 8.1, 9.1, 9.2, 9.3

Description: Plug-in to XML Parser Framework that parses WAP Binary XML.

A.247 Web Recognizers

Development Name: RECOGNISERS

Build File Location: `/common/generic/application-protocols/recognisers/group/`

System Model Location:

Layer: Application Services

Component collection: Content Handling

License Classification: Optional Replaceable

Exposes Third-Party APIs: YES

Present in OS Releases: 7.0, 7.0s, 8.0, 8.1, 9.1, 9.2, 9.3

Description: Web URL and bookmark recognizers implemented as plug-ins to the MIME Recognizer Framework.

A.248 Window Server

Development Name: WSERV8.1

Build File Location: `/common/generic/graphics/wserv/group/`

System Model Location:

Layer: OS Services

Block: Multimedia and Graphics Services

Component collection: Windowing Framework

License Classification: Common Symbian

Exposes Third-Party APIs: YES

Present in OS Releases: 7.0, 7.0s, 8.0, 8.1, 9.1, 9.2, 9.3

Description: Server that owns and manages access to the screen as a drawable resource, making it available to applications through the abstraction of windowed screen areas, and access to the keyboard and pointer or digitizer for GUI applications. Includes the keyclick reference plug-in that produces key or pointer clicks.

A.249 Wireless LAN

Development Name: WIFI_802_11

Build File Location: `/common/generic/networking/802.11/group/`

System Model Location:

Layer: OS Services

Block: Comms Services

Sub-block: Networking Services

Component collection: Link Layer Control

License Classification: Optional Replaceable

Exposes Third-Party APIs: NO

Present in OS Releases: 9.3

Description: Support for wireless LAN based on the IEEE 802.11 specifications.

A.250 WMA 1.1

Development Name: WMA

Build File Location: `/common/generic/ME/group/`

System Model Location:

Layer: Java ME

Component collection: MIDP 2.0 Packages

License Classification: Optional Replaceable

Exposes Third-Party APIs: YES

Present in OS Releases: 8.0, 8.1, 9.1, 9.2, 9.3

Description: MIDP Wireless Messaging 1.1 (JSR-120) APIs supporting SMS and MMS, including Push support.

A.251 WMA 1.1 Push Plug-in

Development Name: WMA

Build File Location: `/common/generic/ME/group/`

System Model Location:

Layer: Java ME

Component collection: Bluetooth and SMS Push

License Classification: Optional Replaceable

Exposes Third-Party APIs: YES

Present in OS Releases: 8.0, 8.1, 9.1, 9.2, 9.3

Description: Plug-in that binds the WMA 1.1 package to the underlying system.

A.252 Word Engine

Development Name: WPENG

Build File Location: `/common/generic/app-engines/wpeng/group/`

System Model Location:

Layer: Application Services

Component collection: Office Application Engines

License Classification: Optional Symbian

Exposes Third-Party APIs: YES

Present in OS Releases: 7.0, 7.0s, 8.0, 8.1, 9.1, 9.2, 9.3

Description: Legacy application engine supporting a word-processor application, originating from the early EPOC built-in applications suite.

A.253 World Server

Development Name: WORLDSERVER

Build File Location: `/common/generic/app-services/worldserver/group/`

System Model Location:

Layer: Application Services

Component collection: Other Application Services

License Classification: Optional Replaceable

Exposes Third-Party APIs: NO

Present in OS Releases: 7.0, 7.0s, 8.0, 8.1, 9.1, 9.2, 9.3

Description: Legacy server providing application access to country and city information, including country, capital city, international and city dialing codes, latitude, longitude, and UTC offset.

A.254 XML Framework

Development Name: XML
Build File Location: `/common/generic/syslibs/xml/group/`
System Model Location:
 Layer: Base Services
 Component collection: Low-Level Libraries and Frameworks
License Classification: Optional Replaceable
Exposes Third-Party APIs: YES
Present in OS Releases: 8.0, 8.1, 9.1, 9.2, 9.3
Description: Extensible framework for XML parsing based on a SAX-2.0-like parser model and supporting DTD and processing plug-ins (e.g. validators and auto correctors) as well as parser plug-ins.

A.255 XML Parser

Development Name: XMLPARSERPLUGIN
Build File Location: `/common/generic/syslibs/xml/group/`
System Model Location:
 Layer: Base Services
 Component collection: Low-Level Libraries and Frameworks
License Classification: Optional Replaceable
Exposes Third-Party APIs: YES
Present in OS Releases: 8.0, 8.1, 9.1, 9.2, 9.3
Description: Non-validating parser plug-in for XML 1.0.

A.256 Zip Compression Library

Development Name: EZLIB
Build File Location: `/common/generic/syslibs/ezlib/group/`
System Model Location:
 Layer: Base Services
 Component collection: Low-Level Libraries and Frameworks
License Classification: Optional Symbian
Exposes Third-Party APIs: YES
Present in OS Releases: 7.0, 7.0s, 8.0, 8.1, 9.1, 9.2, 9.3
Description: Symbian OS port of the zlib compression library, implementing ZLIB, DEFLATE and GZIP for ZIP compression and decompression.

Appendix B: Interviewee Biographies

Geert Bollen

Before joining Symbian, Geert was involved in a startup in the then emerging Electronic Document Management field, where he integrated databases with optical storage technology and made them sit up and perform tricks

At Symbian, Geert initially led the design and implementation of persistent data services in Symbian OS and went on to create its first Java implementation – possibly the first for mobile phones. Following that, he held a wide range of engineering management roles. He is currently VP System Management with overall technical responsibility for Symbian OS.

Martin Budden

Before joining Symbian (then Psion), Martin worked in a variety of computer-related jobs. His first, in 1980, was as a 'Student Scientist' at the Royal Signals and Radar Establishment (now QinetiQ) where he worked on MASCOT (a parallel-processing system) and was exposed to email and the ARPANET. Martin holds an MA in Mathematics from the University of Cambridge.

Martin joined Symbian (then Psion) in 1990 and started working on the MC400 laptop. He worked on a variety of Psion products, including the Series 3 and Series 5 palmtops and then was Technical Lead on an early Philips project – the first mobile phone project to earn Symbian money. He spent a year in Symbian's Swedish offices in Ronneby helping architect the 'Quartz UI' (which later became UIQ). He is currently Symbian's Chief System Architect.

Outside work, Martin is a keen cyclist and has cycled extensively in Europe. In 2002, he spent 26 days cycling 2600 miles across America.

Andy Cloke

Andrew joined Symbian (then Psion) in 1995 from Philips Research Labs where he had been working on hardware and software for WCDMA prototype phones. He worked initially as an engineer and then as an engineering manager for the fledgling Communications group.

From the foundation of Symbian, he worked on the communications subsystems for the first Symbian OS phones, the Nokia 9210 and Ericsson R380. In 2002, he became the company's first Chief Technology Architect and now has ownership of the Technology Strategy.

Charles Davies

Charles became CTO of Symbian in March 2003, after a long career at Psion as a software technologist and as a director since 1982. Charles has been a technology pioneer in software for handheld computers and a lead contributor to many of the architectural concepts underpinning Symbian OS.

As CTO, he provides technology leadership at the executive level, contributes heavily to Symbian's business, product and technology strategies and is the executive sponsor of Symbian's Technology Committee.

Charles holds a first-class degree in Physics from Imperial College, London, and a PhD.

Morgan Henry

Before joining Symbian (then Psion) in 1995, Morgan dabbled in graphic design and wrote software for fun. He stayed in the software arm of the company through its transition from Psion into Symbian. He worked on the Psion Series 5 and was responsible for the kernel port for the Nokia 9210 ('the world's first open Symbian OS phone'). During his time at Symbian, he has also worked as a Technical Lead and System Architect working on many projects during the platform's evolution.

Morgan holds a BSc in Mathematics and Computer Science from Queen Mary and Westfield College, London, and maintains an active interest in drawing, painting and animation.

Ian Hutton

Ian joined Symbian (or Psion, as it then was) in 1995, straight from university. He initially wrote test code for the Text Handling (EText) component of what was still EPOC32. As one of Charles Davies's gang, he played a lead role in designing and implementing the Application Architecture and Printing Services. He later spent two years in Ronneby, Sweden, as Technical Lead seconded to UIQ working on UIQ releases based on Symbian OS v7.0, moving into the Techinical Consulting organization on his return. He now works with Charles Davies in Product Marketing, as an OS Product Planner, helping to shape Symbian's technology roadmap and planning future releases of Symbian OS.

Peter Jackson

Peter joined Symbian (then Psion) in 1994, wishing to apply his mainframe expertise, gained working with the VAX/VMS operating system, to smaller devices including handhelds. He was an early fan of the Psion operating systems created for the Organiser family of devices, and later of the SIBO operating system, the 16-bit precursor of what became Symbian OS.
He designed and implemented the first versions of the Symbian OS localization and internationalization components, led the team that created the FreeType implementation for Symbian OS, and later became responsible for Symbian's source configuration management framework.
He now works with Ravenbrook as a consultant specializing in configuration management.

Keith de Mendonca

Keith was awarded a first-class honors degree in Computer Systems Engineering followed by a DPhil by Sussex University. He joined Psion in 1994 and wrote SDKs and applications for the Psion PDA range.
On joining Symbian, Keith originally worked in the messaging team, initially specializing in email protocols before becoming the Messaging Technology Architect. More recently, Keith was the Chief Technology Architect for Symbian's Application Technology Development group in the UK. He is currently the Chief Technology Architect for Symbian India and is based in Bangalore.

Will Palmer

Will studied electronic engineering at Oxford Polytechnic, before training as a C++ programmer. Prior to joining Symbian, he worked on various leading-edge projects such as vehicle-tracking and automated precision measurement.

Will is a System Architect in Symbian. He joined the company in June 2000 and has worked as an engineer on synchronization technologies, PC- and device-side networking and device management. He is currently trusted with the development and integrity of Symbian's security architecture.

At the moment, Will is traveling an inspiring road with his young family but can quite often be found on a football pitch.

Howard Price

Howard grew up in South Africa. After conscription into the army as a cook for 10 months, he studied for two years towards a six-year degree in architecture before finding that it wasn't what he wanted to do with his life. After six years traveling and working around Europe, he decided to settle down, did a degree in physics and mathematics (including some courses on programming) and, aged 28, joined Symbian (then Psion) as a programmer.

At Psion, Howard developed software for all Psion's PDAs from the Organiser I to the Series 5, including developing the low-level arithmetic operators and the mathematical functions, developing the Series 3/3a OPL runtime, leading the OVAL (Visual Basic for SIBO) run-time and debugger team, leading the Series5 OPL team and designing the OPX framework, and leading the initial Java team. He was working as Engineering Manager of the Applications team at the time Psion Software became Symbian.

At Symbian, after a few more years in management, Howard decided to follow his interests and move back to technical work, joining the System Management Group (SMG) as a senior systems architect. There he designed and developed the Depmodel v1 suite of tools to analyze the structure of Symbian OS automatically. This enabled him to write the System Architecture Overview Documentation for which an understanding of dependencies was essential. The dependency model helped to define the initial system model. More recently, he has joined SMG's Technology Strategy and Analysis team, analyzed the impact of SMP on Symbian OS, researched the technical implications of Moore's Law on Symbian OS and contributed to Symbian's Technology Strategy document.

Murray Read

Murray graduated in Artificial Intelligence and Computer Science at the University of Edinburgh. He worked with NCR and Fortronic on embedded software for cash machines and credit-card terminals. He then joined Origin, where he worked on radio pagers and the Philips smartphone project with Psion. After that, he worked with STNC on the Psion web-browser project.

Murray started working with Psion/Symbian as a contractor on a web project in 1998. Then he worked on the user interface library for the Nokia 7650, which evolved into the S60 platform, focusing on application architecture and the user interface layout system.

Murray is a Chartered IT Professional, a member of the BCS and a regular competitor in Topcoder programming competitions.

Martin Tasker

Martin joined Psion in 1995 after 13 years in the system-software industry. His first commercial software products were a graphics package and debugger for the BBC Micro in its early 1980s heyday, produced while studying Natural Sciences and Computer Science at Cambridge University. On graduation, he joined IBM where he worked in networking and storage management for eight years, programming mainframes in assembler, working on product development, and delivering performance, routing and management improvements in IBM's global VNET network. He learned C++, object orientation and artificial intelligence during a two-year transport research project at Imperial College, London.

Martin then joined the Protea project at Psion, whose architectural decisions form much of the subject of this book. He became responsible for Protea's documentation and SDKs, along with contributions to its architecture. His output included many technical papers on the distinguishing features of Symbian OS, which form the heart of this book. In 2000, his *Professional Symbian Programming* became the contemporary guide to Symbian OS programming which helped Symbian and its customers to grow their engineering teams to achieve the success we see in the marketplace today. *Professional Symbian Programming* also contained the definitive history of the Protea project and, until this book, the most detailed insight publicly available into the design decisions on the project. Martin then served as Product Manager with responsibility for licensing, SDKs and tools. He now works in a technology-strategy role.

Martin is married with four children. He occasionally relaxes with classical music.

Andrew Thoelke

Andrew joined Symbian (then Psion) in 1994 shortly after graduating from Sidney Sussex college, Cambridge with an MA in Mathematics. Within Symbian he became one of the key developers of OVAL, a rapid application development language similar to Visual Basic, for the Psion Series 3a computers. He has since worked as developer, designer and architect on projects throughout the lifetime of Symbian OS, and spanning many of its technology areas such as kernel, data storage, messaging, Java and platform security.

Today, he has one of the most senior technical roles within Symbian, influencing both the technical strategy of the organization and the ongoing architectural development of Symbian OS.

David Wood

David spent eight years at Cambridge University, studying mathematics and then philosophy of science. He drifted into teaching and became head of the mathematics department at Ashbourne College in Kensington, where he specialized in helping A-level-retake students move from, for example, a D-grade pass to an A-grade pass in four months. Towards the end of that period, he taught himself C in his spare time, on an Amstrad PCW8512 word processor running CP/M, and was lucky enough to pick up a job as a junior software engineer in Psion's Harcourt Street offices.

At Psion/Symbian he has headed, at various times, the Development, Technical Consulting, Partnering and Research departments. During the formative stages of Symbian OS, he was the primary integrator of application-level and UI framework code into the ROMs of what was called 'Protea', the Psion Series 5 PDA. This experience is described in more detail in David's 2005 book *Symbian for software leaders*.

References

Aho, A., Sethi, R. and Ullman, J. (1986) *Compilers: Principles, techniques and tools*. Addison-Wesley

Alexander, C. (1979) *The Timeless Way of Building*. Oxford University Press

Alexandrescu, A. (2001) *Modern C++ Design*. Addison-Wesley

Allin, M., Turfus, C., *et al.* (2001) *Wireless Java for Symbian Devices*. John Wiley & Sons

Ambler, S. (2004) *The Object Primer: Agile model-driven development with UML 2*, Third Edition. Cambridge University Press

Appel, A. (1992) *Compiling with Continuations*. Cambridge University Press

Appel, A. (1998) *Modern Compiler Implementation in C*. Cambridge University Press

Assmann, U. (2003) *Invasive Software Composition*. Springer-Verlag

Bar-David, T. (1993) *Object Oriented Design for C++*. Prentice Hall

Beaudouin-Lafon, M. (1994) *Object-Oriented Languages*. Chapman & Hall

Beck, K. (1999) *Guide to Better Smalltalk*. Cambridge University Press

Bishop, J. (1986) *Data Abstraction in Programming Languages*. Addison-Wesley

Bordwell, D., Staiger, J. and Thompson, K. (1985) *The Classical Hollywood Cinema*. Columbia University Press.

Briand, L., Devanbu, P. and Melo, W. (1997) 'An Investigation into Coupling Measures for C++' in *Proceedings of the 19th International Conference on Software Engineering*, 412–21

Brooks, F. (1976) *The Mythical Man-Month*. Addison-Wesley

Buschmann, F., Meunier, R., *et al.* (1998) *Pattern-Oriented Software Architecture*. John Wiley & Sons

Christensen, C. (1997) *The Innovator's Dilemma.* Collins

Craig, I. (2000) *The Interpretation of Object-Oriented Languages.* Springer-Verlag

Davila, A., Epstein, M. and Shelton, R. (2006) *Making Innovation Work.* Wharton

Edwards, L. (2004) *Developing Series 60 Applications.* Addison-Wesley

Funk, J. (2004) *Mobile Phone Disruption.* Hoboken, NJ: John Wiley & Sons

Furber, S. (2000) *ARM System-on-Chip Architecture.* Addison-Wesley

Gabriel, R. (1996) *Patterns of Software.* Oxford University Press

Goldberg, A. and Robson, D. (1989) *Smalltalk-80: The language.* Addison-Wesley

Haikio, M. (2002) *Nokia: The inside story.* Pearson Education

Hansen, P. B. (2001) *Classic Operating Systems* (editor). Springer

Harrison, R. (2003) *Symbian OS C++ for Mobile Phones.* John Wiley & Sons

Harrison, R. (2004) *Symbian OS C++ for Mobile Phones*, Volume 2. Symbian Press

Heath, C. (2006) *Symbian OS Platform Security.* Chichester: John Wiley & Sons

Henderson-Sellers, B. (1996) *Object-Oriented Metrics.* Prentice Hall

Hildebrand, J. D. (1994) *Object Magazine* 3:6 (February 1994). NY USA: SIGS Publications

Himanen, P., Torvalds, L. and Castells, M. (2002) *The Hacker Ethic.* Random House

Johnson, R. (1998) 'Patterns and Frameworks', in Rising, L., *The Patterns Handbook*, Cambridge

Kamin and Samuel (1990) *Programming Languages: An interpreter-based approach.* Addison-Wesley

Kivimaki, J. (2001) *MITA: Mobile Phone Internet Technical Architecture* (editor). Finland: IT Press

Koenig, A. and Moo, B. (1997) *Ruminations on C++.* Addison-Wesley

Lewis, M. (1999) *The New New Thing.* Coronet

Lindholm, C. *et al.* (2003) *Mobile Usability.* McGraw-Hill

Ling, R. (2004) *The Mobile Phone Connection.* San Francisco, CA: Morgan Kaufmann

Lippman, S. (1996) *Inside the C++ Object Model.* Addison-Wesley

MacDowell, I. (2005) *Programming PC Connectivity Applications for Symbian OS.* Symbian Press

Madsen, O., Moller-Pedersen, B. and Nygaard, K. (1993) *Object-Oriented Programming in the Beta Programming Language.* Addison-Wesley

McCarthy, J. (1995) *Dynamics of Software Development.* Microsoft Press

Mével, A. and Guéguen, T. (1987) *Smalltalk-80.* Macmillan

Meyers, S. (1998) *Effective C++*, Second Edition. Addison-Wesley

Myerson, G. (2001) *Heidegger, Habermas, and the Mobile Phone*. Icon Books

Niemeyer, P. and Knudsen, J. (2002) *Learning Java*. O'Reilly

Nonaka, I. and Takeuchi, H. (1995) *The Knowledge-Creating Company*. USA: Oxford University Press

Petzold, C. (1992) *Programming Windows 3.1, Third Edition*. Microsoft Press

Raymond, E. (2004) *The Art of UNIX Programming*. Addison-Wesley

Rising, L. (1998) *The Patterns Handbook*. Cambridge University Press

Sales, J. (2005) *Symbian OS Internals*. John Wiley & Sons

Shepard, S. (2002) *Telecom Crash Course*. McGraw-Hill

Spence, E. (2005) *Rapid Mobile Enterprise Development for Symbian OS: An introduction to OPL application design and programming*. John Wiley & Sons

Stichbury, J. (2005) *Symbian OS Explained*. John Wiley & Sons

Stroustrup, B. (1993) *The C++ Programming Language*, Addison-Wesley

Stroustrup, B. (1994) *The Design and Evolution of C++*. Addison-Wesley

Tasker, M. (2000) *Professional Symbian Programming*. Wrox Press

Tidd, J., *et al*. (2005) *Managing Innovation*. Chichester: John Wiley & Sons

Warren, N. and Bishop, P. (1999) *Java in Practice*. Addison-Wesley

Wilkinson, N. (2002) *Next Generation Network Services*. Chichester: John Wiley & Sons

Wolf, W. (2001) *Computers as Components*. Morgan Kaufmann

Index